THE POLITICS OF PAN-ISLAM

The Politics of Pan-Islam

Ideology and Organization

JACOB M. LANDAU

CLARENDON PRESS · OXFORD

Oxford University Press, Walton Street, Oxford OX2 6DP
Oxford New York Toronto
Delhi Bombay Calcutta Madras Karachi
Petaling Jaya Singapore Hong Kong Tokyo
Nairobi Dar es Salaam Cape Town
Melbourne Auckland
and associated companies in
Berlin Ibadan

Oxford is a trade mark of Oxford University Press

Published in the United States
by Oxford University Press, New York

First published 1990
Reprinted 1992

British Library Cataloguing in Publication Data
Landau, Jacob M. 1924–
The politics of Pan-Islam: ideology and organization.
1. Islamic political ideologies
I. Title
320.5′5
ISBN 0–19–827709–1

Library of Congress Cataloging in Publication Data
Landau, Jacob M.
The politics of Pan-Islam: ideology and organization/Jacob M. Landau.
p. cm.
Includes bibliographical references.
1. Panislamism—History. 2. Islamic countries—Politics and government. I. Title.
DS35.7L36 1989. 320.5′49′0917671—dc20 89-23007
ISBN 0–19–827709–1

Printed in Great Britain by
Antony Rowe Ltd, Chippenham, Wiltshire

There is a dry wind blowing through the East, and the parched grasses wait the spark . . . Islam is the only thing to knit up such a scattered empire.

(John Buchan, *Greenmantle*)

Acknowledgements

Most books owe a great deal to the advice and co-operation offered to their authors. This is certainly true of the present work, which attempts to comprise a large span, timewise and spacewise.

Among those who have generously given me their time and counsel are the following: Professors Bernard Lewis of Philadelphia, Elie Kedourie and Anthony D. Smith of London, Albert H. Hourani of Oxford, Farhad Kazemi of New York City, Zekaî Ertuğrul Ökte of Istanbul, and Ercümend Kuran and Orhan Koloğlu of Ankara. My assistant at the Hebrew University of Jerusalem, Gershon Lokay, has demonstrated no less resourcefulness in locating material than zeal in photocopying it. I am greatly indebted to them all, but the final responsibility is of course mine.

Grants from the Truman Research Institute, Jerusalem; St Antony's College, Oxford; and the Sherman Foundation, London, have assisted me in travel to and research of various archives and libraries, to all of which I would like to express my gratitude.

Finally, very special thanks are due to Oxford University Press, most particularly to Dr Henry Hardy, Nina Curtis, Janet Moth, Elizabeth Sweeney, Jane Robson, and David Neuhaus.

J.M.L.

November 1988

Contents

List of Abbreviations viii
Introduction 1
I. The Hamidian Era: An Imperial Ideology 9
II. The Young Turk Era: Pan-Islam in Peace and War 73
III. Pan-Islam Clashes with the Russian and Soviet Authorities 143
IV. Turkey Opts Out, while India's Muslims Get Involved 176
V. Between Two World Wars: The Convention Age 216
VI. Pan-Islam in Recent Years: New Ideologies and Formal Organization 248
Conclusion 304
Appendices 313
Selected Bibliography 382
Index 427

List of Abbreviations

AE	Quai d'Orsay archives (Paris), Affaires Etrangères
AAT	Château de Vincennes (Paris), Archives de l'Armée de Terre
EI	*Encyclopaedia of Islam*, 1st edn.
EI^2	*Encyclopaedia of Islam*, 2nd edn.
FO	Public Record Office (London), Foreign Office files
IO	India Office Records and Library (London)
MBZ	Algemeene Rijksarchief (The Hague), Ministerie van Buitenlandse Zaken
NA	National Archives (Washington, DC), Department of State Papers
PA	Politisches Archiv (Bonn), Auswärtiges Amt

Introduction

Over the years, few ideas have excited such passions as Pan-Islam. Few have been subject to so many, divergent—even contradictory—interpretations.[1] As early as 1902, two of the best-known Orientalists of that time, E.G. Browne and C.A. Nallino, gave their expert estimates of Pan-Islam. The former considered it non-existent,[2] while the latter saw it as a major trend in modern Islam.[3] The controversy has continued unabated to the present. One recent example is that of two Turkish studies on the Sultan Abdülhamid II, published, respectively, in 1985 and 1987; both touch on Pan-Islam but, while the first maintains that it was one of the corner-stones of the Sultan's policies,[4] the second argues that it was entirely absent from them.[5]

At least some of the divergence in the treatment of Pan-Islam by historians, journalists, and others has been caused by the confusion in the use and misuse of the term, which was extensively employed in the languages of Western Europe. To our knowledge, its first mention in these languages occurred simultaneously in German and English, in Murad Efendi's *Türkische Skizzen*[6] and in a public letter dated 31 December 1877 by the traveller and Turcologist Arminius Vambéry.[7]

[1] For a list of the more far-fetched of these interpretations, see N.R. Farooqi, 'Pan-Islamism in the Nineteenth Century', 284.

[2] E.G. Browne, 'Pan-Islamism', 306 ff., 330.

[3] C.A. Nallino, *Le odierne tendenze dell'Islamismo*, quoted by C.H. Becker, 'Panislamismus', 169 n. 2.

[4] İ. S. Sırma, *II. Abdül Hamid'in İslam birliği siyaseti.*

[5] O. Koloğlu, *Abdülhamid gerçeği.*

[6] This was the pseudonym of Franz von Werner; *Türkische Skizzen*, i. 95, mentions a Young Turk group, 'die den "Pan-Islamismus" in ihr Programm aufgenommen hat'. Quoted by D.E. Lee, 'The Origins of Pan-Islamism', 280.

[7] The letter, published in the *Daily Telegraph* (London) of 12 Jan. 1878, 3, speaks of 'the Moslem population of India, amongst whom Panislamic ideas are spreading from day to day, will not remain inactive in the future should the Christian West continue to indulge in the sport of modern crusades'. Quoted by Lee, 'Origins', 280.

These were however brief, casual mentions only. The first extensive use of the term was made by Gabriel Charmes, a prolific French journalist, who developed an interest in the Ottoman Empire, visited parts of it, and wrote about it repeatedly in the 1880s. Some of his most detailed, lengthy articles were published in the *Revue des Deux Mondes*, one of the best Parisian periodicals of literature and politics of that time, published twice a month. Two of his articles dealt with the Ottoman Empire in a comprehensive approach, and were entitled 'La Situation de la Turquie'.[8] The first of these, subtitled 'La Politique du Califat et ses conséquences', was particularly significant. In it, Charmes described in detail Muslim reactions to France's takeover of Tunisia and official Ottoman activities in order to mobilize Muslim opinion everywhere against it. He labelled the Islamic agitation calling for a Muslim union against Christian Powers 'Pan-Islam'.[9] The prestige of the *Revue des Deux Mondes*, on the one hand, and the fact that Charmes soon incorporated these chapters into a book (with the subtitle 'Le Panislamisme'),[10] which became a bestseller, on the other hand, were probably responsible for the popularization of the term. 'Pan-Islam' was then frequently mentioned in the British press and elsewhere from the early 1880s.[11]

It has been widely accepted by now that the term was of non-Muslim origin and that it was probably employed as a variant of 'Pan-Slavism', 'Pan-Germanism', or 'Pan-Hellenism', which were in frequent use at that time. The chronology is not, however, strictly accurate, for a secret society set up in 1865 and calling itself the Young Ottomans had independently been employing the term *İttihad-ı İslam* or 'Union of Islam' since the late 1860s. It seems that they coined the term on the

[8] G. Charmes, 'La Situation de la Turquie'.

[9] In French, 'Panislamisme', cf. ibid. 47/4 (15 Oct. 1881), 741.

[10] G. Charmes, *L'Avenir de la Turquie.*

[11] Perhaps starting in *The Times* (London), 29 Dec. 1881 ('The Organ of the Pan-Islamist Party at the Palace') and continuing in an article on 'Panislamism and the Caliphate', *The Times*, 19 Jan. 1882, and another, similarly entitled, in *The Contemporary Review*, Jan. 1883, 57 (demanding: 'Panislam must be crushed by a new crusade') and 62 (referring to 'The phantom of a Panislamic League'). See also *The Echo* (London daily), 29 Aug. 1882, 1 ('The Advancing Aggression of Pan-Islamism').

model of *İttihad-ı anasır*, or 'Union of Elements', referring to Ottomanism (a term which they had been formerly using).[12] *İttihad-ı İslam* appears to have first been used in an anonymous article—possibly authored by Ziya Pasha—in the Istanbul *Hürriyet* of 9 November 1868, p. 1,[13] and it surfaced again in other articles in the same periodical. The said articles stated that the Ottoman sultans, at the time of the empire's expansion, had planned a union of all the Muslim peoples. A while later, the Istanbul *Basiret* of 8 April 1872 mentioned *İttihad-ı İslam* as an appropriate antidote to Pan-Slavism and Pan-Germanism. In reply, the well-known poet and writer Namık Kemal (1840–88), in an article, entitled 'İttihad-ı İslam' and published in the Istanbul *İbret* of 27 June 1872, argued that only a union of all Muslims, led by the Ottomans, could save the Ottoman Empire.[14] He later added that the new mass media and means of communication made such a union feasible.[15] The term was used at the same time, to the same effect, by another young Ottoman, Mustafa Nuri Manapir-Zade (1844–1906).[16]

The term *İttihad-ı İslam*, which was to become *İslam birliği* in Republican Turkish, was generally rendered as *Waḥdat al-Islām* (or *al-Waḥda al-Islāmiyya*) in Arabic,[17] later as *Jāmiʿat al-Islām*. All these terms may mean 'Islamic unity' as well as 'Islamic union', which of course has added to the ambiguity of the term and its varied employment. It is obviously not surprising that one of the earliest uses of the Arabic term was in one of the articles of the magazine *al-ʿUrwa al-wuthqā*,[18] edited and published by Jamāl al-Dīn al-Afghānī and Muḥammad ʿAbduh in 1884. Nor was it a coincidence, perhaps, that

[12] İ. Kara, ed., *Türkiye'de İslâmcılık düşüncesi. Metinler/kişiler*, i, p. xxviii. E. Kuran, 'Panislâmizm'in doğuşu ve gelişmesi', 397.

[13] Ş. Mardin, *The Genesis of Young Ottoman Thought*, 60.

[14] Ibid, 60, 332. N. Berkes, *The Development of Secularism in Turkey*, 267–8. Namık Kemal frequently defended Islam in his articles; see e.g. his 'Ḥubb al-waṭan min al-īmān', *Hürriyet*, 20 June 1868.

[15] *İbret*, 6 July 1872; see Mardin, *Genesis*, 60.

[16] *İbret*, 20 June 1872; see Mardin, *Genesis*, 60.

[17] This term, or *waḥda*, is not mentioned in H. Rebhan's *Geschichte und Funktion einiger politischer Termini im Arabischen des 19. Jahrhunderts (1798–1882)* (Wiesbaden, 1986), probably because it came to be employed only later.

[18] *al-ʿUrwa al-wuthqā*, in the reprint of the entire periodical (1957), 67–73.

the editors wrote in Paris at the very time that the term Pan-Islam was increasingly employed there by Charmes and others.

It has already been pointed out that the term Pan-Islam is not only ungainly—in adding a Greek prefix to an Arabic verbal noun[19]—but also artificial and superfluous, since universality and indivisibility between religion and politics are well-established in Islam. Such well-known Orientalists as Louis Massignon in 1920[20] and C.A. Nallino in 1935,[21] among others, have argued forcefully that the idea of political unity is inherent in Islam, whose character is a priori international, no less than a complete moral, cultural, legal, social, and political system.[22] One may add that it is this universalist nature of the religion of Islam (which has enjoined the Believers to continue the war into the *Dār al-ḥarb*, the domain of war, and has forbidden them to give up any territory in their own *Dār al-Islām*, the domain of Islam) that has deeply influenced the politics of Pan-Islam. The universality of Islam has thus been emphasized—and rightly so—by scholars, time and again. However, while the principle of Islamic identity, as pointed out by Bernard Lewis, has long been in existence,[23] and very probably still is, carrying it into some form of union has long manifested itself as an aspiration rather than a consistent activity, an idea more than an organized movement. Muslim activism,[24] most particularly in the cause of achieving unity (just like many a Western evaluation of this activism) has all too often failed to consider the divisive factors in Islam,[25] such as local interests, the impact of nationalism,[26] and the basic divergence between Muslim majority and Muslim minority

[19] E. Michaux-Bellaire, 'La Souveraineté et le califat au Maroc', 118.

[20] L. Massignon, 'Introduction à l'étude des revendications musulmanes', 3–4.

[21] C.A.N. (= C.A. Nallino), 'Panislamismo', 196.

[22] Much has been written on the comprehensive character of Islam. For the methodological issues this has raised, cf. e.g. A. Dawisha, ed., *Islam in Foreign Policy*, esp. 2 ff.

[23] B. Lewis, 'The Ottoman Empire and its Aftermath', 27 ff.

[24] See F. Rahman, 'Internal Religious Developments in the Present Century Islam', 866.

[25] H.A.R. Gibb, 'Unitive and Divisive Factors in Islam', 507–14.

[26] N.H. Aruri, 'Nationalism and Religion in the Arab World: Allies or Enemies', 266.

situations.[27] These have hampered, in various degrees, the politics of Muslim solidarity[28] and unity; the fact that these politics were, at times, defensive, at others, offensive, has characterized them as ambiguous in no little measure.[29] Their being addressed, alternatively, to the Muslim masses, political leaders, or intellectual élites,[30] failed to give them a clear sense of direction. None the less, thinking about the promotion of Muslim political unity—and investing efforts to achieve it, in some concrete manner—has taken place since the late nineteenth century, that is, at a time when the majority of Muslims (both numerically and territorially) were under non-Muslim domination.[31] Conversely, their gradual liberation was bound to weaken Pan-Islam.

We shall refrain from examining previous definitions of Pan-Islam, some of them rather strange.[32] For purposes of this study, we shall consider Pan-Islamic ideology during the last hundred years, approximately, as the corpus of writings (and speeches) which focuses on the importance of overall Muslim unity—less from a religious standpoint and with greater emphasis on political or economic aspects—and proposes ways and means to achieve this end. Pan-Islamic movements will be examined in the context of organized activity to carry out, in practical terms, the above ideology toward a union of Muslims everywhere. Pan-Islamic ideology and movements seem to have presupposed, at various times, a part of all of the following premises for the achievement of unity and union:

a. The need for a strong central authority to lead Pan-Islam and to impose its ideology—generally vested in the Caliph. b. The rallying of the entire Muslim world to the cause. c. Obedience of Muslims everywhere to their leader, the Caliph. d. Total solidarity with the cause, even at the risk of sacrificing personal or local interests. e. Readiness for common

[27] R. Israeli, 'Muslim Minorities under Non-Islamic Rule', 159 ff.
[28] To borrow a term from M. Kramer, *Political Islam*, 80.
[29] T.C. Young, 'Pan-Islamism in the Modern World', 195.
[30] H. Halid, *The Crescent versus the Cross*, esp. 208.
[31] Cf. data in K. Vollers, 'Ueber Panislamismus', 21 ff.
[32] One example is B. Wahby's 'Pan-Islamism,' 863, which asserts that 'the aim of Pan-Islamism ... is the awakening of the Islamic conscience, struggling against the aggressor, be he Pope or Khalifa'.

action—political, economic, even military (in the Holy War, or *Jihad*). f. The establishment of a state comprising all Muslims everywhere.[33]

All these are based, first and foremost, on the commonality of religious sentiment,[34] which one can take for granted while devoting the attention in this study to politics and economics as perceived and employed by Pan-Islam. After all, for Muslims religion has been, and remains for most, the main social and cultural fact of life. However, religion defines their politics as well, for Islam is also a means to articulate political and economic attitudes, too. As a result, relations between Islam as ideology and political relations are complex and multi-varied, one of their expressions being in Pan-Islam.

This study of Pan-Islam, as an exercise in religiopolitics, does not claim to be exhaustive to the last detail; rather, it will examine both the ideology and the movement for Muslim union. In some respects, Pan-Islam is *sui generis*. While some pan-ideologies and movements[35] have a common geographical connection (Pan-Americanism, Pan-Europeanism, or Pan-Africanism) and others an ethnic or linguistic tie, often with an irredentist content (Pan-Slavism, Pan-Germanism, Pan-Turkism, or Pan-Hellenism), Pan-Islam is the only significant one, so it seems, to base itself strictly on religious commonality. Its fortunes will be considered over time and space and, it is hoped, a typology or rather a set of typologies arrived at, on a comparative basis, to explain Pan-Islam's failure in achieving its objectives and to estimate its prospects for the future.

Numerous perceptive works on Islamic resurgence and reassertion have been recently published; hence I shall not discuss them. By contrast, although there exists a vast literature on Pan-Islam, it is less than conclusive. Most of it is outdated or discusses a certain aspect only. One book on Pan-Islam, by G.W. Bury, was published in 1919;[36] since it was based on the

[33] Partly based on L.L. Snyder, *Macro-nationalisms*, 131–2.

[34] C.H. Becker, 'Panislamismus', 181–2.

[35] Snyder, *Macro-nationalisms, passim*. F. Kazemzadeh, 'Pan Movements', 365–70.

[36] G.W. Bury, *Pan-Islam*.

author's own experience as a British intelligence officer in World War I, this work (like many others) is nowadays more in the nature of source material than research. A more recent one, by B.K. Narayan,[37] is mainly a survey of early Islam, with the modern chapters focusing on Islam in politics rather than on Pan-Islam. An article on the origins of Pan-Islam, written almost fifty years ago, has remarked that every specialist dealt with this subject within the geographical section with which he was most familiar.[38] This observation is still correct today. One may add to it, also, that each researcher has investigated the period or the topic which interested him most. Moreover, even two new studies, both published in 1986, suffer from these limitations. Martin Kramer's *Islam Assembled*[39] deals solely with Muslim congresses in the pre-World War I and the inter-war periods, devoting only limited attention to Pan-Islam;[40] while J.P. Piscatori's *Islam in a World of Nation-States*[41] touches on Pan-Islam only incidentally,[42] to assess its role in the formation and evolution of Muslim nation-states. An even more recent book, Reinhard Schulze's *Islamischer Internationalismus*,[43] focuses on one organization, the Muslim World League.[44]

By contrast, the present work intends to concentrate on Pan-Islam and attempts to describe and analyse its ideologies and movements during the last hundred years, approximately, over an area stretching from Morocco to Turkey and India. Islamic values determine the politics—and, later, the economics—of Pan-Islam. Conversely, both Islam and Pan-Islam appear to be used for political and economic ends. One of our objectives will be to try to point out these phenomena, as reflected in Pan-Islam. Another will be to try to consider the situation in modern life, in which religious and political institutions

[37] B.K. Narayan, *Pan-Islamism Background and Prospects.*

[38] Lee, 'Origins', 278–9.

[39] M. Kramer, *Islam Assembled.*

[40] The term is not even listed in his index as a separate item.

[41] J.P. Piscatori, *Islam in a World of Nation-States.*

[42] pp. 38–42, 48 ff., 77–83, 85–9, 104, 109–11.

[43] R. Schulze, *Islamischer Internationalismus im 20. Jahrhundert.*

[44] I will avoid the temptation of dealing with related subjects, such as the revival of Islam, fundamentalism, and the like.

compete with one another, the medieval Islamic theory of unity notwithstanding. Chapters will examine Pan-Islam in the Hamidian and Young Turk eras, then in Central Asia under the Tsarist Empire (and more briefly in the Soviet Union), and in the Indian subcontinent (but not in South-east Asia); and conclude with discussions of the inter-war period and the post-World War II generations, hopefully leading to some comparative observations. Appendices will present several original documents.

The sources of this study are archival materials in London, Paris, Bonn, the Hague, and Washington, as well as books, pamphlets, and press articles in Arabic, Turkish, Russian, and several West European languages. These were collected in Jerusalem, Cairo, Istanbul, Ankara, and from libraries in Western Europe and the United States. An alphabetical list of these materials by author may be found in the selected bibliography, which comprises items listed in this book only. An attempt has been made to keep footnotes as brief as possible, with the full data given in the bibliography. Because so many names of different Asian and African origins are included, no attempt at consistency in transliteration has been made; the tendency has been to keep to generally accepted usage or to the forms preferred by the persons mentioned themselves. In general, Arabic and Persian terms and names have been transliterated following accepted scholarly practice, while Turkish ones (both Ottoman and Republican) have mostly been rendered according to current usage. Those Turks, discussed in an Ottoman context, who later adopted family names (following the 1934 law), have their new family names mentioned in square brackets.

I. *The Hamidian Era: An Imperial Ideology*

1. *Early Developments*

During the nineteenth century, Muslims living in territories which formed an almost continuous land mass, increasingly found themselves wedged between foreign powers and caught between the old and the new. It was their response, largely politicized,[1] to these two challenges that provided the background for Pan-Islam. To use a term of Baber Johansen, this response had (at least, initially), a character of religious anti-imperialism.[2]

In the Ottoman Empire, the largest independent Muslim state at the time, several major ideologies were competing with one another. During the first two-thirds of the century, the empire's élites, while searching for identity, were arguing the relative merits of modernization (meaning the introduction of Westernization) and traditionalism (implying the entrenchment of conservatism, chiefly of Turkish and Islamic values), with the associated conceptions in the areas of government and social life. In the last third of the century, other arguments were added that propounded, firstly, Ottomanism (advocating equality of opportunities for all subjects of the empire, irrespective of ethnic origin or religion, in order to ensure their loyalty); secondly, Islamism (favouring Islam and the Muslims at the expense of other religions in the empire); and thirdly, towards the end of the nineteenth century and the beginning of the twentieth, Pan-Turkism (emphasizing early Turkish values and campaigning for an alliance of all Turkish/Turkic groups, within the empire and outside it). Each school argued that its way was the best, indeed the only one, to save

[1] Cf. R. Schultze, 'Die Politisierung des Islam im 19. Jahrhundert'.

[2] B. Johansen, *Islam und Staat: Abhängige Entwicklung, Verwaltung des Elends und religiöser Antiimperialismus.*

the Ottoman Empire. Within this intellectual turmoil, Pan-Islamists found themselves firmly anchored in the camp propounding Islamism, with a definitely anti-Western orientation.[3] While their original approach was to consider Pan-Islam as a means, an antidote to perilous centrifugal forces, it was also gradually seen by some of them as a centripetal *end* in itself.[4]

While Pan-Islamic ideology and organization in the Ottoman Empire dated mainly from Sultan Abdülhamid II's reign (1876–1909), certain signs point to some earlier developments and preparatory conditions.

The origins of political Pan-Islam can be traced, perhaps, to the Ottoman–Russian treaty of Küçük Kaynarca, in 1774, in which a clause was inserted,[5] asserting the Sultan's spiritual jurisdiction over Muslims (*ehl-i İslam*) outside the Ottoman Empire.[6] At the time, this was probably a face-saving compensation for the Sultan's defeat and the establishment of the Tsar's right to intervene on behalf of Christians within the Ottoman Empire. Attributed to the French Ambassador at the Porte, François Emmanuel Guignard, Comte de Saint-Priest,[7] it is not surprising that this approach, reminiscent of a pontifical concept, should be suggested by a Catholic.[8] Ottoman diplomats were quick, however, to grasp its possible advantages and had it inserted in several subsequent treaties[9] concluded upon Ottoman military defeats and the loss of territory. Understandably, each side was to interpret the treaty differently, either as an Ottoman renunciation of temporal authority or as a recognition of its spiritual one.

Some years later, the legend that Sultan Selim I had inherited the Caliphate from the last Abbasid Caliph in 1517 was

[3] For details, see B. Lewis, 'The Ottoman Empire and its Aftermath', 27ff.

[4] Cf. H. Marchand, 'Un coup d'œil sur l'Islam', esp. 149, 152.

[5] *Muahedat mecmuası* (Istanbul, 1297 H), iii. 256, quoted by E. Kuran, 'Panislâmizm'in doğuşu ve gelişmesi', 396.

[6] B. Lewis, 'The Ottoman Empire in the Mid-Nineteenth Century', 291.

[7] For details on Saint-Priest, see J.M. Landau, 'Saint-Priest and His *Mémoire sur les Turcs*', 127–33.

[8] L. Massignon, 'Introduction', 3. A.-M. Goichon, 'Le Panislamisme d'hier et d'aujourd'hui', 20.

[9] Massignon, 'Introduction', 3. George Young, 'Pan-Islamism', 542. 'Panislamism' (*Asiatic Review*, 1927), 211.

put forward, in order to bolster this spiritual authority. Although controversial, this claim was increasingly accepted,[10] and gained credit, at least among many Muslims, by such inspired-from-above symbolic acts as the girding of nineteenth-century sultans with Caliph ʿUmar's sword on their accession to the throne: Mahmud II in 1808, Abdülmecid in 1839, Abdülaziz in 1861, and Abdülhamid in 1876.[11] Official sanction was inserted in the 1876 Ottoman Constitution as follows: 'His Majesty, the Sultan, as Supreme Caliph, is the Protector of the Muslim religion'.[12]

Muslim religious sentiment in the Ottoman Empire, always a potent force, was fed by the conservative approach of Abdülaziz,[13] in part a reaction to the modernizing trends of Mahmud II. Some informed observers even perceived and described an Islamic revival in the nineteenth century.[14] In the second half of that century, there seems to have been an increasing interest among many Muslims in the wars of the Ottoman Empire, expressed, for instance, by talk in Syria in 1853 of a *Jihad*, or Holy War, during the Crimean War.[15] More particularly, one notices a series of Muslim uprisings in the 1850s (in India and the Caucasus), the 1860s (the Caucasus, Central Asia, and Yunan in China), and the 1880s (Egypt and the Sudan).[16] That these revolts were practically always against foreign domination or intervention was an indication of the

[10] One of the best discussions of the issue is still V. Bartol'd's 1912 study, 'Khalif i Sultan', 203–26, 345–400. See also 'Panislamism and the Caliphate', *The Contemporary Review*, 43 (1883), 57–68; A. Vambéry, 'Pan-Islamism and the Sultan of Turkey', 1–11; T.W. Arnold, *The Caliphate*; L. Saunders, 'The Sultan and the Caliphate', esp. 552–3.

[11] B. Lewis, 'The Ottoman Empire in the Mid-Nineteenth Century', 292–3, for details.

[12] Text in *Düstur*, iv, part 2. See A. Hourani, *Arabic Thought in the Liberal Age*, 106.

[13] B. Lewis, *The Emergence of Modern Turkey*, 123–4.

[14] The *Basiret* article was summarized by Gökhan Çetinsaya, 'II. Abdülhamid'in döneminin ilk yıllarında "İslâm birliği" hareketi (1876–1878)'. On Namık Kemal's article, see e.g. the Orientalist traveller W.G. Palgrave, in his 1872 *Essays on Eastern Questions*, 111–41 ('The Mahometan Revival'), who wrote about Muslim 'fervour'.

[15] F. Steppat, 'Khalifat, *Dār al-Islām* und die Loyalität der Araber zum Osmanischen Reich', 461.

[16] N.R. Farooqi, 'Pan-Islamism in the Nineteenth Century', 283. Goichon, 'Le Panislamisme', 21. K. Singh, 'Pax Islamica', 36.

possible use of Islam as a vehicle of protest.[17] That they all failed to achieve results, after an initial success, was a sign of the lack of co-ordination with other Muslim groups—a conclusion reached soon afterwards. A growing number of Muslims were then simultaneously reasoning in preponderantly religious terms: what European intervention amounted to for many of them was that the Cross had triumphed over the Crescent and that a new version of the Crusades was in the process of succeeding where the old version had failed.[18] The call of 'Crescent versus Cross' was an appealing one;[19] a logical step for a part of the Ottoman press in the early 1870s was to speak of—if not to preach directly—Crescentades.[20]

In addition to press articles, there were indications that interest was growing, during Abdülaziz's reign, in Islamic solidarity—even up to policy decisions. Although there may well have been isolated instances of private deliberations about Pan-Islam before (one such case was noted by Arminius Vambéry),[21] it gained significance during this reign.[22] The Ottoman Government not only started to mobilize Muslim support, for example of the Islamic Fraternities,[23] but also considered—again, on Islamic-defined grounds—intervention in Algeria in 1871, to assist the anti-French revolt there,[24] and then sent military help to East Turkestan in 1874[25] and a military mission to Tunisia in the following year.[26] Although

[17] To use an expression from J. Waardenburg, 'Islam as a Vehicle of Protest', 22 ff.

[18] Goichon, 'Le Panislamisme', 20–1. H. Kohn, *A History of Nationalism in the East*, 38.

[19] P. Schmitz, *All-Islam!*, 17.

[20] Lee, 'Orgins', 282.

[21] Quoted in L. Stoddard, *The New World of Islam*, 65.

[22] A French general, René Brémond, maintained that Abdülaziz had promoted Pan-Islam on the model of Pan-Slavism, due to the impact which the Russian Ambassador Ignatiyev had had on him; but no proof was adduced. See Brémond's *L'Islam et les questions musulmanes*, 56.

[23] B. Lewis, 'The Return of Islam', 21. On the fraternities and Pan-Islam, mainly in North Africa, see also A. Ghirelli, *El renacimiento musulmán*, 46–50; this work has an entire chapter (pp. 40–65) on Pan-Islam.

[24] This plan did not materialize, however. See E. Kuran, 'Panislâmizm'in doğuşu . . .', 396–7.

[25] Ibid. 397.

[26] É.P. Engelhardt, *La Turquie et le Tanzimat*, ii. 94–5, quoted by N.M. Streater, 'Pan-Islamism and Pan-Turanianism', 4.

the above produced only very modest results, they increased the sense of Muslim identity, even solidarity, which was further heightened due to the growth in the number of newspapers (and of their readers) and in the development of communications. All these, however, were to evolve during Abdülhamid II's long reign, when another factor was to have a powerful impact on Islamic—and Pan-Islamic—sentiment. This was the cumulative loss of territories, accompanied by the suffering of the Muslim population at the hands of some of their new rulers.[27]

2. *The Formation of an Ideology—al-Afghānī's Contribution*

Among Muslim writers on Pan-Islam, one perceives variations in arguments and conclusions, but it would be difficult to point out schools of thought. We shall start by examining the Pan-Islamic ideas of the best-known and most powerful personality, Jamāl al-Dīn al-Afghānī (1838–97). al-Afghānī may have become even more famous after his death than in his lifetime.[28] There is a great deal of material written about him—memoirs, books, and articles—which have justified even the publication of a special bibliography of most of those works.[29] There is much less written by himself, at least concerning Pan-Islam.

Born in Asadabad (near Hamadan, in Iran), al-Afghānī toured the Islamic countries, then Europe, with lengthy spells of residence in several of them. His travels included Iran, Afghanistan, India, Turkey, Hijaz, Egypt, France, England, and Russia.[30] Thanks to his intelligence, charismatic oratory, powers

[27] An edifying example is the 102-page long *Appel des Musulmans opprimés au Congrès de Berlin. Leur situation en Europe et en Asie depuis le traité de San-Stéfano.* Addressed in 1878 to Bismarck, as president of the congress of Berlin, this is a collection of authentic complaints, sent to Istanbul by Muslims in Rumelia, Western Thrace, Bulgaria, and Serbia.

[28] See E. Kedourie, *Afghani and ʿAbduh*, 3 ff.

[29] A.A. Kudsi-Zadeh, *Sayyid Jamāl al-Dīn al-Afghānī: An Annotated Bibliography.*

[30] Cf. S. Muṣṭafā, 'Jamāl al-Dīn al-Afghānī al-thā'ir al-sharīd', 16–20. The financing of these travels is still a mystery. A recent article has suggested that it was borne by the Nizam of Hyderabad, where al-Afghānī had stayed for a while. See A. Abdullah, 'Syed Jamaluddin Afghani's Ideas Blaze the Trail', 38.

of persuasion, and writing ability, he earned many admirers, some of whom considered themselves his disciples and continued to spread his message. He seems to have been equally successful in conversing with small élitist groups and in haranguing large crowds in the manner of a populist leader.[31] It was this very success that made al-Afghānī suspect to the ruling circles and he was exiled from—or persuaded to leave—Iran, Afghanistan, India, and Egypt.[32]

al-Afghānī's ideas are not always consistent, particularly in his attitude to religion. However, his concept of politics is coherent, despite changes in emphasis (but in this he is not alone amongst politicians, past and present). From speeches recorded in part by his disciples and, even more so, from his own writings, a general notion of al-Afghānī's political ideology can be obtained. Its recurrent theme was the mobilization of Muslims (and especially their leaders and intellectuals), simultaneously against European aggression and corrupt tyrannical rule at home.[33] Both appeals were revolutionary and their combination made them even more so. Aware of the particularism characteristic of the populations he had visited, he incited each to demand reforms and, even more insistently, to seek means and arms for resisting European intervention or domination.[34] His approach was directed, in the early years of his political preaching, to each Muslim country separately, emphasizing not solely the glory of Islam, but also the country's non-Islamic local greatness.[35] Only later did he reach

[31] al-Ṭāhir Muḥammad ʿAlī al-Bashīr, *al-Waḥda al-Islāmiyya wa-ʾl-ḥarakāt al-dīniyya*, 10–11.

[32] In addition to his own writings, among the the most useful works about al-Afghānī are the following: E.G. Browne, *The Persian Revolution of 1905–1909*; N.R. Keddie, *Sayyid Jamāl ad-Dīn 'al-Afghānī': A Political Biography*; H. Pakadaman, *Djamal-ed-Din Assad Abadi dit Afghani*; Kedourie, *Afghani and ʿAbduh*; H. Srour, *Die Staats- und Gesellschaftstheorie bei Sayyid Ǧamaladdīn 'al-Afghānī'*; M.B. al-Makhzūmī, *Khāṭirāt Jamāl al-Dīn al-Afghānī al-Ḥusaynī*; ʿAbd al-Qādir al-Maghribī, *Jamāl al-Dīn al-Afghānī: dhikrayāt wa-aḥādīth*; M. ʿAmāra, *al-Aʿmāl al-kāmila li-Jamāl al-Dīn al-Afghānī maʿa dirāsa ʿan al-Afghānī*; ʿAlī ʿAbd al-Ḥalīm Maḥmūd, *Jamāl al-Dīn al-Afghānī*; A. Amīn, *Zuʿamāʾ al-iṣlāḥ fī al-ʿaṣr al-ḥadīth*,[3] 63–128, M. Abū Rayya, *Jamāl al-Dīn al-Afghānī 1838–1897*[2]; ʿAbd al-Bāsiṭ Muḥammad Ḥasan, *Jamāl al-Dīn al-Afghānī wa-atharuh fī al-ʿālam al-Islāmī al-ḥadīth*.

[33] N.H. Aruri, 'Nationalism and Religion in the Arab World', 266–7.

[34] M. Colombe, 'Islam et nationalisme arabe à la veille de la première guerre mondiale', 88–9.

[35] F. Rahman, 'Internal Religious Developments', 876.

the conclusion that such was the physical force of each of the European powers that only a united Islamic world could be a match for any of them. He himself, however, perceived no contradiction in terms between his earlier patriotic approach and his later Pan-Islamic one. As H.A.R. Gibb saw it, this was al-Afghānī's 'Pan-Islamic nationalism'[36] or, as N.R. Keddie put it, his 'Proto-nationalism'.[37] For him, nationalism and Pan-Islam complemented one another in their 'liberationist' aspect:[38] The introduction of more liberal regimes into the Muslim countries and popular resistance to European aggression were to be links in an all-Muslim union,[39] the only way to respond properly to the new Christian Crusades.[40]

al-Afghānī's own words are, of course, the best indication of his understanding of Pan-Islam. He was born and bred in a Shiite environment and was familiar, due to his travels and residence, with the Sunnites. One of al-Afghānī's pet ambitions was to bridge their differences. Well-acquainted with the writings of both groups, he argued, again and again, that their differences were a matter of past relevance and that a *modus vivendi* between them could—indeed, should—be found. Thus, in his confidential discussions with his disciple and friend, Muḥammad al-Makhzūmī, he repeatedly stressed the modern irrelevance of these differences,[41] emphasizing that they were 'destroying the bonds of Islamic unity'.[42]

Not surprisingly, al-Afghānī reverted to the Shiite–Sunnite differences in his newspaper articles.[43] After having been exiled from Egypt, he not only contributed to various European newspapers,[44] but also set up several periodicals, in

[36] H.A.R. Gibb, *Modern Trends in Islam*, 119.

[37] N.R. Keddie, 'Pan-Islam as Proto-nationalism', 17 ff.

[38] In Arabic, *taḥarrurī*, as employed by ʿAmāra, *al-Aʿmāl*, 34–5.

[39] Cf. H. Saab, *The Arab Federalists of the Ottoman Empire*, 189.

[40] X, 'Le Panislamisme et le Panturquisme', 183–5. Naṣr al-Dīn ʿAbd al-Ḥamīd Naṣr, *Miṣr wa-ḥarakat al-Jāmiʿa al-Islāmiyya*, esp. ch. 2.

[41] al-Makhzūmī, *Khāṭirāt*, 112–14.

[42] Ibid. 113: 'Tafkīk ʿurā al-waḥda al-Islāmiyya'.

[43] See also K.K. Key, 'Jamal al-Din al-Afghani and the Muslim Reform Movement', 545.

[44] N.R. Keddie, *Sayyid Jamāl ad-Dīn 'al-Afghānī'*, and H. Pakadaman, *passim*, for examples. For others, cf. A. Abdullah, 'Syed Jamaluddin', 42–3. O. Koloğlu, *Abdülhamid gerçeği*, 200, refers to an article by al-Afghānī in the *Figaro* (Paris) of 8 Nov. 1883, in which he warned that, should the Mahdi be victorious in the Sudan, Muslims would rise everywhere.

Arabic, of which the best-known was *al-ʿUrwa al-wuthqā* in Paris.[45] This one, published during 1884, was of particular importance: both he and his disciple Muḥammad ʿAbduh (who is discussed below) wrote its articles, but it was his own views that apparently determined the contents, particularly since ʿAbduh was deeply influenced by him. So *al-ʿUrwa al-wuthqā* repeated al-Afghānī's wishes to reconcile Shiites and Sunnites, and thus lead to a union of the Iranians with the Afghans.[46] This periodical also expressed his views on Pan-Islam in general, at that time, in an article entitled 'al-Waḥda al-Islāmiyya' ('Islamic unity' or 'Islamic union').[47] While one cannot be absolutely certain whether al-Afghānī himself wrote it, or whether ʿAbduh did (if so, probably under his mentor's inspiration), the style seems to point to the former's authorship. Moreover, no less an Islamic scholar than Muṣṭafā ʿAbd al-Rāziq (1886–1947), then Minister of *Waqf*s in Egypt, re-published this article in 1938, with an introduction of his own, maintaining that he recognized it as a product of al-Afghānī's thought.[48] The article[49] is a clear, succinct exposition of al-Afghānī's ideas and may well be their earliest presentation in such detail. The main argument is as follows.

The rule of Islam stretched from the furthest point west, the Maghrib, to Tonkin on the borders of China, and from Fezzan in the north to Sarandib, south of the Equator. The lands inhabited by Muslims were contiguous and formed one single sequence; their government was undefeatable. They had great kings, who ruled most of the globe. Nature was generous to them in plants and trees. Their cities were abundantly populated and competed successfully with other towns in architecture and crafts. They were deservedly proud of their scient-

[45] See e.g. X, 'Les Courants politiques dans la Turquie', 170. C.C. Adams, *Islam and Modernism in Egypt*, 9.

[46] *al-ʿUrwa al-wuthqā* (repr. 1957), 106–10, article entitled 'Daʿwat al-Furs ilā al-ittiḥād maʿa al-Afghān'.

[47] Ibid. 67–73.

[48] Jamāl al-Dīn al-Afghānī, *al-Waḥda al-Islāmiyya wa-'l-waḥda wa-'l-siyāda*, introduced by Muṣṭafā ʿAbd al-Rāziq. This is also the opinion of ʿAmāra, *al-Aʿmāl* . . ., who reprinted this article, 339–46. However, Ṭāhir al-Ṭanāḥī, in his edition of ʿAbduh's *al-Muslimūn wa-'l-Islām*, 32–40, included this article among ʿAbduh's writings.

[49] See below, app. B for an English version (my translation).

ists, men of letters, philosophers, physicians, architects, and scholars. The kings of China and Europe trembled before a word of their caliphs and princes. Until recently, no navy could compete with theirs. The Muslims numbered no less than 400 million, and their hearts were stout and ready to die as martyrs in war. The Koran had made them enlightened and moral. They did not wish to be governed by non-Muslims, even when the rule of others was compassionate. Due to their brotherhood-in-faith, each considered himself subjugated if a Muslim community fell under foreign domination.

However, Muslims lagged in knowledge and industry, after having been the world's teachers. Their lands had been torn and taken over by others, despite the injunctions of their religion not to submit to foreign domination. But they had not forgotten that Allah promised they would inherit the Earth. More importantly, however, than their lagging behind in science and power were their rivalries and internal discord, which brought about poverty, weakness, and disorganization, followed by disunity, corruption, and greed. Muslims had preferred foreign rule, despite the injunctions of their religion, and had requested foreign assistance against their own co-religionists. This occurred since their rule in Spain. The seeking of pleasure ruined their industry, commerce, and agriculture.

Agreeing upon and co-operating in strengthening Muslim government is a basic creed in Islam. However, but for domination-crazy rulers, Muslims east and west, north and south, would unite and work together against the dangers facing them. The Russians were poor, retarded, and had no natural resources, but their spirit of co-operation enabled them to buy the weapons which they could not produce and hire the officers whom they could not train themselves—and all Europe feared them. Muslims could do the same and defend from all attacks the union which joined them.[50] The only ones opposing this union were those local rulers who were steeped in their own daily pleasures and honours; these were like chains around the necks of Muslims. The heirs of the heroes should not let themselves despair, for there was an unbroken sequence of Muslim lands, from Edirne[51] to

[50] In Arabic, *ṣawn al-waḥda al-jāmiʿa lanā.*

[51] Adrinopli.

Peshawar, inhabited by no less than 50 million Muslims, long distinguished by their courage. If these agreed between themselves, showed regard for the needs of other Muslims, and united, they could dam the floods imperilling them from all sides. Melancholy and despair help no cause, but hope and action do. Uniting in the name of the Koran and Islam would guarantee success; Muslim rulers, too, might perceive this before they were done away with.

Characteristic of al-Afghānī's style, this article, flowery in wording and embellished by quotes from the Koran and reminders about the significance of religion, is nothing less than outspoken criticism of particularist tendencies and a resonant appeal for Muslim unity and union, based on communal memory and on factors of territorial and numerical size. It is, also, a revolutionary call in the name of Islam against those local rulers who could prevent the achievement of unity—while leaving them an option to join, for al-Afghānī was, most probably, well aware of their political, economic, and military power. Indeed, some of the few other authentic documents one has of al-Afghānī's projects for Pan-Islam are addressed to the powerful. At some unknown point in his career, al-Afghānī must have reached the conclusion that he would have to convert the Muslim rulers, or at least one of them, to his Pan-Islamic views in order to carry out his plans. The concept of a united Muslim community with a spiritual and political leader at its head was essential to late nineteenth-century Pan-Islam.[52] Although al-Afghānī was probably not its originator, he adopted this concept and markedly toned down his attacks on the Ottoman Sultan, whom he selected as the most likely personality to direct a Pan-Islamic campaign.[53]

Of the two documents relevant to this matter the first was uncovered by Nikki R. Keddie.[54] It is undated, but she thinks it is likely to be from *c.*1877–8,[55] and her arguments are

[52] M. Fakhry, 'The Theocratic Idea of the Islamic State in Recent Controversies', 451.

[53] al-Bashīr, *al-Waḥda*, 18–19, based on al-Makhzūmī's memoirs.

[54] N.R. Keddie, 'The Pan-Islamic Appeal: Afghani and Abdülhamid II', 52–60. Id., *Sayyid Jamāl ad-Dīn 'al-Afghānī'*, 133–8. For a photocopy of the original, see Abdul Hakim Tabibi, *The Political Struggle of Sayid Jamal ad-Din al-Afghani*, App. 1.

[55] Keddie, *Sayyid Jamāl ad-Dīn 'al-Afghānī'*, 131–3.

persuasive. This letter, written in Persian, was sent to a highly placed Ottoman official, very probably with the intention that it should reach the Sultan himself. In the letter, al-Afghānī expressed his worry at the state of the Islamic nation[56] and said that he had devoted considerable time to thought about its salvation. His first suggestion was that, since India's Muslims were numerous, wealthy, generous, and firm Believers, he (al-Afghānī) should go there as a delegate of the Ottomans; explain to them the disadvantages of discord and the advantages of unity; indicate to them the standing of the Ottoman Sultanate in Islam; and collect from them money for the Sultanate—all in the interest of the Complete Unity of Islam.[57] His second was to go subsequently to Afghanistan and appeal to its brave population to beware of Russian designs and throw in their lot with the Ottoman Government. His third was to travel afterwards to Baluchistan and to the Turcoman areas, to persuade them to join a general war in the name of religion, for the sake of the unity of Islam.

The letter is a mixture of Messianic vision and *realpolitik* in the cause of Pan-Islam. In so far as the latter is concerned, one has here a plan for a *rapprochement* between the Ottoman Empire, the Muslims of India, and those of Afghanistan, leading to future co-operation and, perhaps, eventually a union. However, his letter is indicative of even more ambitious plans for the achievement of Pan-Islam. He offers to send Ulema to remote parts of India, others from Afghanistan to Kokand and Bukhara, and yet others from Baluchistan to Kashgar and Yarkand—all as secret emissaries to mobilize support for the cause. Here one has the embryo of a master-plan, bold in its conception and execution. It is likely that this very boldness caused the Sultan (or his advisers) neither to respond favorably, nor to invite al-Afghānī to Istanbul at that time.

If the above letter was written, as Keddie surmises, on an *ad hoc* basis during the Turco-Russian War of 1877–8, in the following years al-Afghānī's plans for achieving Pan-Islam evolved further. Again, in 1892, he offered them to Sultan Abdülhamid, and this time he was invited to Istanbul, to share

[56] In Persian, *Millet-i Islāmiyye.*
[57] In Persian, *Ittiḥād-i tāmm-i Islāmiyye.*

in the Sultan's Pan-Islamic policies. He remained there for the last five years of his life, apparently an involuntary captive, whose every effort to leave was thwarted.[58] It seems that, while al-Afghānī and the Sultan did find common ground on Pan-Islam (and al-Afghānī wrote numerous letters to Iran, to persuade his correspondents of Pan-Islam's merits), they differed in their views on reform and constitutionalism.[59] In addition, various people at the palace worked to circumscribe al-Afghānī's activities [60] and to warn the Sultan of his personal ambitions.[61] Here we are more concerned with his 1892 letter to the Sultan, which I uncovered.[62] The manuscript letter was intercepted by the British and a French translation forwarded to the Foreign Office.[63]

The letter suggests that the agitation in Afghanistan could be turned to the profit of the Ottoman Empire by appealing to the strong religious sentiments of the Afghans. An experienced diplomat should be able secretly to secure the orientation of Afghan policies towards the Ottoman Empire. This ought to compel Iran to take a place, willy-nilly, in an alliance with the Ottoman Empire and Afghanistan, as it would be hemmed in between them. Such a powerful alliance would have to be taken into consideration by both Great Britain and Russia. This alliance would be able not only to choose its partners among the European powers and blocs as it wished, but also to extend the influence of the Caliphate among the Muslims in India and force Great Britain to evacuate Egypt.

Here one has a grand design of Pan-Islamic politics: the establishment of an important Muslim bloc, made up of the Ottoman Empire, Persia, and Afghanistan[64]—the only independent Muslim states at that time—as a stepping-stone towards attracting to a Pan-Islamic union the Muslims in

[58] J.M. Landau, 'An Egyptian Petition to Abdül Hamīd II on Behalf of al-Afghānī', 209 ff.

[59] See Saab, *Arab Federalists*, 189.

[60] Cf. Kedourie, *Afghani and 'Abduh*, 62.

[61] al-Bashīr, *al-Waḥda*, 20–1.

[62] J.M. Landau, 'al-Afghānī's Panislamic Project', 50–4. See also below, app. D.

[63] Ibid. 53–4. This document is located in FO 78/4452, enc. in Hardinge's no. 144, marked 'Secret', to Rosebery, dated in Ramleh, 3 Sept. 1892.

[64] al-Afghānī reverted to this idea, time and again, during his last years in Istanbul, in his discussions with al-Makhzūmī, see A.A. Maḥmūd, *Jamāl al-Dīn*, 318 ff.

India and probably in Egypt. The union, as he confided to close friends in Istanbul, ought to be headed by the Ottoman Empire. However, he seems to have perceived the union in the future only; the more immediate task was resistance to foreign aggression.[65] No longer (as requested in the pages of *al-ʿUrwa al-wuthqā*), would efforts be made to institute study circles near the mosques and *medreses*, or religious schools, to serve as links in a Pan-Islamic union and mobilize public opinion;[66] no more (as envisaged in the earlier document) would activities be focused on the population, using emissaries to rally support in remote locations. Rather, emphasis would be laid on diplomatic efforts (secret, of course), on pressure at the highest levels (compelling Iran to join the alliance), and on negotiating with the European powers on an equal footing (due to the power vested in the Muslim alliance). al-Afghānī had frequently (albeit not consistently) adopted this anti-British stand; he considered Great Britain as the most serious threat to Islam, because of its advanced economic and technological system and its rule over extensive lands inhabited by Muslims.[67] Probably as a result of this letter, al-Afghānī was invited to Istanbul in 1892,[68] to participate in Pan-Islamic propaganda.

al-Afghānī's greatest merit, in our context, was to have shown that, in his days at least, Pan-Islam and nationalism could be mutually complementary (he even toyed with the idea of a confederation of semi-autonomous Muslim states, with the Ottoman Sultan as their suzerain);[69] and to politicize Islam within a Pan-Islamic context or, otherwise said, to transform Islam into a political ideology. However, he was not the only Pan-Islamic ideologue, nor even the first one.

3. *The Formation of an Ideology—Other Contributions*

At least a part of the great impact which al-Afghānī had on his listeners and readers was due to preparatory work by others,

65 B. Tibi, *Nationalismus in der dritten Welt*, 153.

66 See H. Srour, *Die Staats- und Gesellschaftstheorie*, 183.

67 M. Ruthven, *Islam in the World*, 304.

68 Abū Rayya, *Jamāl al-Dīn*, 62–8, for al-Afghānī's years in Istanbul.

69 al-Afghānī's discussions with al-Makhzūmī, reported by M. Colombe, 'Islam et nationalisme', 91–3.

which had made interest in Islam and an Islamic union, if not yet a byword in every household, at least current within certain élites in the Ottoman Empire. While Islam had long been, of course, a potent religious and social force in the empire, with an extensive bureaucracy serving it, the last third of the nineteenth century witnessed increasing concern with Islam among circles besides the religious establishment and the common people. Termed 'Islamism', or *İslamcılık* in Turkish, this trend seems to have focused, at first, on returning to fundamental Islamic values and on traditional matters, such as education[70] and public morals. Gradually, however, Islam assumed increasing political significance among the Young Ottomans, those Turks who were striving for political reform in the 1860s and 1870s, in the teeth of the conservative policies of the Sultans Abdülaziz and Abdülhamid. As they became increasingly politically minded, Islam assumed for some of them an added significance. The Russian conquest of formerly autonomous Muslim Khanates in Central Asia and the subsequent loss of Ottoman territories increased political awareness. Press reports about the persecution of Muslims by the new rulers of those territories, chiefly in the Crimea and the Balkans, served to politicize still further, in the cause of Islam, a good number of public figures.[71] Later, separatist activities by some of the empire's non-Muslim minorities aroused several Young Ottomans and others to the possible merits of a politically minded Islamist policy, one of whose most forceful expressions was Pan-Islam.

One does not know precisely when the idea started getting around. The contention of the French author Bertrand Bareilles, who lived in Turkey, that Halil Pasha had been the first to advocate, in 1865, the joining of the forces of Islam, so as to provide the Caliph with a weapon with which to fend off European pressures,[72] is not backed by proof and seems rather doubtful. Anyway, by 1871, Pan-Islam seems to have been well-known enough in the Ottoman Empire to bring about

[70] Cf. T.Z. Tunaya, *İslâmcılık cereyanı*, 91–3.

[71] S.J. Shaw and E.K. Shaw, *History of the Ottoman Empire and Modern Turkey*, ii. 259.

[72] B. Bareilles, *Les Turcs*, 207–8.

at least one strong condemnation. Written in Istanbul on 16 August and published in Paris in that year, the 31-page anonymous pamphlet *Unité Islamique*[73] was forthright in its opposition to Pan-Islam. Its author (possibly a Christian in Istanbul) favoured Ottomanism, even though he had doubts about its being immediately practicable. He attacked Pan-Islam which, he maintained, was an insidious policy fomented by fanatical agitation and well regarded at the Sublime Porte. It was particularly dangerous as a political school,[74] which would not only be unable to save the empire, but rather would destroy it: on the religious level, it would be impossible to unite the disparate groups and sects; on the political level, the Sultan–Caliph would fail to impose himself on all Muslims, only provoking international opposition, especially of the powers ruling Muslim populations.[75]

This policy seemed to advance with the progress of foreign conquest.[76] It meant, on the one hand, strengthening the bonds with Muslims in the Ottoman Empire, including those in the lost territories, such as Algeria, the Crimea, Cyprus, Tunisia, and Egypt; and on the other, forging ties with Muslims from across the empire's boundaries—in a spirit of solidarity or even union (*İttihad-ı İslam*).[77] Such ideas were fostered not solely by the Opposition, but also by the Establishment,[78] quick to sense the advantages which could accrue to the Sultan, in both his own rule of the empire and his prestige outside it. A number of government-sponsored newspapers increasingly supported a version of internal and external Pan-Islam, the main objectives of which were, respectively, to assure the loyalty of non-Turkish Muslims (Arabs, Albanians, and others); and to attract political and financial support from

[73] A copy is available in the library of the Institut National des Langues et Civilisations Orientales, Paris. I am grateful to Dr François Georgeon for supplying me with a photocopy.

[74] *Unité Islamique*, 7: 'Une école politique, un drapeau'. Also p. 16: 'Nouvelle école politique'.

[75] See selections from this pamphlet below, app. A.

[76] As noted by Keddie, 'Pan-Islam as Proto-nationalism', 19–21.

[77] P. Schmitz, *All-Islam!*, 59.

[78] Examples in B. Lewis, 'The Ottoman Empire in the Mid-Nineteenth Century', 293–4, and id., *The Middle East and the West*, 101 ff.

those in India and elsewhere, as a first step, and their union with the empire, as a future one—modelled perhaps on the recently successful unifications of Italy and Germany.

It should be emphasized at this point that Pan-Islamic propaganda became more widespread and insistent in the later years of Abdülhamid's reign and under the Young Turks (as we shall see below). However, a few examples of the main themes and arguments in the Hamidian period are relevant here.

As early as 1881,[79] a Turkish manuscript on the subject was presented to Abdülhamid and dedicated to him; it was then incorporated into the Sultan's library. Written by Süleyman Hasbi (died 1327/1909), author of several printed works,[80] it was entitled 'A Treatise about Union for the Happiness of the Islamic Millet' and comprised fifteen leaves.[81] While generally starting from religious premisses, it discussed *ittifak* (co-operation) and *ittihad* (union or unity) as essential for the future of the entire Islamic nation—attacking divisiveness (*tafarruk*). It pointed at the Pilgrimage as an important unifying factor for Muslims everywhere and reminded these of the obedience they owed the Caliph who still combined spiritual with temporal authority. In the 1870s and 1880s, Namık Kemal,[82] already mentioned as a prominent poet and writer in Young Ottoman circles, was so revolted by Ernest Renan's (1823–92) claim, in a famous lecture entitled *L'Islamisme et la science*, that Islam was hostile to science, that he not only replied in some detail, defending traditional Islamic and Ottoman values, but also discussed the option of Pan-Islamic unity under Ottoman leadership, to better resist Europe's designs. However, Namık Kemal viewed Pan-Islam as a mainly cultural phenomenon[83] and had already expressed these opinions in an 1872 article entitled 'İttihad-ı İslam' (one of the

[79] The MS was dated 25 Dhī al-Ḥijja 1289 H (equivalent to 23 Feb. 1873).

[80] On whom see B.M. Tahir, *Osmanlı müellifleri*, i. 288–9, which also mentioned this MS.

[81] S. Hasbi, 'Risale-yi ittihadiye li-saadet-i millet-i el-İslamiye'.

[82] On Namık Kemal and his ideas see, *inter alia*, H.Z. Ülken, *Türkiye'de çağdaş düşünce tarihi*, index; M.N. Özön, *Namık Kemal ve İbret gazetesi*.

[83] B. Lewis, *Emergence*, 142, 341. Keddie, 'The Pan-Islamic Appeal', 48. E. Kuran, 'Panislâmizm'in doğuşu ...', 397. Shaw and Shaw, *History of the Ottoman Empire*, ii. 259.

first times this term was used) in the Istanbul daily *İbret*.[84] In it, he maintained that 100–200 million Muslims, with a common religion and culture, were a force, if their agreeing[85] transcended the frontiers of the Ottoman Empire.[86] Another contributor to *İbret*, Nuri, in an article entitled 'Strengthening the Ties',[87] argued that, since Prussia had been striving to unite all Germany, Russia all Slavs, Sardinia all Italians, while some stateless peoples, also, had been thinking of uniting, strengthening inter-Muslim ties was imperative.

Muḥammad ʿAbduh (1849–1905)[88] has already been mentioned as al-Afghānī's disciple and close collaborator in *al-ʿUrwa al-wuthqā*, in Paris in 1884. After his return to Egypt—from where he had been expelled for his share in the ʿUrābī uprising of 1881–2—ʿAbduh became a central figure in the teaching and interpretation of Islam, rising to the post of Grand Mufti. His active interest in politics diminished over the years, perhaps due to the need to remain on good terms with the British in Egypt,[89] but more probably because of the all-absorbing interest he developed in Islam. In his view, Islamic values had been eroded, but the Muslims themselves were at fault; a regeneration of the early unadulterated Islam would restore its pristine character and enable it to compete successfully with European values.[90]

In his later years, ʿAbduh seems to have realized that a united Muslim state was politically unfeasible, even if he did call on Muslims to unite against their enemies.[91] He increasingly favoured what he believed to be the more successful, albeit slower, process of education in religious Pan-Islam,

[84] *İbret*, 1/11 (27 June 1872), 1. The article has been reprinted several times, e.g. by M.N. Özön, *Namık Kemal*, 74–8.

[85] In Turkish, *ittifak*.

[86] The article ends with a promise by Namık Kemal to indicate, at some future time, the ways and means of this *ittifak*.

[87] Nuri, 'Teşyid-i revabit'.

[88] ʿAbduh has received considerable attention. See, in addition to his own works, M.R. Riḍā, *Taʾrīkh al-ustādh al-imām*, i–iii; Adams, *Islam and Modernism*; Kedourie, *Afghani and ʿAbduh*; M. H. Kerr, *Islamic Reform*; Naṣr, *Miṣr wa-ḥarakat al-Jāmiʿa*, ch. 3. Aḥmad Amīn, *Zuʿamāʾ al-iṣlāḥ*, 302–69.

[89] M. Ruthven, *Islam in the World*, 304–5.

[90] B. Lewis, *Middle East*, 103–4. M. Fakhry, 'The Theocratic Idea', 454.

[91] Cf. K. Singh, 'Pax Islamica', 38.

within the framework of Islamic religious and social reforms.[92] When, in *c.*1886, he sent two messages to the Şeyhülislam (Shaykh al-Islām), the chief religious dignitary in Istanbul, and the Vali, or Governor, of Beirut, respectively, arguing for the preservation of the Ottoman Empire, he considered it an article of faith.[93] In May 1900, he published, in the Egyptian newspaper *al-Mu'ayyad*,[94] a reply to the critical remarks about Islam in two newspaper articles in *Le Journal* (Paris) of March 1900 by Gabriel Hanotaux (1853–1944), French historian and former Minister for Foreign Affairs[95] (who had previously served in the French Embassy in Istanbul briefly, in 1885–6). The gist of ʿAbduh's reply, entitled 'Unifying the Muslims and the Unity of the Religious and Political Powers',[96] is as follows. After praising the virtues of Islam as a moral religion and social way of life, he acknowledged that some innovations had been added to it over the years. These innovations were responsible for the rare expressions of fanaticism of which Hanotaux accused the Muslims—and, anyway, this was not to be compared to the blatant fanaticism of Christians, east and west. He went on to assert that, even though Muslims in various lands traded with, and lived off, one another, none of them thought of uniting in action (that is, political action). Even the Pilgrimage was intended solely for religious ends—for Muslims to learn Islam from each other and to devise among themselves how to defend their own interests; something which the Europeans, too, were wont to do. Their overall loyalty to the Ottoman state and to its sultan was nothing to wonder about, as—united by the Koranic revelation and common history—they expected the leader of the largest Muslim state to defend them and improve their

[92] Cf. P.J. Vatikiotis, 'Muḥammad ʿAbduh and the Quest for a Muslim Humanism', 58–9. Acc. to Vatikiotis, ʿAbduh's disillusionment with political Pan-Islam started much earlier, even while he was editing *al-ʿUrwa al-wuthqā* conjointly with al-Afghānī.

[93] Riḍā, *Ta'rīkh* ..., ii. 339, quoted in Adams, *Islam and Modernism*, 61–2.

[94] 6 and 13 May 1900. Repr. in M. ʿAmāra, *al-Islām wa-'l-ʿUrūba wa-'l-ʿalmāniyya*, 109–14. See also M.A.Z. Badawi, *The Reformers of Egypt*, 45.

[95] M.T. Harb, ed., *L'Europe et l'Islam: M.G. Hanotaux et le cheikh Mohammed Abdou*, reprinted Hanotaux's articles, a French translation of ʿAbduh's replies, and further articles by Hanotaux in the same matter.

[96] 'Tawḥīd al-Muslimīn wa-waḥdat al-sulṭatayn al-dīniyya wa-'l-siyāsiyya'.

belief and their morals. Again, this was nothing that should worry Hanotaux, nor should he fret over the Sultanate and Caliphate being joined in one person; for in Islam the Sultan ruled, but all religious decisions were in the hands of the Ulema.[97]

These articles, characteristic of ʿAbduh's writing at the time, have an obviously apologetic tenor and represent a visible retreat from his attitude in former days concerning political Pan-Islam.[98] This may be explained, in part, as a wish to dispel suspicions abroad of Muslim fanaticism and the danger of Pan-Islam. Among other examples of Islamic apologetical literature are a book by a Turk, who had studied at the Universities of Istanbul and Cambridge, named Halil Halid (died 1934),[99] *The Crescent versus the Cross*, published in 1907,[100] and a letter from Prince Sabaheddine. The latter (1877–1948), grandson of Sultan Abdülmecid and nephew of Abdülhamid II, had left the Ottoman Empire in 1900 for Europe, to continue there his opposition to the Hamidian regime and his struggle for reforms.[101] As soon as he arrived in Europe, Sabaheddine and his brother Lutfullah wrote and printed a *Lettre au Sultan Hamid II*,[102] calling for immediate and other administrative reforms in the Ottoman Empire. On 4 August 1906, Sir Edward Grey, British Foreign Secretary, commenting in the House of Commons on the Danishway Incident in Egypt,[103] asserted that the incident had happened

at a time we were one of four Powers in occupation of certain islands belonging to the Sultan because of difficulties which had arisen with Turkey over the Macedonian question. The fact had begun to have its

[97] At the time, this was also the view of Snouck Hurgronje, a leading Dutch Orientalist, on whom see G.S. van Krieken, *Snouck Hurgronje en het Panislamisme*. See also M.R. Buheiry, 'Colonial Scholarship and Muslim Revivalism in 1900', 7, and Detlev Khālid, 'Aḥmad Amīn', 60.

[98] For an overview of ʿAbduh's political ideas, see ʿAbd al-ʿĀṭī Muḥammad Ahmad, *al-Fikr al-siyāsī li-'l-imām Muḥammad ʿAbduh*, esp. 289–301.

[99] See *Türk Ansiklopedisi*, xviii. 383–4, s.v. Cf. N. Berkes, *Development of Secularism in Turkey*, 360.

[100] For a review, cf. A. Fevret, 'Le Croissant contre la croix', 421–5.

[101] On Sabaheddine and his views, see Berkes, *Development*, 309 ff. Şerif Mardin, *Jön Türklerin siyasî fikirleri*, 215–24.

[102] A copy can be consulted in PA, Türkei no. 198, vol. 2.

[103] Parliamentary Debates, 4th ser., First Session of the 28th Parliament, vol. 152 (= vol. 11 of the 1906 session), col. 1832.

> effect in Egypt, and when you have conflict of that kind with the Turkish Government ... you may be quite sure it is bound to have a certain effect on the Mahomedan races who are under British rule or in countries under British occupation.

In another part of his speech, Grey used the term 'fanaticism'.

Prince Sabaheddine, writing from Paris, responded in a letter to the London *Times*,[104] simultaneously sending a translated version to *Le Siècle*.[105] While denying fanaticism, Sabaheddine did confirm that a part of the East, assuming that it had been imposed upon, could be brought to sympathize with the Pan-Islamic doctrine, in which 'might be found the elements of an understanding between all Mussulman countries, whereby the constituent unities of the nation might be preserved'. He acknowledged further that Pan-Islamic policies would obviously be influenced by religious feeling, 'the connecting link which binds together a most heterogeneous mass numbering several hundreds of millions'. In European political circles, added Sabaheddine, Istanbul was regarded as the centre of Pan-Islamic activity. Yet, it had been the seat of the Caliphate for four centuries, without doing anything for the political union of Muslim nations. Only in the second half of his reign had Abdülhamid been favouring such a policy—on a religious level, in order to strengthen his hold on the Muslims within his own empire.

Although, in typical apologetic style, the emphasis in Sabaheddine's letter was laid on the religious character of Pan-Islam and the blame for the whole movement was squarely put on the shoulders of the European powers which imposed (their rule, most probably) on Muslims, a careful reader would notice that the Prince neither denied the existence of Pan-Islam, nor did he attempt to minimize its potential.[106] His

[104] *The Times*, 13 Aug. 1906, 6. The letter was written on 9 Aug. 1906. It was considered important enough for the German Ambassador to London to send the *Times* extract to Reichskanzler von Bülow, see PA, Türkei no. 198, vol. 4.

[105] *Le Siècle* (Paris), 15 Aug. 1906, repr. in P. Fesch, *Constantinople aux derniers jours d'Abdul Hamid*, 401–4.

[106] For Sabaheddine's further views on Pan-Islam, see Y.H. Bayur, *Türk inkilâbı tarihi*, ii. part 4, esp. 33, 94–7. For his general views, cf. H.Z. Ülken, *Çağdaş Türkiye düşünce tarihi*, index. Later, he was to head a liberal opposition party in the Ottoman parliament of 1908.

views were echoed by other Young Turks in exile, notably by one of their most articulate spokesmen, Ahmad Rıza. In 1906, in *Mechveret*, the French monthly supplement which he edited in addition to the Turkish *Meşveret* (both in Paris), replying to accusations about Islamic fanaticism in *Le Temps*, Rıza argued that, if existent, Islamic sentiment was a reaction to European domination and exploitation.[107] A year later, he maintained that Pan-Islam as a sentiment of solidarity and a desire for union was praiseworthy and considerably less dangerous than either Pan-Slavism or Pan-Germanism.[108]

Evidently, when writing for their own readership, Muslim supporters of Pan-Islam were much more frank about its objectives, political and otherwise. We shall limit ourselves to three examples—a Turk, an Arab, and a Persian.[109]

Abdürreşid İbrahim (Ibragimov) (1853–1944), a disciple of al-Afghānī and a bitter foe of the West and, even more so, of Russia, was of Tatar origins. Educated in Kazan and Medina, he spent some time in Istanbul, where he was influenced in 1897 by al-Afghānī.[110] Two years earlier, he had published there an important political pamphlet, in Tatar, *Çolpan Yıldızı* (The Morning Star); in it he strongly denounced the harsh treatment of Muslims in the Russian Empire. Back in Russia, he was appointed *qadi* in Orenburg and led a campaign of journalistic propaganda for Pan-Islam among Russia's Muslims.[111] In 1910, he left Russia, and travelled extensively in India and Japan, settling afterwards in Turkey, where he continued his Pan-Islamic work. He spent a good part of World War I in Berlin, where he was active in general Muslim affairs.[112] One of his most important works is a two-volume book, largely based on his travels in the Muslim lands and in Japan (he was the foremost Ottoman expert on Japan at the time). In this work, published afterwards,[113] he discussed

[107] A.R. (= Ahmad Rıza), 'Le Panislamisme', 3–4.
[108] Id., *La Crise de l'Orient*, esp. 32–3.
[109] More examples, in some detail, will be given below, in ch. II.
[110] al-Maghribī, *Jamāl al-Dīn*, 123.
[111] His role is discussed below, in ch. III.
[112] PA, Orientalia Generalia no. 9, vols. 7 and 8, *passim*.
[113] A. İbrahim, *20. asrın başlarında İslâm dünyası ve Japonya'da İslâmiyet*. I am quoting from a later edition. On the author, see K. Kreiser, 'Vom Untergang der *Ertoghrul* bis zur Mission Abdurrashid Efendis', esp. 235–8. H.

Pan-Islam (*İttihad-ı İslam*),[114] as follows. Pan-Islam was a matter of life and death for all Muslims, yet it had been but little explained. While the foreign press, serving the enemies of Islam, wrote extensively about it, accusing Pan-Islam of being fanatic and destructive (and even daring to threaten), the Muslim press was content with publishing short, infrequent articles about it. Foreigners accused Pan-Islam of being a union of terrorists[115] which, unlike Pan-Slavism and Pan-Germanism, it was not; at no time had any Muslim held such a view. Rather, Pan-Islam was a fraternity born 1,300 years before and which was going to last for ever; its expression was co-operation.[116] A catastrophe, past and present, that afflicted Muslims in India, would affect Muslims in Siberia, Tukestan, China, and Afghanistan. The Believers were like the human body, which suffers when one of its organs is unwell, or like a building, the stones of which support one another. Pan-Islam was, truly, one of the main duties of religion. Pan-Islam meant unity.[117] If Muslims were true Believers, then the Foreigners indeed had cause to worry. The 250,000 Muslim soldiers in the Russian army would not only refuse to attack other Muslims, but would also refuse to defend their own lands.[118] Muslims would refuse to fight in inhuman wars; the Egyptian army would not fight against Muslims in the Sudan, nor Indian Muslims against Afghanistan. The two states most pre-occupied with Pan-Islam were Russia and the British Empire, due to their large Muslim populations.

These are strong words, especially interesting in the references to the military implications of Pan-Islam. It should be noted that, according to Abdürreşid İbrahim himself, he was making these comments to an audience of Muslims in Central Asia. Verbal propaganda by such travelling emissaries was supplemented by written works. One of the most prolific in this respect was Shaykh Abū al-Hudā al-Ṣayyādī, one of the

Bräker, *Kommunismus und Weltreligionen Asiens*, i. part 1, 59–60. E. Edib, 'Meşhur İslâm seyyahı Abdürreşid İbrahim Efendi', 3–4, who dates his life as 1850–1944. Mahmud Tahir, 'Abdurrashid Ibragim, 1857–1944', 135–40.

114 A. İbrahim, *20. asrın başlarında İslâm dünyası*, i, part 1, 205–7.

115 The author puns here, ibid. 206: 'İttihad-ı Islam ... yeni bir "ittihad-ı eşkiya" değildir'.

116 Literally, 'mutual assistance'.

117 Or 'union'.

118 In the case of a Muslim war.

highest-placed Arabs at the Court of Abdülhamid II.[119] Abū al-Hudā had serious differences of opinion with al-Afghānī, in Istanbul, on personal and other grounds and, according to Ibrāhīm al-Muwayliḥī (1846–1906), an Egyptian man of letters who was very close to al-Afghānī in his last years, was largely responsible for al-Afghānī's falling out of grace with the Sultan.[120] One wonders whether there was not a difference of views between al-Afghānī and Abū al-Hudā concerning Pan-Islam. While the former was attempting to convince Abdülhamid of the merits of action among Muslims living beyond the borders of the Ottoman Empire, the latter was steadily propagating Pan-Islam within the empire, most particularly among its Arab subjects. Abū al-Hudā wrote (probably with some assistance) some 212 books and pamphlets,[121] most of which were published in the years he spent at the Court, between 1878 and 1908.[122] A large proportion was devoted to defending the legitimacy of Abdülhamid's assumption of the Caliphate and calling on all Muslims to obey him. This was Pan-Islamic propaganda, chiefly in the empire's Arab lands, through which Abū al-Hudā sought to disseminate the concepts that obedience to the authority of the Caliph was binding upon all Muslims and that union, too, was an act of faith. Since Abdülhamid possessed all virtues and was defending both religion and faith, submission to him and union under him were imperative. Although these appeals were generally intended for the empire's Muslims, especially the Arabs (whose own solidarity on ethnic or linguistic grounds Abū al-Hudā was striving to prevent), the call for universal Muslim solidarity behind the Caliph probably served Pan-Islamic propaganda abroad, too, in a more limited measure.

A third instance is that of a Persian prince, Abū al-Ḥasan Mīrzā Qājār (1846/7–1917/18), writing under the pseudonym of Shaykh al-Ra'īs. He left Iran in 1884 and soon afterwards settled in Istanbul, where he was active in promoting, in both

[119] See B. Abu Manneh, 'Sultan Abdulhamid II and Shaikh Abulhuda al-Sayyadi', esp. 140–8.

[120] J.M. Landau, 'An Insider's View of Istanbul', 76 and *passim*.

[121] Many of them can be found in the library of the Middle East Centre, St Antony's College, Oxford.

[122] B. Abu Manneh, 'Sultan Abdulhamid II', 140 and 151, n. 101.

his writing and personal contacts, the proposition of a Shiite–Sunnite *rapprochement,* aiming at a union headed by the Ottoman Sultan. In 1894, he published in Bombay a 96-page tract in Persian, entitled *Ittiḥād-i Islām,*[123] which was an appeal for the spread of Islam, on the one hand, and the unity of Muslims in a Pan-Islamic union, on the other hand. The author maintained that the salvation and success of the *umma* or Islamic community depended on brotherhood and unity.[124] The *modus operandi* he suggested was a treaty between the Ottoman Empire and Iran, aiming at a union comprising all Muslim territories which had once been a part of these two empires; proclamations to this effect should be distributed at Mecca, during the Pilgrimage.[125] Sunnite Ulema and Shiite Mojtahids ought to increase their efforts to enlighten the people about the merits of such a *rapprochement.* Since the Shah of Iran could not combine temporal with spiritual power,[126] while the Ottoman Sultan could and did, the latter should be obeyed; the central authority of the Ottoman Sultanate, both spiritually and temporally, would ensure it the leadership of a Pan-Islamic union.[127] Although he cautiously avoided calling Abdülhamid a caliph, which would have antagonized Shiite readers, and referred to him consistently as a sultan, the author's appeal to Sunnites and Shiites alike to assist Abdülhamid in uniting the Muslim world and to acknowledge him as their leader showed great courage.

The Islamic press occasionally published articles in support of Pan-Islam, rarely opposing it at that time. It should be remembered that the Turkish and Arabic press had initially been sponsored and financed from above, so that non-official newspapers and periodicals were somewhat of a novelty and they, too, tried to get government subsidies (which usually meant government control). While during Abdülaziz's reign—chiefly

[123] Shaykh al-Ra'īs (i.e. Abū al-Ḥasan Mīrzā), *Ittiḥād-i Islām* (Bombay, 1894), 74–5. A copy of this edition is available in Istanbul University's Central Library. A 2nd edition, edited by S. Sajjādī, was published in Tehran in 1984. See also H. Enayat, *Islamic Modern Political Thought,* 122, and Kramer, *Islam Assembled,* 22–4.

[124] Shaykh al-Ra'īs, *Ittiḥād-i Islām,* 1894 edn., 49–52.

[125] Ibid. 58–60.

[126] Ibid. 76 ff.

[127] Ibid. 71–5.

toward its end—infrequent articles supporting Pan-Islam were published,[128] these seem to have become somewhat more plentiful in the later years of Abdülhamid's reign.[129] A characteristic example is that of the weekly *Türk*, published in Cairo, which in 1904 carried a debate about the merits of a Pan-Islamic union, both sides agreeing about its importance for Muslims, but differing in their evaluation of its impact on the European powers.[130] Another instance is the Pan-Islamic weekly *Mizan* (Balance or Scales), edited in Turkish by Mehmed Murad (1853–1912). A Daghestani who had made his way to Istanbul as a young man of 20, due to his feeling that, as a Muslim, he would be better off there, Murad became a teacher in philosophy at the prestigious *Mülkiye*, the author of history books, and the editor of *Mizan*. Published as a weekly in 1886–90, then in Egypt in 1896, and yet again in Istanbul in 1908, the periodical was committed to constitutional reform and Pan-Islam. In the former respect, it criticized Abdülhamid's arbitrary rule, but in the latter its line was not so different from the Sultan's—to see all Muslims saved from foreign rule through the Caliphate and united in a large Muslim Empire. A supreme council, made up of spiritual representatives from Muslim countries in Asia and Africa, would ensure the Caliphate's leadership over the world of Islam. Further, Murad attacked the foreign colonialists in North Africa and proposed to encourage Muslim commerce everywhere, to set up Muslim schools in Black Africa, in the Caliph's name, and to send the Caliph's representatives far and near.[131] In 1911, Murad published a book, entitled *Sweet Hopes—Bitter Truths*,[132] suggesting the establishment of a permanent nine-member committee of Muslims from foreign-ruled lands, to deliberate the

[128] Details in 'Panislamism' (*Asiatic Review*, 1927), 212, mentioning the Arabic *al-Jawā'ib*, and Berkes, *Development*, 268.

[129] Becker, 'Panislamismus', 181, mentioning *al-Ma'lūmāt* of Istanbul and *al-Mu'ayyad* of Cairo, in 1900. See also ibid. 187.

[130] Turgut, 'İttihad-ı İslam meselesi'; Orhan, 'İttihad-ı İslam'.

[131] A.B. Kuran, *İnkılâp tarihimiz ve Jön Türkler*, 40–62. H.Z. Ülken, *Çağdaş Türkiye düşünce tarihi*, 120–4. E.E. Ramsaur, *The Young Turks*, 27–8, 38; see also 48–50. O. Koloğlu, 210–12. Mardin, *Jön Türklerin ...*, 46–92. Berkes, *Development*, 307–9. F.A. Tansel, 'Mizancı Mehmed Murad Bey', esp. 75–7; she maintains that he died in 1914, see ibid. 88.

[132] Mehmed Murad, *Tatlı emeller, acı hakikatler* (1911).

sharīʿa (Turkish, *şeriat*), or Islamic law, and even select the Caliph.[133]

Lastly, something may be said about Pan-Islamic congresses. They are discussed in the section on ideology, rather than on organization, for the simple reason that such congresses never met during Abdülhamid's reign, very probably due to his opposition, as he wished to keep decision-making in his own hands. However, the idea was put forward and may well have had a powerful impact in subsequent years.

The most renowned case was that of the fictional plan in Arabic of ʿAbd al-Raḥmān al-Kawākibī (1854–1902), a Syrian Arab journalist and public official. In 1900, he published in Cairo his *Umm al-Qurā* (Mother of the Villages or Head Village), an imagined congress of Islamic revival, attended by Muslim participants from various countries—along with detailed organizational and procedural guide-lines. Its serialization in the widely circulated, influential Muslim monthly *al-Manār*, in Cairo, popularized the idea of such a congress, Pan-Islamic in its composition more than in its objectives, perhaps.[134]

Another project, much less known than al-Kawākibī's, was proposed, in Abdülhamid's last year, by a Persian named Mīrzā ʿAlī Agā Shīrāzī, who campaigned early in 1909 for an international Muslim parliament.[135] Having founded and published *Moẓafferī*, the first Persian newspaper in the south of Iran, in Bandar Bushir, in about 1899, he used it as a platform for demanding constitutional and other reforms. In 1908 and early 1909, he travelled to Bombay, to Egypt, and to the Arabian Peninsula, from where he intended to go to Istanbul. In Mecca, he had a special issue of *Moẓafferī* printed, comprising unequivocal Pan-Islamic ideas. These are to be found in an appeal to all Muslims, everywhere, without distinction of rite or sect, to unite. He emphatically declared that this was feasible and as his proof, somewhat curiously, cited the fact that

133 See Kramer, *Islam Assembled*, 49–50, 205.

134 The most recent discussion of al-Kawākibī's book and plans is by Kramer, ibid., ch. 3. Kramer also provides an excellent bibliography—to which one may add, however, B.B. Ghali's 'Le Grand Dessein de al-Kawakibi', 93–102; A. al-Jundī's *al-Yaqẓa al-Islāmiyya*, 141–6; and A. Amīn's *Zuʿamāʾ al-islāḥ fī al-ʿaṣr al-ḥadīth*, 267–301.

135 Details in L. Bouvat, 'Un projet de parlement musulman international', 321–2.

Christian sects, despite mutual hatred and endless wars, did unite against Islam. Shīrāzī considered the annual Muslim pilgrimage as most fitting for such a reconciliation and appealed to the Ulema and enlightened persons to set up a parliament there (which he called an *Enjumān*, or society, of Muslim alliance). This body would convene annually at Mecca and consist of delegates officially empowered by their governments and of Ulema. Its terms of reference were to be strictly religious and moral, not political, in order not to arouse European apprehensions of Pan-Islam and thereby justify further foreign intervention in the Islamic states.

4. *Organization of Pan-Islamic Activity*

If the initial stages of Pan-Islamic ideology can be examined on the basis of available writings, the organization and implementation of Pan-Islamic activities are much less known. In 1906, Valentine Chirol, a British journalist interested in Ottoman affairs, asserted in a lecture to the Central Asian Society in London, 'The channels through which Pan-Islamism works are so tortuous, its ramifications so subtle . . . '.[136] These comments are very apt, although some patterns are evident nevertheless. Abdülhamid transferred the decision-making centre of state policies, in both major and numerous minor matters, from the government departments to the palace.[137] This was certainly so in the case of Pan-Islam. Since most of this clandestine activity started from the Sultan's palace, or led directly to it, available information is scanty, sketchy, and often contradictory. It appears that little was committed to writing and even this has yet to surface.

Abdülhamid II's own memoirs (assuming that they are authentic), published in Turkish[138] and later in an Arabic

[136] V. Chirol, 'Pan-Islamism', 6.

[137] Cf. A.M. Broadley, *How We Defended Arábi*, 455. Broadley, 'Urābī's lawyer in Egypt, was usually well-informed about Ottoman affairs at the time. See also in this matter 'Panislamism and the Caliphate', *The Contemporary Review*, 63. E.D. Akarlı, 'Abdulhamid's Islamic Policy in the Arab Provinces', 45–6. Koloğlu, *Abdülhamid gerçeği*, 199–200.

[138] First as Sultan Abdülhamit, *Siyasî hatıratım*, then as *Abdülhamid'in hatıra defteri*.

translation,[139] provide little information on this matter. Dictated by the deposed Sultan, mainly during March and April 1917, these were more of an *apologia pro vita sua* and a means of settling accounts with his rivals and detractors than bona fide memoirs. There were no more than a few casual, brief remarks about Pan-Islam, mentioning, *en passant*, the importance of a Muslim *rapprochement* as an antidote to foreign designs on his empire and to British intrigues aiming at abolishing the Ottoman Caliphate—but no details about his own Pan-Islamic projects, organization and moves. Another edition mentioned that he had encouraged Islamic propaganda everywhere, at least partly responding to Christian propaganda;[140] this, again, offered no information on ways and means. The Ottoman state archives, which should have been the main depository of this information, have been delved into by several historians, such as İ.S. Sırma, with meagre results as far as Pan-Islam is concerned. It seems that these archives contain more information on what Abdülhamid was preventing others from doing than about what he was achieving himself. None the less, from other archives, contemporary memoirs, press reports, and various bits and pieces of information, a fascinating, if incomplete, picture emerges. We shall discuss the Hamidian organization under several headings: Broader concepts, official moves, Pan-Islamic societies, and non-institutionalized activities.

5. Broader Concepts

The basic issue which Abdülhamid's Pan-Islamic policies had to face up to was his assertion that he was the Caliph of all Muslims, able to unite them in obedience to him. Support for his claim seemed more likely in his own times than previously, as a result of foreign aggression against Muslim lands. After the French conquest of the Maghrib started in 1830, parts of the Muslim populations in Algeria, Tunisia, Tripolitania, and Egypt reversed an earlier trend of weakening their ties with

[139] Translated by Muḥammad Ḥarb ʻAbd al-Ḥamīd as *Mudhakkirāt al-Sulṭān ʻAbd al-Ḥamīd*.

[140] Sultan Abdülhamit, *Siyasî hatıratım*, 165 ff.

the Ottoman Empire for the sake of increased autonomy; some started to see it, indeed, as their best and only protector from European aggression. Muslim delegations from Central Asia, threatened by Russian advance, even groups from South-east Asia, also came to the Court of the Sultan–Caliph to request assistance.[141]

Abdülhamid's claim to the Caliphate was hotly contested as early as 1877, soon after his accession,[142] and as warmly defended,[143] and not only abroad; the argument went on within the Ottoman Empire, too.[144] He had to assert himself, therefore, not only as the Caliph of Ottoman Muslims (who were more amenable to persuasion), but also of those who were not his subjects (which was more difficult), and to convince the European powers that his own spiritual leadership of Muslims everywhere was a significant contribution to his temporal power (thus offsetting, at least in part, the empire's military and economic weakness).[145]

The first of these three objectives was largely taken care of by a policy of Islamism, which increasingly favoured the central government against the periphery, and the Ottoman Empire's Muslims at the expense of others—chiefly in public office, education, and economic opportunities. Its main signs, over the years, were the following: an increasing number of devout Muslims were appointed to the upper levels of the state bureaucracy and to the Sultan's Court; the Sultan himself appointed *qadis*, teachers, and other Ulema in both the empire and territories which it had lost; pensions and salaries were increased, first and foremost for the Ulema; religious institutions were repaired and new ones built; religious schools

[141] A.J. Toynbee, 'The Ineffectiveness of Panislamism', 692–3.

[142] See e.g. G.P. Badger, 'The Precedents and Usages Regulating the Muslim Khalifate', 280 ff.

[143] J.W. Redhouse, *A Vindication of the Ottoman Sultan's Title of 'Caliph'*, 1–19.

[144] e.g. N. Azoury, *Le Réveil de la nation arabe dans l'Asie turque*, 246–8. One instance is an appeal favouring the Grand Sharīf of Mecca as the sole legitimate caliph, published in *al-Jihād*, Nov. 1901, then in a French translation in a supplement to *El Centinela del Estrecho de Gibraltar* (Gibraltar), 20 Nov. 1901.

[145] See also Tawfīq ʿAlī Barrū, *al-ʿArab wa-ʾl-Turk fī al-ʿahd al-dustūrī al-ʿUthmānī 1908–1914*, 34–7.

were inaugurated and lessons in Islam were introduced into other schools too; contributions to religious fraternities and charity were increased; basic books on Islam were printed and distributed free of charge or at low cost; and free schools started for Muslim families coming to Istanbul.[146] Special attention in all this was given to the Arabs,[147] the most important Muslim element in the empire, besides the Turks, and even a Turkish–Arab empire was briefly considered.[148] Abdülhamid suspected Egypt's Khedive, ʿAbbās Ḥilmī,[149] and then the Sharīf of Mecca, of intriguing for the establishment of an Arab Caliphate, to replace the Turkish one, and naturally strived to prevent this. In addition to intensive propaganda, financial investments were made (for example, in Syria and Iraq), favouring Muslims economically. In a parallel manner, efforts were made to appeal to the sentiments of Islamic solidarity of Albanians and Bosnians and to encourage them to co-operate (chiefly the former) against the Montenegrins, for example, in 1878.[150] In some instances, this may well have been the embryo of a plan to bring closer to one another neighbouring Muslim groups, as a step towards more general measures. Moreover, by emphasizing the Christian threat to Islam and combining it with the dangers for the empire, this approach succeeded in attracting the support not only of conservative elements among the empire's Muslims, but also of certain circles of liberal Turks, concerned about the future of their religion and state.[151]

[146] *Vossische Zeitung* (Berlin), 8 Sept. 1898. İ. Kara, *Türkiye'de İslamcılık düşüncesi*, i, pp. xxviii–xxix. Shaw and Shaw, *History of the Ottoman Empire*, 259–60. F.B. al-Ṣawwāf, *al-ʿAlāqāt bayn al-dawla al-ʿUthmāniyya wa-iqlīm al-Ḥijāz*, 126–8.

[147] Cf. PA, Orientalia Generalia, no. 1, vol. 5, von Oppeheim's no. 133, to von Bülow, dated in Oberkassel, 26 Sept. 1901.

[148] E.Z. Karal, *Osmanlı tarihi*, viii. 544–5. Anwar al-Jundī, *al-Yaqẓa al-Islāmiyya*, 94–6, exaggerates in asserting that the powers made strenuous efforts, at that time, to disrupt Turco-Arab solidarity.

[149] ʿAbbās Ḥilmī himself asserted in his memoirs, without adducing any proof, however, that the British had planted these suspicions. See *al-Miṣrī* (Cairo daily), 7 May 1951, 12.

[150] B. Samardžiev, 'Traits dominants de la politique d'Abdülhamid II relative au problème des nationalités (1876–1885)', partly based on Albanian sources.

[151] X, 'Les Courants politiques dans la Turquie', 170–1. Keddie, 'Pan-Islamic Appeal', 50–1.

The second objective required the major effort of the Pan-Islamic thrust, particularly since it was to be an Ottoman and Muslim response to the moves of the powers on behalf of the Christians in the Ottoman Empire; this meant an intervention of the Caliph on behalf of the Muslims in *their* empires.[152] This level of Pan-Islam was often interlocked with activities in the Ottoman Empire's recently lost territories (the Pan-Islamic policy being intended to compensate, in part, for the loss of these lands), as well as in more remote areas. Our examination of these activities will not always be able to distinguish between these two levels.

The third objective, it was hoped, would take care of itself: when the powers were persuaded that the Sultan–Caliph's appeal to their own Muslim subjects was effective they would be deterred from attacking the Ottoman Empire.[153] Pan-Islamic propaganda, as directed from Istanbul, threatened Great Britain, France, and (to a lesser extent) Russia and the Netherlands implicitly, while explicitly denying any such intent.[154]

6. *Official Moves*

These were not numerous and involved such activities as the opening of Ottoman consulates in China, India, North Africa, Japan, and elsewhere[155]—to which Islamic and Pan-Islamic activity was delegated. As Abdülhamid wrote in his memoirs, some of these consuls, or *şehbender*s, were considered overzealous in their propaganda and were expelled, as in the case of Sadık Bey in Java.[156] Either because Abdülhamid feared countermoves by the European powers or, more simply, because his favoured *modus operandi* was in clandestine (or

[152] See 'Exploiting the Crescent', 7. E.Z. Karal, *Osmanlı tarihi*, viii. 540–2. Evidently, the Ottoman Empire was not powerful enough to carry this out. Cf. D. Kushner, *The Rise of Turkish Nationalism 1876–1908*, 47.

[153] For a recent article on this, see C. Farah, 'The Islamic Caliphate and the Great Powers: 1904–1919', esp. 44 ff.

[154] In an interesting article, recently published, K.H. Karpat focuses solely on the first objective, Islamism. See his 'Pan-İslamizm ve ikinci Abdülhamid', esp. 28 ff.

[155] Ahmed Emin, *Turkey in the World War*, 180.

[156] Sultan Abdülhamit, *Siyasî hatıratım*, 162–3.

semi-clandestine) channels, he generally opted for the latter.

In 1874, Sultan Abdülaziz had assisted the Muslims headed by Yakub Beg in Kashgar by sending him 200 rifles, three cannons, and several Turkish officers to train his troops for a confrontation with the Russians.[157] Abdülhamid followed this up with a Turkish mission to Afghanistan in June 1877, soon after his accession to the Sultanate,[158] that is, just at the time when we have our first indications of Pan-Islamic ideas beginning to take formal shape.[159] Although *raisons d'état* concerned with the Ottoman Empire's foreign relations prompted this mission, a Pan-Islamic element entered into it as well. The war with Russia had started two months previously and Abdülhamid was responding to the proclamations of the Tsar and Grand Duke Nikolay about the Crusading character of this war. Moreover, he seems to have had hopes, reflected in the Turkish press, of Muslim volunteers ariving from outside the Ottoman Empire to fight for it.[160] The new British Ambassador to Istanbul, A.A. Layard (who had arrived there in April, just before war broke out), reported that Abdülhamid was considering the establishment of an international 'Muslim League' to assist the Ottoman Empire militarily and was attempting to unite all Asiatic Muslim states against Russia.[161] An official mission from Istanbul was sent to Afghanistan via Egypt and India. Although it failed in its objectives, it did at least provide a first instance of political and diplomatic activity, initiated in Istanbul, to further Pan-Islamic solidarity. The fact that a mullah, or Muslim cleric, rather than a diplomat or a military officer, was selected to go supports the view that the basic approach of Abdülhamid in this matter was Pan-Islamic.

Thwarted in Central Asia and defeated in war, Abdülhamid

[157] L.E. Frechtling, 'Anglo-Russian Rivalry in Eastern Turkistan, 1863–1881', 479. M. Saray, '1874 'de Kaşgar'a gönderilen Türk subayları', 244–51. Id., *Rus işgalı devrinde Osmanlı devleti ile Türkistan hanlıkları arasındaki siyasi münasebetler (1775–1875)*, 98–109. Cf. ibid. 109–12, for other contacts of the Ottoman Government with the Afghan and the Turkmens. See also Y. Halaçoğlu, 'Binbaşı İsmail Hakkı Bey'in Kaşgar'a dâir eseri', 521–4. G. Çetinsaya, 'II. Abdülhamid döneminin', 57 ff.

[158] D.E. Lee, 'A Turkish Mission to Afghanistan, 1877', 335–6.

[159] See above, in my introduction.

[160] Lee, 'A Turkish Mission', 336, and the sources in the footnotes.

[161] Quoted ibid. 339–42.

seems to have turned his attention to Africa's Muslims. While one has no corroboration of P. Frémont's story (the author claims that he was 'formerly in Ottoman service')[162] that, early in his reign, the Sultan was preparing to secure the loyalty of Chad's Muslims, but gave up the project, due to lack of funds,[163] there is more substance about Abdülhamid's plans in North Africa, somewhat later. Some attempts may have been made to mobilize Islamic sentiment in Algeria, along with fomenting trouble against the French.[164] If so, these must have been secret moves. However, there was little that was clandestine in the Pan-Islamic response from Istanbul to France's occupation of Tunisia in 1881. Wilfrid Scawen Blunt, the champion of Arab nationalism, argued at the time[165] that the French had played into Abdülhamid's hands in Tunisia, by enabling him to gain the sympathy of North African Muslims, many of whom had been opposing the Ottoman sultans for centuries.[166]

In the Tunisian crisis, the Sultan did not merely attempt to arouse the European powers against France. He briefly considered sending the Ottoman navy against it, but dispatched instead an Ottoman general, Hüseyin, to Tripolitania in order to raise and train an army there against the French in neighbouring Tunisia; symptomatically, Hüseyin was accompanied by sheikhs whose duty was to arouse Islamic sentiment.[167] However, the Sultan's main reaction was an orchestrated press campaign, with easily detectable Pan-Islamic overtones, chiefly in the Istanbul Turkish *Vakıt* (Time) and Arabic *al-Jawā'ib* (Circulating News), both dedicated to the cause of Pan-Islam. Most particularly, *al-Jawā'ib*,[168] since it appeared in Arabic, was transported to Tripoli (Libya) during the Tunisian crisis in large quantities, for circulation among the tribes.[169] In fairly

162 'Un ancien fonctionnaire ottoman'.

163 P. Frémont, *Abd Ul-Hamid et son règne*, 70–2.

164 Ibid. 72–9.

165 W.S. Blunt, *The Future of Islam*, based on his 1881–2 articles in *The Fortnightly Review* (London).

166 Id., *The Future of Islam*, 87. See also 'Panislamism and the Caliphate' (1882), 8.

167 Details in Charmes, 'La Situation de la Turquie', 721–2.

168 See below, in this chapter.

169 See the report of *The Times* correspondent in Istanbul, published on 29 Dec. 1881, with examples from *al-Jawā'ib*.

violent terms, the Tunisians were encouraged to continue their struggle,[170] with assurances of future help, for the Sultan was willing to shed his last drop of blood and to spend the last piastre of his Treasury to defend the sacred religion of Islam.[171] One of the other frequently repeated themes was that, when a Muslim people fell under the domination of *gâvur*s (infidels), the sole remedy available was to establish a league of all Muslims in Africa. Other newspapers used this opportunity to appeal to all Muslim rulers and peoples to enter into relations with the Caliphate of Islam, to entrust to it the direction of their politics and to obey its orders. The oft-repeated argument of the press then was that the Muslim world, although divided into many states and innumerable groups, each with its own particular objectives, was united by a common leadership and working, in diverse ways, towards the same ends.[172]

In 1881, Abdülhamid II had, therefore, his first opportunity of semi-officially proclaiming Muslim unity within the Ottoman Empire and outside it. Consequently, it was no coincidence that Pan-Islam started then to be evaluated in world politics and in the international press. While he failed to dislodge France from Tunisia, Abdülhamid made an important step forward in the foreign relations of Pan-Islam. When, later in that year, the ʿUrābī uprising occurred in Egypt, public opinion in Europe immediately suspected the Sultan of instigating it, as a part of his Pan-Islamic campaign. The London *Times* wrote at the time that most people considered the Sultan responsible and were apprehensive about his mysterious influence over Muslims.[173] In this case, however, nothing could have been further from the actual facts, because Abdülhamid was unhappy about the start of the uprising and even more so about its results—the 1882 British Occupation of Egypt. True enough, Aḥmad ʿUrābī and other leaders of the movement in Egypt, most particularly Sāmī al-Bārūdī, wrote

[170] See also *The Times*, 19 Jan. 1882, 5.

[171] *al-Jawāʾib*, 27 Dec. 1881, 1, editorial by Aḥmad Fāris [al-Shidyāq]. It is interesting that the 'sacred religion of Islam', rather than the right of the Ottoman provinces to hold on to its provinces, is mentioned.

[172] Charmes 'La Situation', 723–4.

[173] *The Times*, 22 Dec. 1881. The French had very much the same suspicions.

to the Sultan, reporting to him and asking for his protection.[174] The Sultan, however, would have none of this and in order to persuade the Egyptian Muslims to support the Khedive (the ruler officially appointed by him), rather than ʿUrābī, had his representative in Egypt, Derviş Pasha, publish in Cairo a message enjoining on all Muslims obedience to the Khedive, by the Sultan's order, emphasizing that Islam, being religion and nation in one, had no place for tribalism and nationalism.[175] So here one has Abdülhamid not necessarily using Pan-Islam against the British (that is, *before* they occupied Egypt), but, rather, openly attempting to check nascent nationalism and substitute Islam—and Pan-Islam—for it, at home and abroad. This was probably the utmost this sultan could do in his own foreign relations, based on his attempt to maintain the status quo by preserving a policy of neutrality and peace-keeping (since he well knew he could not vanquish in war any of the Great Powers).[176]

The dispatch of the Ottoman warship *Ertuğrul* for a visit to Japan was symptomatic of the interest aroused in Muslim communities by Japanese progress.[177] The ship went via Suez, Aden, Bombay, Ceylon, Singapore, Saigon, and Hong Kong, in the years 1888–9, stirring much interest and excitement in local populations, especially among the Muslims, particularly as the ship's officers and sailors went ashore to pray in the mosques and the name of the Sultan–Caliph was proclaimed in the Friday prayers.[178] In 1901, a mission, led by General Hasan Enver Celâlettin Pasha,[179] was sent to the Muslims

[174] Details, with quotes from the relevant documents, in the Ottoman archives, in S. Deringil, 'The "Residual Imperial Mentality" and the ʿUrabi Paşa Uprising in Egypt', esp. 33 ff.

[175] Text in the Ottoman archives, quoted ibid. 35. The text about Islam reads as follows: *din ve millet bir olduğu gibi*.

[176] S. Deringil, 'Aspects of Continuity in Turkish Foreign Policy', esp. 39–46.

[177] K. Kreiser, 'Der japanische Sieg über Russland (1905) and sein Echo unter den Muslimen', 209 ff.

[178] V. Bérard, *Le Sultan, l'Islam et les puissances*, 36–7. Kreiser, 'Vom Untergang der *Ertoghrul*', 239–48. Koloğlu, *Abdülhamid gerçeği*, 204–5.

[179] See his biography in *Tarih ve Toplum* (Istanbul), 1 (Jan. 1984), 6. Acc. to A.A. Rey, *La Question d'Orient devant l'Europe*, 2, 16, and *Le Réveil de l'Islam est-il possible?*, 7–9, who discussed this mission, this Enver Pasha belonged to the Polish family Boginskiy (which had converted). Several

of China,[180] visiting on the way numerous Muslim communities and greeting them in the Sultan's name. The mission seems to have been most successful in delivering its Islamic and Pan-Islamic messages, both *en route* and in China itself.[181] At about the same time, the Ottoman Consul in Java, Sadık Bey, was trying to defend the rights of Muslims there, to the displeasure of the Dutch authorities,[182] while the Ottoman Consul-General in Batavia, Kâmil Bey, was sent home because of his Pan-Islamic propaganda.[183] However, the Sultan's interest in the Muslims of East and South-east Asia persisted, as apparently did his hope of attaching them to his Pan-Islamic projects. In 1908, indeed, an envoy from Abdülhamid, Ali Rıza, arrived in China to request official recognition for Ottoman consuls in China to look after Muslim interests, a demand which in this case also was rejected.[184]

Meanwhile, official moves were being made to establish closer relations with Iran and Afghanistan. The former was Shiite and the latter rather wary of Ottoman intentions; both their rulers were very reluctant to recognize the Ottoman Sultan as their Caliph. However, both states, although independent, were hard pressed by the Russian–British rivalry in the area. The Emir of Afghanistan, ʿAbdurrahman, sent envoys to Abdülhamid, but little came out of it.[185] Then, just before the assassination of Nāṣir al-Dīn Shah in 1896, Abdülhamid sent an official mission to him, to pave the way for a Muslim *rapprochement*, which was meant to lead to union at some future date. The mission was considered so important at the Ottoman palace that one of Turkey's ablest diplomats and a known scholar in Islamic theology, Münif Pasha, Minister for Education, was nominated to head it. According to an

contemporary historians have wrongly attributed the leadership of the mission to the more famous Enver Pasha (later, one of the triumvirate who ruled the empire after the Young Turk revolution). However, at the time, the latter was just graduating from the military academy and was going to spend the next few years fighting guerrillas in Macedonia.

180 E. Kuran, 'Panislâmizm'in doğuşu . . .', 398.

181 İ.S. Sırma, 'Sultan II. Abdülhamid ve Çin müslümanları', 199–205.

182 E. Kuran, 'Panislâmizm'in doğuşu . . .', 399.

183 *Vossische Zeitung*, 12 Feb. 1901.

184 W.W. Cash, *The Moslem World in Revolution*, 26–7.

185 A. Vambéry, *Western Culture in Eastern Lands*, 353.

earlier report, written in 1878 by the Russian Ambassador in Istanbul, this Münif Pasha, an Arab born in Syria, had been the initiator of the policy to strengthen the international authority of the Sultan–Caliph via Pan-Islam.[186] The assassination of the Shah, just before the mission's arrival, was a serious setback, but the Sultan appointed Münif Pasha as Ambassador to Tehran and various attempts at reaching co-operation continued.[187] Shortly afterwards, in 1900, there were press rumours that the new Shah would be visiting Istanbul in the cause of Islamic unity.[188] Moẓaffar al-Dīn Shah came in early October,[189] but the visit did not advance Pan-Islam's interests. So the Sultan had to be content with inviting young Persians to study in Istanbul, particularly in its military schools.[190]

Lastly, a major effort was invested by Abdülhamid in a very special relationship with the new German state, well noted at the time by both Ottoman[191] and foreign observers. Although the Ottoman Empire's *realpolitik* determined the various steps made, Pan-Islam had a share in this as well. Germany's economic interests in the Ottoman Empire are too well-known to require a special mention.[192] As for the Ottoman leadership, it had very good reasons for this *rapprochement*. There had never been a war with Germany, which nowhere had even a common frontier with the empire. German investments in building new railroads were much appreciated by an Ottoman Government which was less than sure of the loyalty of its non-Turkish Muslims, chiefly the Arabs, and desired speedy access to their areas for its troops. However, it is likely that the fact that Germany had very few Muslims in its lands and no ambitions on territories inhabited by Muslims—contrary to Great Britain, France, Russia, even Austro-Hungary—

[186] Russian Ambassador A.B. Lobanov (1824–96) to Minister for Foreign Affairs N.K. Girs (1820–95), acc. to I.L. Fadyeyeva, *Ofitsial'niye doktrini i idyeologii v politikye Osmanskoy Impyerii*, 135–6.

[187] R. Ahmad, 'A Moslem's View of the Panislamic Revival', 523–4.

[188] Cf. Vollers, 'Ueber Panislamismus', 30–1.

[189] Becker, 'Panislamismus', 189. *The Times*, 1, 2, 3, and 9 Oct. 1900.

[190] XX, 'La Solidarité islamique et l'Angleterre', part 3 (May 1922), 68.

[191] Such as Ahmed Refik [Altınay], a well-known man of letters, in his ('A.R.') *Abdülhamid-i sani ve devr-i saltanatı*, iii. 1043–61.

[192] See, among others, *L'Europe et l'Empire ottoman*, 387–9; İ. Ortaylı, *İkinci Abdülhamit dönеminde Osmanlı İmparatorluğunda Alman nüfuzu*.

weighed heavily in the Sultan's policy decisions, too.[193] Germany started training a small number of Turkish officers, while its own officers assisted, and partly conducted, the Turkish campaign in Thessalia;[194] the fact that this was practically the only Ottoman military victory during Abdülhamid's reign did not pass unnoticed. But the most significant act, as far as Pan-Islam was concerned, occurred during Kaiser Wilhelm II's visits to the East.

The Kaiser visited Istanbul three times—a record for any European monarch, past or present[195]—in November 1889, October 1898, and October 1917. The third visit, being after Abdülhamid's dismissal, need not concern us.[196] The first was to Istanbul only, upon Abdülhamid's invitation which spoke explicitly of 'a new agreeable opportunity to strengthen and consolidate the harmony and amity between our two empires'.[197] This visit was beneficial to both parties. According to a letter from Istanbul,[198] the Kaiser had been impressed by the quality of the Ottoman army (partly trained by Germans) and, in particular, by its officers; the Sultan's influence and prestige could not but grow in consequence of this visit. However, the second visit was more important, since bilateral relations had been improving meanwhile, the Kaiser supporting the Ottomans politically (as in the Turco-Greek conflict over Crete, in 1897–8) and militarily (with training, weapons, and munitions)[199] and getting in return economic concessions (for German business). It was also a lengthier visit, covering Istan-

[193] Cf. the evidence of a high palace official, Tahsin Paşa, *Abdülhamit Yıldız hatıraları*, 52–4, 230 ff. See also E.Z. Karal, *Osmanlı tarihi*, viii. 548. E.M. Earle, *Turkey, the Great Powers and the Baghdad Railway*, 64–5.

[194] R. Herly, 'L'Influence allemande dans le Panislamisme contemporain', ch. 1, p. 594.

[195] PA, Preußen 1, nos. 1–4v, 'Geheim', comprises 13 vols., largely devoted to these visits.

[196] See ibid., vol. 13.

[197] Ibid. vol. 3, cable dated in Yıldız Palace, Istanbul, 4 Oct. 1889; Wilhelm II's reply, cable dated in Berlin, 5 Oct. 1889.

[198] The letter, dated 8 Nov. 1889, was published as 'Le Séjour de Guillaume II à Constantinople', in *La Turquie* (Paris), 13 Nov. 1889.

[199] Rafiüddin Ahmad, an Indian writing on 'The Kaiser, the Caliph, and the Khedive', in *Pall Mall Gazette* (London), 31 Aug. 1898, accurately predicted, 16 years before the event, that, 'when Armaggedon would begin Germany could count upon the assistance of the brave Turkish army'.

bul, Haifa, Jaffa, Jerusalem, Damascus, and Beirut[200] (the Khedive ʿAbbās Ḥilmī invited him to Egypt, too,[201] but the visit was later cancelled by the Kaiser. In this instance, there were more than the quid pro quo calculations mentioned above. On one hand, the needs of the Ottoman Empire had become more pressing. On the other hand, German foreign and economic policies, dictated by the Kaiser, had become more global, especially after Bismarck's dismissal in 1890. Hence, in his second visit, Wilhelm II wished to become not only the Protector of all Christians in the Holy Land but, simultaneously, the Friend of Muslims everywhere as well—no mean achievement.

Rumours had been spread that the Kaiser was going to convert to Islam; these had no basis whatsoever, but were never denied. The Kaiser's garment, when he entered Damascus in 1898, was a curious combination of a medieval Lohengrin uniform (in front) and an Arab–Muslim coat (from behind)—which, again, only reinforced the rumours about his attitude to Islam. Excitement reached its peak when, after having laid flowers on Saladin's grave, the Kaiser attended a huge banquet at which Shaykh ʿAbd Allāh, speaking for the Ulema of Damascus, asserted that the Kaiser had earned the gratitude of all the world's Muslims. The Kaiser thereupon proclaimed, 'The Sultan and the three hundred million Muslims who revere him as their leader should know that the German Emperor is their friend forever'![202] Here was international legitimation of Pan-Islam and of Abdülhamid's efforts to promote it politically. Some observers thought then that, thanks to German support,[203] Abdülhamid's Pan-Islamic policies had become more daring.[204] German–Ottoman co-operation also bore other fruits, some of them curious enough. In 1908, several months after the Chinese had refused the request to have

[200] For the visit, see PA, Preußen 1, nos. 1–4v, 'Geheim', vols. 6–9.

[201] Cf. ibid., vol. 4, ʿAbbās Ḥilmī's official invitation, dated 24 June 1898.

[202] Ibid., vol. 8*a*. R. Herly, 'L'Influence allemande', ch. 1, p. 594. Vambéry, 'Pan-Islamism', 553. F. Cataluccio, *Storia del nazionalismo arabo*, 20.

[203] e.g. in a cable, dated in Berlin, 28 Feb. 1901, Wilhelm complimented Abdülhamid on his firmness in maintaining 'the dignity of his throne and the prestige of Islam'. Cf. PA, ibid., vol. 11.

[204] See editorial in *The Morning Post* (London), 6 Aug. 1906.

Ottoman consuls look after the interests of all Muslims in China, the German Minister to Peking informed the Chinese Government that Germany had been requested by the Sultan to undertake the protection of Ottoman—and, by implication, Muslim—subjects in China![205] The protection of Ottomans in China had formerly been in French hands[206] and the change was another step in the Ottoman–German *rapprochement*.

7. *Associations and Fraternities*

While organization seems to have been unmethodical (not to say haphazard),[207] some forms of it did exist. Pan-Islamic associations were few and small in the Hamidian era—and information is proportionately scanty. Most were short-lived, such as a society set up in Istanbul near the end of Abdülaziz's reign, in 1872, to promote Ottoman leadership in defending Islam against the Western menace.[208] Another, allegedly set up by al-Afghānī and ʿAbduh, in Paris, in 1884, to strive for the reform and unity of Islam, had several branches, for the short span of its life.[209] It is very likely that there existed other similar associations (probably made up, in part, of Muslims who had immigrated into the Ottoman Empire),[210] of which no definite information has reached us.

A longer-lasting association was the League of Prizrend, sometimes called the Albanian League.[211] It was set up by a meeting of local notables on 10 June 1878, primarily to strive for the autonomy of Albania. However, a more immediate task before its founders was to prevent the partition

[205] PA, Orientalia Generalia, 9, vol. 4, German Ambassador to Great Britain, Graf Wolff-Metternich's no. 689, to von Bülow, dated in London, 18 July 1908. Cf. W.W. Cash, *The Moslem World*, 27, where the date, 1909, should be corrected to 1908.

[206] PA, Orientalia Generalia, 9, vol. 4, German Ambassador to China, Graf von Rex's no. A212, to von Bülow, dated in Peking, 31 Oct. 1908.

[207] Leading such close observers as C.S. Hurgronje to doubt whether any organization existed. See his *Holy War 'Made in Germany'*, 25, 29.

[208] Berkes, *Development*, 268.

[209] Riḍā, *Taʾrīkh* . . ., i. 284. Hourani, *Arabic Thought*, 109. Hassan Saab, *The Arab Federalists*, 189.

[210] R.H. Davison, *Reform*, 274–5.

[211] For which see S. Ballvora, 'La "Ligue de Prizrend" et la question de l'autonomie de l'Albanie', 49–55.

of Albania at the Congress of Berlin, so that the two basic documents establishing the League, soon after its foundation, emphasized loyalty to the Ottoman Empire and the Sultan, referring also to the *sharīʿa*.[212] There was, in 1878, frequent communication between the League of Prizrend and the government in Istanbul.[213] This stress on Muslim and Ottoman identity was only natural for one of the empire's Muslim groups, finding assurance at a time of grave crisis in the common religion.[214] Abdülhamid was well aware of this: when, three years later, there were signs of an anti-Ottoman movement among the Albanians, Derviş Pasha was ordered to appeal to them for loyalty to the Sultan–Caliph, in the name of the *sharīʿa* and Pan-Islamic interest.[215] Notably, a generation later, in October 1911, there occurred in Albania, again, demonstrations of support for the Sultan–Caliph, expressing readiness to join the fighting against the infidels who had invaded Libya.[216]

Pan-Islamic centres were reportedly set up as far afield as Tunisia,[217] Shanghai, and Java,[218] but no detailed information about them is available. Vambéry tells us about a

> Pan-Islamic Society established in London, in 1886, under the protection of the Sultan of Turkey, the Khedive [of Egypt], the Amir of Afghanistan, the Sultan of Morocco, and others, with the professed object of bringing about a fraternization of all Moslems all over the world, but which so far has only a very limited circle of activity. The society has not the necessary means, nor is London the place from which a sufficient influence can be exercised upon the Islamic world.[219]

Contrary to Vambéry's view, however, the Sultan may have valued London for its central position, for, in 1903, yet another

[212] Ibid. 50.

[213] B. Kodaman, *Sultan II. Abdulhamid devri Doğu Anadolu politikası*, 84.

[214] As it is natural for present accounts in Communist Albania to gloss over the Pan-Islamic sentiment of the league.

[215] I.L. Fadyeyeva, *Ofitsia'lniye doktrini*, 137, based on the Russian archives.

[216] Türkei no. 198, vol. 7, German Consul Gerhard von Mutius to Reichskanzler Bethmann Hollweg, dated in Salonica, 8 Oct. 1911.

[217] Acc. to a report in the Tunisian archives, quoted by Béchir Tlili, 'Au seuil du nationalisme tunisien', 220–1. Pan-Islamic activity in Tunisia was carried out covertly.

[218] G. Young, 'Pan-Islamism', 542.

[219] Vambéry, *Western Culture*, 351–2.

association, aptly named 'Pan-Islam', was set up. Its president was Abdullah al-Mamun Suhrawardy and its secretary M.H. Kidwai (more about him and his writings below),[220]—both of them barristers from India. Suhrawardy was the elder of the two, a graduate of the Universities of Calcutta and Edinburgh and the author of works on Islamic law. That Abdülhamid may have had something to do with this association can be gathered from the invitation of Suhrawardy and Kidwai to Istanbul and their decoration there.[221] The association, which changed its name from 'Pan-Islam' to 'Islam', a year after its foundation, was active until 1907, when both president and secretary returned to India.[222] It is best known for the journal it published, *Pan-Islam*, which, in addition to Pan-Islamic propaganda, aiming at the union of Sunnites and Shiites, appealed to humanitarian and socialist values.[223] Not much is known about the association's other activities, if any.[224] They must have been connected, however, to the objectives of this association which, thanks to a book by Kidwai[225] and a memorandum drawn up by the historical section of the Foreign Office in London, are known to us.[226] These, in addition to promoting the advancement of the Muslim world and removing misconceptions about Islam, included support for the Ottoman Caliphate and carrying out propaganda in this sense among Muslims and others, while obtaining the Caliph's support for all Muslim communities.

One has hardly any information about such associations in Istanbul, up to the very end of Abdülhamid's reign, when a Muhammadan Union (*İttihad-ı Muhammedi Fırkası*), led by Vahdeti, a Bektashi dervish from Cyprus, was launched on 5 April 1909 (its newspaper, entitled *Volkan*, had been appear-

[220] See ch. IV.

[221] *Revue du Monde Musulman* (Paris), 3/9 (Aug.–Sept. 1907), 136–7. Kohn, 48–9.

[222] *Revue du Monde Musulman*, ibid.

[223] G. Young, 'Pan-Islamism', 542. L.L. Snyder, *Macro-Nationalisms*, 133. I have been unable to consult this periodical.

[224] Vambéry, *Western Culture*, 230. Majid Khadduri, 'Pan-Islamism', 227.

[225] M.H. Kidwai, *Pan-Islamism*. The 'Objects of the Pan-Islamic Society of London' are reproduced ibid., on an unnumbered page (preceding p. 1).

[226] This is entitled *Muhammedan History: The Pan-Islamic Movement* (1920). See esp. 50–1 and extracts in app. Q, below.

ing, however, since 10 November 1908). It combined extremist Islam with militant Pan-Islam, in its emphatic demands for the establishment of an Islamist state.[227] It also attempted a *putsch* to return Abdülhamid to real power, drive away the Young Turks' Committee of Union and Progress, and replace the Constitution with the *sharīʿa*.[228] The failure of the *putsch* also spelt the banning of the Muhammadan Union and the temporary eclipse of organized Pan-Islam.

Judged by its formal organization, the Pan-Islamic movement was not at all impressive.[229] The reason that one does not find many Pan-Islamic associations within the Ottoman Empire in the Hamidian era seems to be that the Sultan suspected organized groups as liable to develop opposition to the regime; moreover, he preferred to attempt to harness for his ends already extant religious associations, rather than create new ones. Of these, the religious fraternity or *tarikat* (Arabic, *ṭarīqa*) was his first choice—not always with success. Indeed, one does not yet have precise information as to how and to what extent the fraternities were drawn into the circle of Pan-Islamic activity.[230] Western researchers have emphasized this aspect but, since they were writing as outsiders, may well have exaggerated. Local sources, on the other hand, offer little reliable information, most probably because Abdülhamid seems to have been even more secretive than usual in his relations with the fraternities.

What is known, however, indicates that Abdülhamid attempted to harness the entire *Sufi* structure to his Pan-Islamic activities, under the supervision of a specially nominated official, the sheikh of the *tarikat*s (in Arabic, *Shaykh al-*

[227] That is, a state based on the *sharīʿa* and governed by Islam, not by parliament. See Berkes, *Development*, 371–2. T.Z. Tunaya, *Türkiye'de siyasal partiler*², i. 182–98, for this group, and 199–205 for its programme.

[228] The best documented work on this *putsch* is still S. Akşın's *31 mart olayı*. See also B. Lewis, *Middle East*, 105–6. Id., *Emergence*, 215. Shaw and Shaw, *History of the Ottoman Empire*, ii. 280–1. M. Sadiq, 'The Ideological Legacy of the Young Turks', 186–90.

[229] See the opinion of Albrecht Wirth, in *Der Deutsche*, 1 Dec. 1906, quoted (in French) by N.S. (= Slousch), 'Une opinion allemande sur le panislamisme', 406–7.

[230] For a summary of the main views, see C.H. Becker, 'Panislamismus', 188–9. To these one may add sources in the following footnotes, as well as M.M. Feduchy, *Panislamismo*, esp. 63 ff.

Ṭuruq).[231] No less indicative of the Sultan's intentions was the fact that among his small circle of advisers and confidants at the palace were several high officials (of whom more below) in the fraternities, highlighting his great interest in the matter. In general, the *Sufis*, or ascetics, who made up the fraternities, were more concerned about religious purity, mystic contemplation, and retreat from material life than about politics, even Pan-Islamic ones; moreover, some of their leaders had reservations about the Sultanate in Istanbul with its emphasis on worldly affairs; in addition, all the fraternities displayed a lack of enthusiasm for co-operating with each other. A keen observer of Turkish affairs, Sir Edwin Pears, was to remark later that the immersion of the fraternities in the religious aspects of Islam was one of the reasons for the failure of political Pan-Islam to spread in the Ottoman Empire.[232]

Abdülhamid, however, had been interested in religious matters before his ascension to the throne and, according to some sources, had even joined the Madaniyya Fraternity,[233] an offshoot of the Shādhiliyya.[234] If this was indeed so, he must have appreciated even then—as he certainly did later—the potential of tightly knit religious associations such as the fraternities, zealous in the cause of Islam and scattered over wide areas in Hijaz, Syria, and Iraq (the Rifāʿiyya) and in the deserts or other inaccessible parts of North Africa (the Shādhiliyya–Madaniyya and the Sanūsiyya), where the non-Muslim authorities, for example in Algeria and Tunisia, found it difficult to control their activities.[235] There are also some indications that the Sultan established relations with fraternities in Anatolia as well (the Tījāniyya).[236]

Abdülhamid failed to mobilize the crucial support of the most potent fraternity of his time in North Africa, the Sanū-

[231] R. Simon, *Libya between Ottomanism and Nationalism*, 18.

[232] E. Pears, *Life of Abdul Hamid*, 151.

[233] O. Depont and I.T. d'Eckhardt, 'Panislamisme et propagande islamique', 12.

[234] C.S. Hurgronje, 'Les Confréries religieuses, la Mecque et le Panislamisme', 266–7. We shall refer to it, henceforth, as Shādhiliyya–Madaniyya.

[235] Cf. O. Depont and X. Coppolani, *Les Confréries religieuses*, pp. xiv, 261 ff. Vollers, 'Ueber Panislamismus', 34 ff. E. Layer, *Notes sur le Panislamisme et la géographie équatoriale*, 3–30.

[2236] J.O. Voll, *Islam Continuity and Change in the Modern World*, 201.

siyya.[237] Important reasons[238] (albeit not the only ones) were the official attempts, in 1882–94, to collect taxes from the tribes in which the Sanūsiyya was based; and the animosity between Muḥammad al-Mahdī, head of the Sanūsiyya (and son of its founder) and Shaykh Ẓāfir in the Sultan's entourage. Abdülhamid, however, seems to have made some impact in favour of political Pan-Islam—presented to the fraternities in the guise of religious activity—within the Shādhiliyya–Madaniyya in North Africa, as well as among the Rifāʿiyya in the Arab East (chiefly in Hijaz). This was largely due to his keeping two central leaders of these fraternities at his Court and entrusting many others with special missions (to be discussed below).

One can get an idea about the Sultan's *modus operandi* in the cause of Pan-Islam from several documents uncovered by Sırma. In the late nineteenth century, the French became so worried about the Pan-Islamic activities of the fraternities in North Africa that they established a special *Service des Affaires Musulmanes et Sahariennes* to supervise them.[239] It would appear that it was mostly the Shādhiliyya–Madaniyya that was busy with Pan-Islamic activity at the time. A secret report, sent by Lacau, the French Consul-General in Tripoli (Libya), to the French Ministry for Foreign Affairs on 12 February 1902,[240] states that the Shādhiliyya–Madaniyya Fraternity

[237] The relations of the Sanūsiyya with Abdülhamid have not yet been definitely ascertained. Even members of this fraternity were not quite sure about them; see the opinion of one of them, S. El Khalidi, 'Pan-Islamism', 256–7. Cf. PA, Orientalia Generalia 9, no. 1, vol. 1, von Oppenheim's no. 12, to Reichskanzler Hohenlohe-Schillingfürst, dated in Cairo, 22 Aug. 1896; ibid., vol. 3, von Oppenheim's nos. 83–6, to Hohenlohe-Schillingfürst, all dated in Cairo, Apr. 1900; ibid., vol. 6, von Oppenheim's no. 168, to von Bülow, dated in Cairo, 23 Apr. 1903; PA, Orientalia Generalia 9, vol. 4, von Oppenheim's (unnumbered) report to von Bülow, dated in Cairo, 27 May 1909. See also M. Le Gall, 'The Ottoman Government and the Sanūsiyya', 91 ff.

[238] For other reasons, see C.H. Becker, 'Panislamismus', 190 ff. E. Ghersi, *I movimenti nazionalistici nel mondo musulmano*, 13. H.A. Wilson, 'The Moslem Menace', esp. 382–7. al-Bashīr, *al-Waḥda*, 38 ff. 'K voprosu o Panislamızm'', 8 ff. See also Caffarel's 1888 report below, in app. C.

[239] Cf. Sırma, 'Quelques documents inédits sur le rôle des confréries Tarîqat dans la politique Panislamique du Sultan Abdulhamid II', esp. 285–91.

[240] Id., *II. Abdül Hamidin İslam birliği siyaseti*, 52–62. For the relevant French text, see Sırma's paper 'Fransa'nın Kuzey Afrika'daki sömürgeciliğine karşı Sultan II. Abdülhamid'in Panislamist faaliyetlerine ait bir kaç vesika'

in Tunisia and Libya was active in Pan-Islamic propaganda. According to this report, Sī Qāsim[241]—the brother of Shaykh Ẓāfir, the head of the Shādhiliyya–Madaniyya, living in Istanbul—had returned to Tunisia from Istanbul, with ample material means at his disposal. Although no overall plan for procuring weapons and using them was known, local Muslims, and most particularly members of the Shādhiliyya–Madaniyya Fraternity, were busy arousing the Arabs and Negroes in Africa to armed struggle, at the behest of the Ottoman Government. Should a revolt occur, the fraternities would have a significant share in it, the report concluded.[242]

8. *The Gathering of Intelligence*

Intelligence activities are shrouded in secrecy everywhere. This was especially true of the late Ottoman Empire. One does know, however, that Abdülhamid's curiosity was insatiable, as witness, for instance, the large collection of photographs and reports from the Ottoman Empire and elsewhere,[243] in the library of the Yıldız Palace.[244] Much of this was concerned with Islam, and it is known that this sultan commissioned various works on Muslims everywhere, written in Turkish, the only language in which he was proficient; at least some of these were intended to serve for Pan-Islamic activities.

However, from an original document in the state archives at the Hague, one gets a somewhat better insight into the workings of Pan-Islamic intelligence gathering.[245] This is a 1907

166–9—based on AE, Direction Politique, Série B, Carton 80, Dossier 3. See also below, app. E.

[241] In another text, he is named Sī Ḥamza; cf. Sırma, 'Fransa'nın Kuzey Afrika'daki', 162, and Caffarel's report, below, app. C.

[242] Yet another French report, found by Sırma in AE, was written in Jedda, on 20 Apr. 1902, but dealt with the general activities of the Madaniyya and the Rifā'iyya, without any reference to Pan-Islam. See Sırma, 'Ondokuzuncu yüzyıl Osmanlı siyâsetinde büyük rol oynayan tarîkatlara dâir bir vesîka', 187–90.

[243] For an example, see J.M. Landau, *Abdul-Hamid's Palestine.*

[244] Most of this collection can be consulted today in Istanbul University's main library.

[245] This document, which I have found in MBZ, is listed as dossier 451 (A190), A 190/B 107 (24534), French Legation to the Netherland's Minister for Foreign affairs, dated in The Hague, 19 Dec. 1907. See also below, app. H.

letter from the French Envoy in the Hague to the Dutch Minister for Foreign Affairs, concerning İsmail Hakkı, a lieutenant-colonel in the Ottoman army, who, a year earlier, had written to the French Resident General in Tunis, requesting statistical data about Tunisia's population; but no data were provided. Then Hakkı wrote to French officials in Dahomey, Senegal, Niger, Mauritania, and the Congo, asking for Muslim statistics in those countries; again his request was turned down. He then addressed himself to the *Société de Géographie d'Alger*, demanding to become a member; he failed, again. Frustrated at these attempts, Hakkı wrote to a Muslim notable in Java, requesting detailed data on Islam in Java and Sumatra—a step which brought about the correspondence between the French and the Dutch officials. What is suggestive in the matter is that this İsmail Hakkı was not only the son of Tevfik Pasha, then Ottoman Minister for Foreign Affairs, but was himself at the age of 23 already an adjutant of Abdülhamid himself.[246]

9. *Financing*

One knows as little about the financing of Pan-Islam as about its gathering of intelligence. One reason may be the Sultan's privy purse, which received gifts and disbursed donations without any record.[247] Another was the disorganization of the state finances and the relevant records, which had become less punctilious in their accounting than in earlier generations. Hence speculation rather than information is forthcoming, such as the assertion in 1909 (and in 1916), that half of the Sultan's revenues was being spent on Pan-Islam.[248]

Some information is available, however, about two state enterprises, which had important Pan-Islamic functions as well, the Red Crescent Society and the Hijaz Railway.[249] The former, set up on a modest scale in 1868, obtained its official name of *Osmanlı Hilâl-i Ahmer Cemiyeti* in 1877. Placed then

[246] On this İsmail Hakkı and his personality, cf. article in *Algemeen Handelsblad*, 17 Oct. 1907.

[247] A.L.C. (= A. Le Chatelier), 'Le Panislamisme et le progrès', 465.

[248] S. El Khalidi, 'Pan-Islamism', 256. S.G. Wilson, *Modern Movements among Muslims*, 68.

[249] G.L. Le. (= G.L. Lewis), 'Pan Islamism', 413.

under the patronage of Abdülhamid himself,[250] it rendered services in Muslim wars both within and without the empire, with its Pan-Islamic character emphasized time and again. During the Ottoman wars, donations and gifts arrived from Muslims in many lands.[251] The society's branches spread to all parts of the empire, and its volunteers not only cared for the wounded, but also assisted the resettlement of Muslim immigrants.[252] Its budget was basically covered by the state, but donations were collected from Muslims everywhere in times of peace and, even more so, of war.[253] It was thus a significant contributing factor to Islamic solidarity.[254] The Hijaz Railway, a much larger and more renowned enterprise, had an even more Pan-Islamic flavour, being advertised as a railroad to be constructed entirely by Muslims for the use of Muslims —which referred, of course, to the Pilgrimage to Mecca and Medina.[255] The construction, from 1901 to 1908, which was continuously used for Pan-Islamic propaganda, involved a huge investment, only a part of which came from the Sultan himself. Donations arrived in sums, large and small, from Muslims all over the world.[256] Not everything is known and accounted for—and even less in the case of liberal donations which the Sultan disbursed in the cause of Pan-Islam[257] or of sums sent to him for the same purpose.[258]

10. *Propaganda*

Sizeable sums of money must have been spent on propaganda, although, again, one does not yet have reliable figures. Islamic

[250] *al-Jawā'ib* (Istanbul), 12 Aug. 1877, 2.

[251] H. Kohn, *A History*, 46.

[252] *Revue du Monde Musulman*, 25 (1913), 358.

[253] This occurred, for instance, in the 1897 war in Thessaly; see some figures in R. Ahmad, 'A Moslem's View', 520.

[254] The weekly *al-Jawā'ib* reported these contributions regularly. For a somewhat later period, see *Revue du Monde Musulman*, 25 (1913), 358–9.

[255] J.M. Landau, *The Hejaz Railway and the Muslim Pilgrimage*, introduction.

[256] The finances of the Hijaz railway have been discussed by W. Ochsenwald, *The Hijaz Railroad.*

[257] Thus, X, 'Les Courants politiques dans le milieu arabe', 277, writes merely of generous donations and large expenses, without facts and figures, as does, also, A.R. Colquhoun, 'Pan-Islam', 909.

[258] Depont and Coppolani, *Les Confréries*, 263, mention a gift to the Sultan of 8,000 Francs, collected in French-ruled North Africa.

and Pan-Islamic ideas were instilled in a variety of ways. The leaders of the Pilgrimage and its other officials were ordered to propagate them[259] among the empire's Muslims and those arriving from abroad. From 1894, Kurds and Albanians were persuaded, separately, to join special military units.[260] These were not only trained in the arts of war, but indoctrinated religiously to range themselves behind the Ottoman Pan-Islamic Government; instruction of other units, too, continuously emphasized the military and political need for Muslim solidarity.[261] No less important, an *aşiret mektebi*, or tribal school, was inaugurated, in 1892, for the sons of Arab tribal chiefs and, to a lesser extent, of Kurdish and Albanian ones. This was evidently intended to strengthen the loyalty of the tribal chiefs and to indoctrinate their sons in Islam and Pan-Islam.[262] It was to provide a five-year course for boys aged 12 to 16: the curriculum was focused on the religious sciences, but also included Turkish, Arabic, and Persian, as well as calligraphy, arithmetic, Islamic history, and Ottoman geography. Fifty students were accepted for the first grade, forty for each of the subsequent ones (making a total of 210).[263] Pupils from abroad, too, were offered grants to come to Istanbul and study there—no doubt in the hope that they would propagate Islam and Pan-Islam in their own countries later.

The press, however, served as the main means of Pan-Islamic propaganda. In 1871–3, towards the end of Abdülaziz's reign, *Basiret*, one of the most widely read Turkish newspapers in Istanbul, printed articles asserting the unity of Islam in the world and declaring that the Ottomans ought to band together all the Muslims of the empire. It also called on the world of Islam for a Muslim campaign to assist both the Algerians and India's Muslims against Europe.[264]

[259] A. Ular and E. Insabato, *Der erlöschende Halbmond*, 182–3.

[260] I.L. Fadyeyeva, *Ofitsia'lniye doktrini*, 182 ff., based on the Russian archives. B. Kodaman, *Sultan II. Abdulhamid devri*, 84–94. S. Duguid, 'The Politics of Unity', 140 ff.

[261] Fadyeyeva, *Ofitsia'lniye doktrini*, 182 ff. B. Samardžiev, 'Traits dominants', 72.

[262] Kodaman, *Sultan II. Abdulhamid devri*, esp. 68–79.

[263] Ibid. 69–74.

[264] R.H. Davison, *Reform*, 275–6, with footnotes. *Basiret* continued its Pan-Islamic propaganda for a while, see G. Çetinsaya, 'II. Abdülhamid döneminin', 67 ff.

Abdülhamid seems to have grasped, early in his reign, the power of the press. One consideration was, of course, its size. Upon his accession to the throne in 1876, there were some 107 newspapers and periodicals in the Ottoman Empire. In all Muslim countries, the total number of newspaper and periodicals rose, reportedly, from about 200 in 1900 to more than 1,000 in 1919.[265] However, a serious obstacle consisted in the many different languages employed by Muslims in and out of the empire. This difficulty seems to have been dealt with by encouraging the publication of as many newspapers as feasible in different languages to support Pan-Islam, and by subsidizing several of them entirely, while donating money to others, whether published in the Ottoman Empire or elsewhere (e.g. *Shamsul Akhbar* in Madras).[266]

Some examples for the early years of Abdülhamid's reign may be found in *Osmanlı* (French title *L'Osmanli*), a Turkish–French bi-weekly, then daily, published in Istanbul from 29 July 1880 to 29 July 1885.[267] Although basically dedicated to serving the Ottoman Empire, *Osmanlı* asserted that it was aiming at defending the fatherland, family, and religion,[268] the last one evidently referring to Islam. The newspaper increasingly rebutted arguments by detractors of Islam[269] and advocated Muslim interests everywhere,[270] most particularly after France's conquest of Tunisia. A Pan-Islamic tone crept in, too, such as the joyful report of 'A New Success of the Muslim Caliphate',[271] referring to the Caliph's banner floating over the walls of Mecca, something not achieved by Abdülhamid's ancestors. While it cautiously disclaimed the planning of a grand Islamic union to dispossess the European powers of their Muslim-populated territories, it argued that the Ottoman

[265] L. Stoddard, *The New World of Islam*, 80. Cf. N.R. Farooqi, 'Pan-Islamism', 288.

[266] See PA, Orientalia Generalia 9, vol. 3, German Acting Consul-General in Calcutta Keller's no. 205, to von Bülow, dated in Calcutta, 16 Dec. 1906.

[267] Incomplete sets are available at Atatürk Kitaplığı and Hakkı Tarıq Üs library, in Istanbul.

[268] *Osmanlı*, 1/1 (29 July 1880), editorial.

[269] Examples in *Osmanlı* 1/5 (12 Aug. 1880) and the following issues (Ahmed Midhat's articles).

[270] e.g. *Osmanlı*, 1/92 (27 June 1881).

[271] *Osmanlı*, 1/83 (26 May 1881).

Government had to protect itself from the designs of these powers to dispossess it of its Arab-populated lands.[272] And, in an article entitled 'İttihad-ı İslam' or, in French, 'L'Union Islamique',[273] *Osmanlı* rejoiced in the heartfelt attachment of Muslims everywhere, especially the support of those in India, for their Caliph.[274]

On 29 December 1881, an article in *Osmanlı* by a well-known man of letters and lecturer in history at Istanbul University, Ahmed Midhat (1844–1912),[275] maintained that, even though Pan-Islam had no aggressive intentions, the union of Muslims was the only way to defend Islamic lands against European would-be colonizers—by ending the isolation of every single Muslim unit (this was evidently directed primarily against France and Great Britain, although Russia probably continued to be suspect, too). Another Pan-Islamically inspired article, on 9 February 1882, emphasized that the Ottoman Empire and Germany were the only powers interested in maintaining the status quo and ought to collaborate in order to prevent others from attempting to annex further Islamic territories.[276]

A characteristic Turkish daily was *Malumat* (subtitled 'Journal politique quotidien illustré'). Edited by Mehmed Tahir, it started publication in Istanbul on 2 June 1897.[277] This was supplemented by a weekly in a lighter vein, bearing the same name and edited in Istanbul by the same Mehmed Tahir, in Turkish (with some Arabic), between 1895 and 1903.[278] The daily, at least, was subsidized; it sold very cheaply and was distributed in large numbers within and without the Ottoman Empire; whenever it was banned, it was smuggled in. In 1897, its propaganda was two-pronged. On the one hand, it strived to reassure the European powers that, while the unity of all Muslims was very real, this was solely a spiritual union, so that all imputations of political fanaticism were purely an

[272] *Osmanlı*, 1/84 (30 May 1881).

[273] *Osmanlı*, 1/53 (3 Feb. 1881), written by Salih.

[274] This will be discussed in more detail below, ch. IV.

[275] On whom see Ş. Rado, *Ahmet Midhat Efendi* (with selections from his articles).

[276] Summarized in Fadyeyeva, *Ofitsia'lniye doktrini*, 155–7.

[277] A set is available in Atatürk Kitaplığı, Istanbul.

[278] A complete set can be consulted in the University Library, Tübingen.

invention of the European press, such as *Le Temps* of Paris.[279] On the other hand, perhaps under the impact of the 1897 Turco–Greek War (for which it claimed warm support among India's Muslims),[280] it started a campaign against the European powers among the Muslims, praising Abdülhamid as the successor of the Prophet Muḥammad and the Protector of all Muslims. Describing the allegedly inhuman domination of Christians over Muslims, *Malumat* exhorted Muslims in India, Africa, and elsewhere not to lose courage and trust the Father of the Believers to come to their help at the right moment, to achieve the triumph of Crescent over Cross and paganism.[281]

In general, Abdülhamid appears to have been more supportive of non-Turkish organs. One knows, for instance, of preparations to publish in Istanbul pamphlets and manifestos directed at India's Muslims—a matter which was taken so seriously by the British that their ambassador, Lord Dufferin, called personally on the Sultan to persuade him to desist.[282] However, an Urdu organ, entitled *Peïk Islām* (The Messenger of Islam), was set up in Istanbul, around 1881, and was subsequently dispatched to India. Its success was minimal, but it made its mark, nevertheless.[283]

The greatest emphasis was laid, for obvious reasons, on Pan-Islamic propaganda in the Arabic periodicals. An example is the government-subsidized *al-Jawā'ib*,[284] already mentioned for its role in the Tunisian crisis of 1881. Published in Istanbul as a weekly (but sometimes less regularly), *al-Jawā'ib* had been appearing since 1860, edited by Aḥmad Fāris al-Shidyāq (1801–87), a well-known Arab man of letters.[285] Soon after Abdülhamid's accession it became the main organ of his Pan-

[279] *Malumat* (Istanbul), 1 Oct. 1897. Cf. translation in *Le Monde Oriental* (Istanbul), 3 Oct. 1897. See also PA, Orientalia Generalia 9, vol. 1, reports from Sept.–Oct. 1897.

[280] *Malumat*, 10 Oct. 1897, editorial, entitled 'Makale-yi mahsusa: cevap-ı müfid'.

[281] 'Der Panislamismus', *Vossische Zeitung*, 8 Sept. 1898.

[282] E. Pears, *Life*, 150. V. Chirol, 'Turkey in the Grip of Germany', 237–8.

[283] *The Times*, 19 Jan. 1882, 8.

[284] Incomplete sets are available at the Université St. Joseph, Beirut, and Atatürk Kitaplığı, Istanbul.

[285] On whom see L. 'Awaḍ, *Ta'rīkh al-fikr al-Miṣrī al-ḥadīth*, ii. ch. 3.

Islamic propaganda in Arabic, distributed in Syria[286] and other Arab provinces, as well as abroad. It was so outspoken in its support of Pan-Islam that the British banned its entry into India,[287] and the French did the same in North Africa several times in the late 1870s and early 1880s.[288] The Sultan decorated al-Shidyāq for the strong support of *al-Jawā'ib* in the Tunisian crisis, which brought about, not unexpectedly, a complaint by the French Ambassador in Istanbul.[289] No wonder—for in a single issue there were no less than five articles encouraging resistance to the French invasion of Tunisia, by inspiring belief that the local Muslims could count on the assistance of the Sultan; if he had not yet intervened, he might well do so in the following summer—and 10 million Muslims were expecting his orders in North Africa alone.

Pan-Islam dictated, to a large degree, both the contents and style of *al-Jawā'ib*, in its news and its articles, especially from early 1877, as the following examples may indicate. When the Ottoman Empire was pressured by Russia for concessions in the Balkans, an editorial[290] agreed only to concessions consistent with the integrity of its territories and with the empire's honour, warning of Muslims thinking of union everywhere:

> The Muslims in all the countries of the world are thinking nowadays that they must unite. This is due to the proximity of their connections with each other, seeing that transportation and communication have become so much easier. This approach leads them to esteem the Government of the Ottoman Empire. Undoubtedly, the Muslims in lands of Central Asia and in the farthest parts of Africa respect the Sultan and consider him the symbol of Islamic unity.[291]

In another editorial, soon afterwards,[292] *al-Jawā'ib* attempted to raise the morale of its readers by reminding them of Russia's enemies in Europe and elsewhere, including among these not

286 On Pan Islamic propaganda in Syria, cf. M.D. Samra, 'Pan Islamism and Arab Nationalism', 5–10, 18 ff.

287 Koloğlu, *Abdülhamid gerçeği*, 94.

288 *al-Jawā'ib*, 4 Jan. 1877, 1; 12 Dec. 1877, 3.

289 *The Times*, 29 Dec. 1881.

290 *al-Jawā'ib*, 6 Jan. 1877.

291 Or: union.

292 *al-Jawā'ib*, 10 Jan. 1877, 1.

only the Muslims in Kashgar, Bukhara, and Afghanistan, but also those in India (both Sunnites and Shiites) and in Africa—all led by the Ottoman Empire. Once war with Russia had started, the weekly called on Muslim Circassians in the Caucasus to join the Ottoman army.[293] It also appealed for donations, later printing frequent and detailed information about financial help for the war effort and the Red Crescent from all parts of the Ottoman Empire and India's Muslims (in Bombay, Calcutta, and Madras), as well as from those in Singapore, Pyonang, and elsewhere.[294] This was repeated subsequently, chiefly in times of crisis. It was then that praise for Abdülhamid, ever present, assumed more unambiguously Pan-Islamic lines, maintaining[295] that he had helped Muslims everywhere, pursuing a policy of peace with foreign states, except with those harming Muslim rights, as in the case of France in Tunisia,[296] 'for our Lord, the Imam of the Muslims,[297] has been entrusted with the religion of Islam, as Caliph of the Muslims'.

Other Arabic organs with similar tendencies were issued in the provinces of the Ottoman Empire. A later one, with definite Pan-Islamic sympathies, *Nibrās al-Mashāriqa wa-'l-Maghāriba* (The Light for Muslims, East and West), was published in Cairo in 1904–7.[298] Edited by Sayyid Qāsim Ibn Sayyid al-Shamākhī al-ʿĀmirī and Sayyid Muṣṭafā Ibn Ismāʿīl al-ʿAmrī al-Fārikhī, it was also read outside Egypt, as is evident from an official report from Muscat which reached the India Office.[299] Major Grey received an issue of the periodical,

[293] *al-Jawā'ib*, 20 May 1877, 2; for a letter from those Circassians, cf. 24 June 1877, 3.

[294] e.g. *al-Jawā'ib*, 22 and 25 July 1877 and many other issues.

[295] *al-Jawā'ib*, 27 Dec. 1881, editorial.

[296] Another editorial by al-Shidyāq, *al-Jawā'ib*, 3 Jan. 1882, 1, chastized France for having humiliated the Muslims.

[297] *Mawlānā imām al-Muslimīn*.

[298] Acc. to A. Ahmed-Bioud *et al.*, eds., *3200 revues et journaux arabes de 1800 à 1965*, 147, no. 2911, it was published three times a month.

[299] IO, 1/P&S/19, FO Confidential Prints, Asiatic Turkey, 1906–8, Major Grey to Major Cox, 'Confidential', dated in Muscat, 24 Dec. 1906, being enc. 2 in India Office dispatch no. 43, to the Foreign Office, dated 25 Mar. 1907. The two editors sent, on 18 Dec. 1905, a letter to the German Emperor, Wilhelm II, praising his pro-Islamic sentiments and his friendship with Abdülhamid II. The manuscript letter is in PA, Orientalia Generalia 9, vol. 2.

containing inflammatory addresses supposed to have been delivered at an imaginary Pan-Islamic conference. These were followed by exhortations addressed individually to prominent Muslim rulers, urging them to resist Christian interference and look to Turkey for assistance—a leitmotif repeated in other issues as well. In the part addressed to the Sultan of Muscat,[300] he was assured that he would retain his own independence ('a holy right which should be untouched') and enjoined to propagate Pan-Islamic views not only in his own territory but in neighbouring ones too.

It is a tribute to Pan-Islamic propaganda that it spread far and wide, with relative rapidity and with adept consideration of the languages required. An example is provided by Muḥammad ʿAbduh's reply to Hanotaux in 1900, discussed above. His reply was printed in Arabic, and translated into Turkish and Persian as well; it was dispatched, with remarkable speed, not solely to Mecca, Syria, Asia Minor, Iran, and Afghanistan, but also to Khartoum and Chad.[301] And, in addition to the press and pamphlets listed above, albums containing photographs of Abdülhamid and his predecessors were sent far and wide.

So far, this section has dealt with the Hamidian Pan-Islamic propaganda in Muslim concentrations, which obviously were its main target. However, propaganda in Europe, although of a much more limited scope, was not entirely neglected. It was largely directed to Muslims living in Europe, as in the instance of the Arabic leaflets distributed in France, in 1906, calling for loyalty and obedience to the Sultan–Caliph.[302] Propaganda for non-Muslims was couched in different terms. In London, there seem to have been irregular press publications subsidized by Abdülhamid, but in Paris, from the 1880s until the Young Turk Revolution of 1908, there regularly appeared a weekly named *L'Orient Journal de Défense des Intérêts de l'Empire Ottoman.*[303] This was published under the direction and with

[300] Ibid., enc. 3. See also below, app. G.

[301] *La Revue* (Paris), 1 Feb. 1905.

[302] PA, Orientalia Generalia 9, no. 1, vol. 9, von Grünau's pol. no. 186, to von Bülow, dated in Cairo, 10 Sept. 1906.

[303] Its name varied, being also entitled *La Turquie: Journal de Défense des Intérêts de l'Empire Ottoman.* Incomplete sets are available at the Newspaper section of the British Library at Colindale, London, and the Library of Congress, Washington, DC.

the funds of the Ottoman Empire, for propaganda purposes. Again and again it emphasized that Pan-Islam was a religious, not a political movement,[304] with the aim of persuading the French and other powers that it was not intended against them (in the event, it failed in this).

11. *Emissaries*

Although due consideration was given to modern means of communication, much of the emphasis (and perhaps most of it) was laid on the traditional, time-hallowed method of sending emissaries, carrying the Sultan's orders and responsible solely to him. These complemented and supplemented the press propaganda by bringing the message to people who were illiterate (or could not be reached regularly by the press). Moreover, a human element was thus added and the emissaries could—and did—distribute many thousands of Korans[305] and report back to Istanbul on the impact of Pan-Islamic propaganda.

A few such emissaries have been mentioned above, in connection with some of Abdülhamid's official moves, but in most cases their missions were personal and clandestine (or semi-clandestine), employing special codes for telegrams and letters.[306] None the less, some information about them and their activities has reached us. Vambéry perhaps best described the situation in 1906:[307]

Messengers under the guise of religious preachers and expounders of the Koran were sent to all quarters of the globe proclaiming the pious feelings of the Khalifa, and exhorting the true believers to persevere in their faith and to unite in a common bond in defence of Islam. These seemingly unofficial missions were from time to time

[304] e.g. *L'Orient*, 20/26–8 (29 June–6/13 July 1906), 5, 'Encore le Panislamisme!'. The non-political behaviour of India's Muslims was adduced as an example.

[305] B. Bareilles, *Les Turcs*, 208–9.

[306] A.-M. Goichon, 'Le Panislamisme', 22.

[307] Vambéry, 'Pan-Islamism', 548. See also G. Roy, *Abdul-Hamid le Sultan rouge*, 73 ff. ʿAbd al-Shāfī Ghunaym ʿAbd al-Qādir and Raʾfat Ghunaymī al-Shaykh, *Qaḍāyā Islāmiyya muʿāṣira*, 98.

answered by delegations from Bukhara and Afghanistan, as well as by learned Mohammedans from India.

Earlier, in 1895, the Sultan was suspected by the Russians of fomenting a Pan-Islamic rebellion in Turkestan.[308] Some further evidence about Abdülhamid's emissaries being active in Turkestan and Bukhara, in the early twentieth century, is available in the writings of Count Pahlen (1861–1923). A Russian aristocrat from a family of officials in state service, Senator Pahlen was entrusted with reporting on the situation in Turkestan in 1906–9. While his 20-volume report is not available, his memoirs are.[309] They speak about small secret nuclei for Pan-Islamic propaganda being set up by Abdülhamid's agents. Also, since would-be teachers in the *medrese*s of Samarkand were constrained to follow higher religious studies in the scholarly institutions of the Ottoman Empire (there being no equivalent ones in their own area), some of them were apparently recruited there for Pan-Islamic activity back home.[310]

From all the above and other available information, one gets a picture of a constant stream of emissaries to and from Istanbul, bearing the Pan-Islamic message. Many were sheikhs, some of whom were associated with fraternities, others Ulema and men of religion.[311] More rarely, they comprised notables, traders, and businessmen.[312] Probably they were not as numerous as is generally assumed,[313] but their activity was dedicated and intensive. At various times, one reads or hears of their having visited southern Russia,[314] Iran,[315] Afghanistan,[316]

[308] E.D. Sokol, *The Revolt of 1916 in Russian Central Asia*, 59–64.

[309] K.K. Pahlen, *Mission to Turkestan.*

[310] Ibid., esp. 46–52, 61.

[311] Charmes, 'La Situation de la Turquie' (1881), *passim*. See also 'Le Panislamisme turc en Afrique et en Arabie et la presse arabe', 59.

[312] L. Hubert, 'Avec ou contre l'Islam', 8; A. Le Chatelier, 'Lettre à un conseiller d'état', 54.

[313] In 1881, Charmes estimated their number at several hundred. See his 'La Situation de la Turquie', 744.

[314] Vambéry, *Western Culture*, 351. 'Pan-Islamic Revival', *The Morning Post* (London), 15 Nov. 1906, 3.

[315] L. Bouvat, 'La Presse anglaise et le Panislamisme', 405.

[316] Ibid. Also Vambéry, *Western Culture*, 351. X, 'Panislamisme et Panturquisme', 185.

Central Asia,[317] India,[318] China,[319] the Malay Archipelago,[320] the Dutch Indies,[321] and particularly Java,[322] Japan,[323] the Philippines[324] (where Abdülhamid was said to have used his influence over the Muslims, in 1898, to persuade them not to fight the US troops);[325] and, naturally, Morocco,[326] Algeria,[327] Tunisia,[328] Egypt,[329] Sudan,[330] and Arabia.[331] Of the Arab populated countries, Morocco seems to have been of particular interest to Abdülhamid, no doubt because it was the only Muslim area in North Africa never to have been incorporated into the Ottoman Empire. Although under French rule, Morocco had its own sultan in 1880, when Abdülhamid sent him an Arab emissary, Ibrāhīm al-Sanūsī (no doubt connected with the Sanūsiyya Fraternity), bearing rich gifts. The fact that this emissary visited the Sultan of Morocco, on Abdülhamid's be-

317 Vambéry, *Western Culture*, 351.

318 Ibid. X, 'Panislamisme et Panturquisme', 185. Bouvat, 'La Presse anglaise', 405. O'Leary, *Islam at the Cross Roads*, 123. E. Kuran, 'Panislâmizm'in doğuşu ...', 399.

319 Vambéry, *Western Culture*, 351. X, 'Panislamisme et Panturquisme', 185. E. Kuran, 'Panislâmizm'in doğuşu ...', 399. R.R. 'Un Uléma de Pékin à Constantinople', 612–7. İ.S. Sırma, 'Sultan II. Abdülhamid'in Uzak-Doğu'ya gönderdiği ajana dair', 6–9. Acc. to French documents, published by Sırma, 'Pekin Hamidiyye Üniversitesi', esp. 161 ff., Abdülhamid's prestige was high among China's Muslims, in 1908. Ali Rıza's mission has been mentioned above; although it ostensibly aimed at promoting Muslim schooling, it had an obviously Pan-Islamic dimension.

320 'Panislamism' (1927), 212. A. Reid, 'Nineteenth Century Pan-Islam in Indonesia and Malaysia', 267 ff.

321 A. Cabaton, 'Panislamisme ou commerce', 281–2. L.W.C. van den Berg, 'Het Panislamisme', esp. 410 ff.

322 Vambéry, *Western Culture*, 351. X, 'Panislamisme et Panturquisme', 185.

323 X, 'Panislamisme et Panturquisme', 185. E. Kuran, 'Panislâmizm'in doğuşu ...', 399.

324 Ahmed Emin, 180.

325 It was alleged that the US Ambassador Oscar S. Straus asked the Sultan to intervene. See PA, Orientalia Generalia 9, no. 1, vol. 6, von Oppenheim's no. 176, to von Bülow, dated in Cairo, 28 May 1903.

326 Le Chatelier, 'Lettre', 54.

327 B. Lewis, *Emergence*, 342.

328 Le Chatelier, 'Lettre', 54.

329 B. Lewis, *Emergence*, 342.

330 X, 'Panislamisme et Panturquisme', 185. Le Chatelier, 'Lettre', 54.

331 'Le Panislamisme turc', 60. An emissary visited Malta briefly, too, in 1906, but was recalled to Istanbul by the Sultan. See PA, Orientalia Generalia 9, vol. 2, German Consul Maximilian von Tucher's cables and reports, dated in Valetta, 22 and 25 Jan. and 5 Feb. 1906, respectively.

half, three times in the same year, raises the possibility that the Ottoman Sultan wished to open a legation there and was planning a Muslim alliance against France.[332] Hence the French were suspected of scotching both projects.[333] Direct Pan-Islamic propaganda in Morocco subsided until Wilhelm II's visit there in 1905.[334] Then it gathered force in both Morocco and Tunisia, according to confident information provided by the Ottoman Ambassador to France for the German military attaché in Paris, Captain von Strempel, late in 1906.[335]

Other emissaries of Abdülhamid went to Central and Eastern Africa,[336] but stories about sizeable Muslim involvement in the Boer War[337] seem fanciful (although the Muslim community in Cape Town did send the Sultan–Caliph a message of their unwavering loyalty).[338] But there were emissaries, so it seems, who reached the ruier of Wadai, the confines of Lake Chad,[339] Tanganyika, and Natal.[340] Zanzibar's Muslims were a special case: Aḥmad Ibn Sumayṭ (1861–1925), a local scholar, had studied with Faḍl Ibn ʿAlawī Ibn Sahl (*c.*1830–1900), who in his later years went to Istanbul, where he became a close collaborator in Abdülhamid's Pan-Islamic activities until his (Faḍl's) death.[341] When Aḥmad Ibn Sumayṭ had to flee Zanzibar, it was natural for him to travel to Istanbul, in order to rejoin Faḍl. The Sultan provided generously for him (as he had done for Faḍl), almost certainly using him for some Pan-Islamic activity and hoping to have him work in Zanzibar for the cause upon his return there as an emissary. And return

[332] Or so the French, at least, suspected. Cf. *Documents Diplomatiques Français*, 1st ser., iii, doc. no. 264, summarized by Samardžiev, 73.

[333] *Express-Orient* (Bucharest), 17/29 April 1889.

[334] A. Stead, 'The Problem of the Near East', 589–90.

[335] PA, Orientalia Generalia 9, vol. 2, von Strempel's no. 846, to von Bülow, 'Vertraulich', dated in Paris, 10 Nov. 1906.

[336] X, 'Panislamisme et Panturquisme', 185. Acc. to J.O. Voll, *Islam Continuity and Change*, 135, Pan-Islamic activism in East Africa was quite pronounced in the late 19th century.

[337] 'Panislamism in Africa', *The Times*, 22 Mar. 1900, based on the Austrian newspaper *Vaterland* of 21 Mar. 1900.

[338] PA, Orientalia Generalia 9, vol. 4, German Consul-General Humboldt-Dachtroeden's no. 290, to von Bülow, dated in Cape Town, 8 Sept. 1908.

[339] Fadyeyeva, *Ofitsia'lniye doktrini*, 152, for details.

[340] Le Chatelier, 'Lettre', 54.

[341] For Faḍl, cf. R.L. Pouwels, *Horn and Crescent*, 156.

Aḥmad did in 1888, and it can be safely assumed that he was instrumental there in propaganda which increased the awareness of Pan-Islam and the popularity of Abdülhamid. The latter was so interested in strengthening his own ties with the Muslims of Zanzibar that, early in 1889, he sent a special emissary, Mehmet Bey, to Sayyid Khalīfa, the Sultan of Zanzibar, to ask whether an Ottoman legation could be established there. In April 1889, the Sultan of Zanzibar granted Abdülhamid's request.[342] Then and there, the Ottomans appointed ʿAbd al-Qādir al-Dānā, an Arab *qadi* from Beirut, as their first representative in Zanzibar. He was to be accompanied, so it seems, by a military attaché.[343] The Zanzibar archives contain a letter (*c.*1907) from Zanzibar's ruler Sayyid ʿAlī Ibn Ḥamūd to Abdülhamid, in which he accepted the latter as Caliph and requested his assistance against British activities in Zanzibar.[344] The Friday prayers were recited in Abdülhamid's name until 1910, a year after his final deposition;[345] the effectiveness of his Pan-Islamic propaganada may be gauged, perhaps, by the public demonstrations and the boycott of Italian goods, late in 1911, after Italy's invasion of Libya.[346]

12. *Istanbul as Centre of Pan-Islamic Activity*

From all of the above, there emerges a picture of Istanbul as the nerve centre of Pan-Islamic activities. This was, however, a centre which was very different from parallel institutions

[342] PA, Türkei no. 174, German Ambassador Joseph Maria von Radowitz's no. 68, to von Bismarck, dated in Istanbul, 15 Apr. 1889. Cf. *Journal des Débats* (Paris), 15 Apr. 1889.

[343] Radowitz, no. 68, in PA, Türkei no. 174, and his no. 80, to von Bismarck, dated in Istanbul, 4 May 1889. Cf. *Wiener Politischer Korrespondenz* (Vienna), 1 May 1889. Contrast *Le Temps* (Paris), 10 May 1889.

[344] R.L. Pouwels, *Horn and Crescent*, 207. B.G. Martin, 'Notes on the Members of the Learned Classes of Zanzibar and East Africa', 542 n. 69. Id., *Muslim Brotherhoods in Nineteenth-Century Africa*, 232, n. 114. The reference given for this document is A.R.C. 122 a. I have obtained from the Zanzibar archives a photocopy with this reference no. but it deals with another matter. Perhaps the document referred to is located elsewhere (in Istanbul?).

[345] For the whole matter, cf. Pouwels, *Horn and Crescent*, 205–7; Martin, 'Notes', esp. 542–3; id., *Muslim Brotherhoods*, 174–6.

[346] PA, Orientalia Generalia 9, vol. 4, German Consul Schmidt's reports to Bethmann Hollweg, dated in Zanzibar, 2 Dec. 1911 and 15 Jan. 1912.

in some other states. In order to grasp its essentials better, a brief exposé of the background, summarized from an article on 'Panislamism and the Caliphate', written by the Istanbul correspondent of the London *Times* in early 1882,[347] may be relevant.

The article stated that Ottoman statesmanship had undergone by then a radical change, best studied in the palace, where the direction of all affairs was in the Sultan's own hands. The Sultan aimed no more at an honourable place in the councils of Europe but, rather, at creating a gigantic Pan-Islamic Confederation capable of resisting aggressive Christendom. He perceived among Muslims in Asia and Africa an immense field of compensation for the empire's losses in Europe. He must have liked, also, the underhand intrigue which would be required for this enterprise. So the Sultan surrounded himself with men who sympathized with the idea of Pan-Islam and who worked for its accomplishment, busily scheming within the empire and out of it. Distinguished persons from all over the Muslim world were invited to visit Istanbul and were received there with marked consideration and liberal hospitality;[348] foreign students were granted generous scholarships; while others were encouraged to bring there their grievances.[349] Simultaneously, agents were dispatched to Muslim lands, far and near, to propagate the idea that the Sultan, as Caliph, was the Protector of all Muslims. The propaganda had relatively more success in outlying regions, which had not experienced Ottoman administration, especially among India's Muslims. In Afghanistan, too, it made some headway, while in Hijaz[350] and Egypt it was resisted for a while by Ḥusayn, Sharīf or Mecca, and Khedive Ismāʿīl, respectively. However, after the former had been assassinated and the latter deposed, Pan-Islam progressed on the eastern shores of the Red Sea and in

[347] *The Times*, 19 Jan. 1882, 8.

[348] For one example, concerning the visit of Muslims from China, returning from the Pilgrimage, PA, Orientalia Generalia 9, no. 1, vol. 10, von Oppenheim's no. 363, to von Bülow, dated in Cairo, 15 May 1907. For other visitors, cf. ibid., von Oppenheim's no. 364, to von Bülow, dated in Cairo, 16 May 1907.

[349] See also Chirol, 'Turkey in the Grip', 237. Rouire, 'La Jeune-Turquie et l'avenir du Panislamisme', esp. 260.

[350] For this Pan-Islamic propaganda in Hijaz, cf. G.S. van Krieken, *Snouck Hurgronje*, 16–17.

the Nile Valley. In the Maghrib there was also, initially, less success, but France's invasion of Tunisia changed all that.

In this apparently well-documented report (which sums up several other matters, which we have discussed above), we would like to focus on one aspect—Abdülhamid's close assistants and collaborators in his Pan-Islamic activities. Be his management as personal and centralized as he liked it, he still had no choice but to delegate authority (at least, some of it), while keeping control. This appears to have been a *sine qua non*, considering the huge size and spread of lands inhabited by Muslims at that time, on the one hand, and the enormous task which Pan-Islam set before him, on the other. Moreover, although he could conceal successfully many of his agents and emissaries, he was unable to do this with his main assistants, about whom some reliable information is available.

Obviously, over the years there were personnel changes among the Sultan's major assistants in Pan-Islamic activity. Khayr al-Dīn (Khayreddin) Pasha (1822–90), a Tunisian of Circassian origin, had arrived in Istanbul in 1871, with the intention of securing support against foreign threats to Tunisia. On arrival, he proclaimed that the Sultan was considered by Tunisian Muslims as their Protector.[351] He ultimately failed to obtain support for Tunisia, but stayed on in Istanbul to become one of the outstanding Arabs in high positions and, indeed, served as Grand Vizir in 1878–9, having a share in the formation of Pan-Islamic policies and continuing to advise Abdülhamid even after his own demotion.[352] However, a small number of advisers assisted the Sultan in his Pan-Islamic plans, and their execution, for much longer periods. Later, another Arab, ʿIzzat Pasha al-ʿĀbid, was responsible for the construction of the Hijaz Railroad and other matters.[353] From a detailed confidential report, prepared in 1888, by Caffarel, the

[351] Davison, *Reform*, 275. On Khayr al-Dīn's career, in Tunisia and Istanbul, see *EI*², iv. 1153–5, s.v. M. ʿAmāra, *Tayyārāt al-yaqẓa al-Islāmiyya al-ḥadītha*, 95–111; acc. to p. 95, Khayr al-Dīn was born in 1810.

[352] P. Bardin, *Algériens et Tunisiens dans l'Empire ottoman de 1848 à 1914*, 91 ff., 108 ff. E.D. Akarlı, 'Abdulhamid's Islamic Policy', 52. Cf. Broadley, 'How We Defended', 460. Mardin, *Genesis*, 385–93. Hassan Saab, *The Arab Federalists*, 193–4. T.Y. Ismael and J.S. Ismael, *Government and Politics in Islam*, 27, 32. Aḥmad Amīn, *Zuʿamāʾ al-iṣlāḥ*, 158–97.

[353] Tahsin Paşa, *Abdülhamit*, 23 and *passim*.

French military attaché in Istanbul,[354] one can get a good idea of who the people in charge were and how their difficult task was shared.[355] The division was geographical, rather than functional, fitting well into the bureaucratic conceptions of the time.

Abdülhamid's main assistants were the following: (i) Shaykh Muḥammad Ẓāfir, a Tripolitanian and one of the leaders of the Shādhiliyya–Madaniyya Fraternity, was responsible for Pan-Islamic propaganda and activities in Egypt and in North Africa, including, of course, those of his own fraternity; it appears that he dealt with other parts of Africa as well. (ii) Sayyid Faḍl, from India, was responsible for the southern shores of the Red Sea and of the Arabian Peninsula, from Aden to Massawa, as well as for India. (iii) Shaykh Aḥmad Asʿad (Esat), born in Hijaz and a wealthy man, was responsible for the Holy Cities of Mecca and Medina and for the Arab tribes in Yemen. (iv) Shaykh Abū al-Hudā al-Ṣayyādī (died 1909), from Syria, one of the leaders of the Rifāʿiyya Fraternity, was responsible for other Arab lands (Syria, Iraq, and to some extent Hijaz, a centre of the Rifāʿiyya) as well as active in the overall propaganda to the Arabs.[356] These four not only advised, but were sent to focal areas: Ẓāfir to Tripolitania in 1881, when France was conquering Tunisia, Asʿad to Egypt, in 1882, when the ʿUrābī uprising was brewing and the British were occupying the country; and Faḍl to India, for a while.

In addition to these, several Persians were commissioned by Abdülhamid to work to attract Iran into a Muslim union; at least some of these believed that this was the one and only

[354] AAT, Archives de la Guerre, 7N1629, E. Caffarel's no. 164 to the French Minister for War, dated in Istanbul, 17 Feb. 1888. See also below, app. C.

[355] Some additional details were found in later sources. See O. Depont and X. Coppolani, *Les Confréries*, 262–3; O. Depont and I.T. Eckhardt, 'Panislamisme', 9–13; I.T. von Eckhardt, 'Panislamismus und islamitische Mission', 62 ff.; Hurgronje, 'Les Confréries religieuses', 266 ff.; Goichon, 'Le Panislamisme', 22; Z.N. Zeine, *Arab–Turkish Relations and the Emergence of Arab Nationalism*, 54; H. Pakadaman. *Djamal-ed-Din*, 107; Akarlı, 'Abdulhamid's Islamic Policy', 52–4; Abu Manneh, 'Sultan Abdulhamid II', 131 ff.; Werner Ende, 'Sayyid Abū l-Hudā, ein vertrauter Abdülhamid II'.

[356] There are many other bits of information on these four. For Abū al-Hudā, see C.E.B., 'Notes sur le Panislamisme'; Abu Manneh; and W. Ende, 'Sayyid Abū l-Hudā'. For the others, cf. B. Bareilles, *Les Turcs*, 208.

way to serve and save Iran, too.[357] The Sharīf of Mecca, when he was on good terms with the Sultan, also employed his elevated position to speak up for Pan-Islam. Further, a French document of 1907[358] gives details about a legal expert at the Court of Abdülhamid, called İsmail Hakkı (not the Sultan's adjutant of the same name, mentioned above), who had visited China and Japan on Pan-Islamic missions and claimed to be president of the Central Islamic Committee—about which I have not found any information (if it existed at all). However, these were, most probably, only the tip of the iceberg floating in Pan-Islamic waters.

[357] Pakadaman, 107, for their names.

[358] MBZ, dossier 451 (A. 190), A/190/B107 (24534), French Legation to Netherlands' Minister for Foreign Affairs, dated in The Hague, 19 Dec. 1907. See also below, app. H.

II. *The Young Turk Era: Pan-Islam in Peace and War*

1. *The Ideological Debate*

With the relegation of Abdülhamid II to a ceremonial role in 1908, by the Committee of Union and Progress ('The Young Turks') and his deposition a year later, Pan-Islam suffered a serious set-back. Although Pan-Islamic activities were continued by the new government, they failed to get the priority treatment which they had rated previously. Moreover, since all the organization of Pan-Islam and much of its ideological propaganda had been centred in the Sultan's palace, the disappearance of Abdülhamid and his trusted advisers from the political scene could not but curtail drastically Pan-Islamic activities.[1]

Whatever the attitudes of the new rulers to Pan-Islam (and we shall have more to say about this later), ideological debates went on and intensified, particularly since censorship had become less severe, for a while, under the new regime: liberalism versus authoritarianism, Ottomanism versus nationalism, Islamism (and Pan-Islam) versus Turkism (and Pan-Turkism).[2] A part of the press and special publications—not necessarily government-inspired—continued to argue about the merits of Pan-Islam and the means for achieving world-wide Muslim solidarity and unity, in order to forge the forces which would save the Ottoman Empire. We shall discuss these publications and their arguments generally, then select a few characteristic examples for a more detailed examination.

As we shall see later, Pan-Islam enjoyed only lukewarm sup-

[1] Cf. Rouire, 'La Jeune-Turquie et l'avenir du Panislamisme', 264 ff.

[2] Shaw and Shaw, *History of the Ottoman Empire and Modern Turkey*, ii. 273.

port from the Young Turk leadership (at least, in the pre-World War years), so that its protagonists had to fend for themselves, more or less. Indeed, leading ideologues of the Ottoman Empire, in its last decade, like Ziya Gökalp (1875/6–1924), were of two minds about the role of Islam inside the empire and out of it, as is evident from his book, *Türkleşmek, İslamlaşmak, muasırlaşmak* (Adopting Turkism, Islamism, and Modernization). Although his poem 'İslam ittihadı'[3] supported an Islamic union—provided each Muslim land had achieved its own independence first—his general inclination was towards Turkism and Pan-Turkism. As to the Pan-Islamists themselves, they paid only lip-service, if that, to the opinions advocating modernization or Ottomanism (generally these were attacked, sometimes violently), but they seemed to express a more conciliatory attitude to Turkism and Pan-Turkism, which were increasingly becoming the official ideology of the core leadership of the Young Turks.[4] Some accommodation with it was proposed, on the lines that Pan-Islam and Pan-Turkism could coexist by supplementing one another.[5] However, the bulk of Pan-Islam literature at the time was devoted to a discussion of the nature of Pan-Islam, its policies, and the ways and means to further its cause. Not surprisingly, most of the discussion was carried out by Muslims—chiefly, Turks and Arabs—who had close ties with Islamist circles and whose opinions were mostly published in the organs of those circles. Since the main interest of these circles was in Islam, the debates concerning Pan-Islam were generally tied up with arguments about Islam and Islamism.[6]

Consequently, while there may have been, in the years of Young Turk rule (1908/9–1918), short-lived periodicals devoted to Pan-Islam alone, many of the press articles concerning it were published in such Turkish periodicals as the weekly *Sırat-ı Müstakim* (The Straight Way), founded on 11 August 1908, continued by *Sebilür-Reşad* (The Right Way); both were

[3] Repr. in e.g. Bayur, *Türk inkilabı tarihi*, 371 n. 53. See also Kara, *Türkiye'de İslamcılık düşüncesi*, p. xliii.
[4] J.M. Landau, *Pan-Turkism in Turkey*, 44 ff.
[5] al-Jundī, *al-Yaqẓa al-Islāmiyya fi muwājahat al-istiʿmār*, 101 ff. for details.
[6] Tunaya, *İslamcılık cereyanı*, 91–3 and *passim*.

published in Istanbul and had an Islamist character. In these, indeed, Pan-Islam was proclaimed boldly.[7] Along with an emphatic request for a return to the ancient Muslim values, demands for the unity—and union—of all Muslims were formulated, deemed necessary in order to resist the further onslaught of Christian Europe.[8] Although Mehmed Âkif [Ersoy] and H. Eşref Edib were editors of the above periodicals for some time (the former an admirer of al-Afghānī),[9] it is well-nigh impossible to speak of a definite editorial policy of the above periodicals concerning Pan-Islam, chiefly because journalists moved from one newspaper to the other, some of them establishing new periodicals, with similar views, such as the Turkish-language fortnightly *İslam Mecmuası* (Turkish Periodical), founded in 1913 in Istanbul. The slogan of *İslam Mecmuası* was *Dinli bir hayat hayatlı bir din*[10] (which can be roughly translated as 'A life full of religion and a religion full of life').

Arabic-language newspapers also discussed Pan-Islam, sometimes indicating full approval. Thus, the Istanbul-based Arabic monthly *al-Hidāya* (The Guidance) of March 1913 called on all Islamic nations to relinquish their racial differences and turn to Islam as their main bond, in order to set up a strong bloc of nations.[11] A year later, the Indian Muslim *al-Balāgh* (The News) cogently distinguished between two types of Pan-Islam —religious and political. The former, demanding religious unity, was mandatory in Islam and should be promoted by Muslims, one and all. The latter, political union, was warmly recommended by a majority of Muslim statesmen, but could only be based on the former. Therefore to shape a political alliance, great efforts must be invested, in buttressing religious unity, via the standardization of curricula in the schools of Muslim lands; making the study of Arabic compulsory; pre-

[7] Kara, p. xxix. Yu. V. Marunov, 'Pantyurkizm i Panislamizm Mladoturok (1908–1911 gg.)', 40.

[8] Shaw and Shaw, ii. 304.

[9] Cf. his 'Cemaleddin Efganî', *Sırat-ı Müstakim*, 4/90 (13 May 1326/1910), 207–8.

[10] Kara, p. xxxi. A set is available in the Beyazıt Library, Istanbul.

[11] *al-Hidāya*, 4/4 (Rabīʿ al-ākhir 1331), 177, quoted by W.L. Cleveland, 'The Role of Islam as Political Ideology in the First World War', 85. An incomplete set of this monthly is available at the Beyazıt Library, Istanbul.

paring new textbooks (conforming to both the spirit of Islam and modern civilization); setting up a religious moral newspaper in every town; printing and distributing, without charge, books on history and morals; and sending emissaries to all villages to propagate the word of Islam.[12]

2. *Some Ottoman Writers on Pan-Islam*

Some characteristic works on Pan-Islam by writers of the Young Turk era have been selected for more detailed examination.

Mehmed Âkif [Ersoy] (1873–1936), a famous poet, writer, and activist for Islamism in the late Ottoman Empire,[13] favoured Pan-Islam, which he arrived at via his strong religious convictions. In a moving *vaaz*, or sermon, delivered at the Beyazıt Mosque in Istabul in 1913[14] and entitled 'Islamic Union and Nationalism',[15] he maintained[16] that, although Islam had commanded the Muslims to be united (*ittihad*), they were divided—despite the Prophet Muḥammad's warning that division was certain to bring about the annihilation of the Muslims by their enemies. As a result, there were several nations (*millets*) in the Ottoman Empire: Albanians, Kurds, Circassians, Arabs, Turks, and Lazes.[17] The bond among them is Islam, which does not acknowledge nationhood (or nationalism, *milliyetçilik*) as Christianity does. In the last four or five years, Muslims had been divided, and revolts followed one another—all because foreigners sowed discord, in order to take over 'our country' (*memleketimiz*), as they had done in Spain,[18] India, Algeria, and Iran. All groups (*fırka*) should unite immediately against the enemies; otherwise, the throne of the Sultanate would be in danger. This could be prevented, if the

[12] Reported in *Revue du Monde Musulman*, 27 (1914), 387–9.

[13] For a good monograph on him, see Fevziye Abdullah Tansel, *Mehmed Akif*.

[14] Printed in *Sebilür-Reşad*, 9/230 (29 Sefer 1331/7 Feb. 1913), repr. in Kara, i. 376–82.

[15] Mehmet Âkif Ersoy, 'Bayezid kürsüsünden vaaz'.

[16] The following is a summary of the main points.

[17] Mehmet Âkif obviously refers here solely to the Muslim groups in the empire.

[18] Refers to the Christian *reconquista* in the 14th and 15th centuries, nothing to do with the Turks—but much with Islam.

Muslims in Russia kept up their faith, those in Algeria did not convert to Christianity, and those in India raised their voices. If they did not preserve this Sultanate, they would be annihilated, which is what their enemies wished. None the less, they had not united in order to set up an energetic Muslim Government (*bir müslüman hükümeti*)—although this Government and this Caliphate were the last hope of Islam.

Şehbenderzade Ahmed Hilmi (1865–1914),[19] a Bulgarian Muslim from Philippopolis,[20] had spent years in Tripoli (Libya) carrying out Pan-Islamic and anti-French propaganda among the Sanūsiyya Fraternity and other Muslims of the area. In addition to lecturing in philosophy at Istanbul University, he edited and published in Istanbul an Islamic and Pan-Islamic weekly in Turkish, suitably named *İttihad-ı İslam*, of which eighteen issues appeared between 4 December 1908 and 23 April 1909.[21] Its subtitle, 'Union Sociale Musulmane', indicated the approach to Islam and Pan-Islam that the periodical would follow. Indeed, the first issue comprised an editorial by Ahmed Hilmi himself, entitled 'What in Our View Does the Union of Islam Mean?'[22] Under the impact of the Young Turk Revolution, with the new rulers having other considerations besides Pan-Islam, this article did not support political Pan-Islam. Hilmi maintained then that a plan to set up a state for all Muslims contradicted the laws of both history and human nature and was a fantasy, just like Pan-Hellenism. Social Pan-Islam should aim, instead, to raise and develop Muslim society, in the spirit of the Committee of Union and Progress. This leitmotif was often repeated in favour of social and cultural development for all Muslims and all Ottomans,[23] thus expressing Ottomanism, then popular with the new regime.[24]

None the less, editorial policy frequently emphasized politics—advocating independence for all Muslims or printing news and commentaries about the political situation of Mus-

[19] 'K' voprosu o Panislamizm'', 11. Ülken, *Türkiye'de çağdaş düşünce tarihi*, 278–87. Kara, i. 3. Landau, *Pan-Turkism in Turkey*, 46.

[20] Which is why he is frequently referred to as Filibeli Ahmet Hilmi.

[21] Incomplete sets are available at Atatürk Kitaplığı and Hakkı Tarık Üs Library, Istanbul.

[22] *İttihad-ı İslam*, 1 (4 Dec. 1908), 1–2: 'Bizce ittihad-ı İslam ne demek?'

[23] e.g. *İttihad-ı İslam*, 2 (24 Dec. 1908), 1.

[24] e.g. *İttihad-ı İslam*, 13 (12 Mar. 1909), 1–2, editorial.

lims everywhere. An 'Islamic policy', it was argued, should encourage a strong tie[25] among all Muslims.[26] A four-part article on 'The Future of Islam' emphasized the huge potential power of 300 million Muslims.[27] Meanwhile, Ahmed Hilmi wrote sympathetically about political Pan-Islam in several other newspapers as well, usually assuming a pen-name.[28] In another Turkish periodical, *Hikmet* (Sagacity), which he edited in Istanbul during 1910–11, Ahmed Hilmi took an even more outspoken line, calling on the world's 300 million Muslims to unite. The extreme agitation against the European powers irked the ruling Committee of Union and Progress, which closed down *Hikmet*.[29] Hilmi's activity was extensive enough to interest Sir Gerard Lowther, the British Ambassador in Istanbul,[30] who in 1910 sent to Sir Edward Grey, at the Foreign Office, a report on Ahmed Hilmi's activities, along with the English translation of a pamphlet by Hilmi signed Mihriddin Arusî.

No less relevant to our discussion is a book by Ahmed Hilmi, again published under the pen-name of Mihriddin Arusî (in 1911), entitled *A Guide to Politics*[31] *for the Twentieth-Century World of Islam and for the Muslims of Europe*.[32] As far as Pan-Islam is concerned,[33] the author considered that the Muslim elements in the Ottoman Empire were the only ones hoping for the empire's survival. Their unity (or union) was the sole remedy for the empire's troubles and the only guarantee for its independence. On the other hand, division would be disastrous. After the 25 million warlike Arabs in Algeria, Tunisia,

25 In Turkish, *rabita*.

26 Hilmi, 'Siyaset-i İslamiye', *İttihad-ı İslam*, 5 (1 Jan. 1909), 1.

27 Id., 'İslam istikbali', *İttihad-ı İslam*, 4–5. Hilmi signed here with his pseudonym, Mihriddin Arusî.

28 Such as Mihriddin Arusî, Coşkun Kalender, Kalender Geda, or Özdemir. See Kara, i. 3. Marunov, 39.

29 On *Hikmet*, see 'K' voprosu o Panislamizm'', and Marunov, 39. This was a daily and a weekly, by turns.

30 IO, L/P&S/19, FO Confidential Prints, Asiatic Turkey and Arabia, 1910, 20–7, Sir Gerard Lowther's no. 715 to Sir Edward Grey, dated in Istanbul, 9 Oct. 1910, and enc. 1. See also below, app. I.

31 Or policy. *Siyaset* means both 'politics' and 'policy', like 'politique' in French.

32 *Yirminci asırda âlem-i İslâm ve Avrupa* (Istanbul, 1327 H).

33 Ibid. 2–11, 63–77, 87–96, repr. or summarized in Kara, i. 29–33.

and Egypt had lost their independence, how could 5 million Yemenites and 3 million Syrians hope to maintain theirs, should the Ottoman Empire break down?[34] The hope of all Muslims everywhere was in the Ottoman Empire and the Caliphate. Although there existed conflicts between various Muslim elements,[35] impeding unity, a union of these elements was imperative; sovereignty implied the decision of the majority, which was Muslim. When united, Muslims had failed to demonstrate their worth to the Great Powers. Should they remain divided, they would be exposed to two dangers—conflicting interests and the co-operation of the powers against them. Against these dangers, united Islam, enjoying the moral support of 400 millions, would not be a force to be taken lightly.

Yet another Turkish work on Pan-Islam, undated but published about the same time as Ahmed Hilmi's book, was a 34-page pamphlet by Esat, entitled *The Union of Islam*.[36] The author was employed at the time as a recorder at the Court of Maritime Commerce in Istanbul. After having described the feats of united Islam in its early history, Esat praised Muslim unity, then advocated an Islamic union[37] as absolutely essential for all states and nations in achieving happiness and maintaining it. Examples for successful unions of states were the United States, the German Empire, and Italy.[38] These ought to set a model for a Pan-Islamic union,[39] to be expressed both internally and externally. Uniting all Muslim peoples would create a Great Power, with the Caliphate at its centre. Divisiveness among Mulims was noxious and the Koran, also, had recommended unity and condemned divisiveness.[40] The union should become 'a state of Islam',[41] capable of repulsing foreign aggression.[42] Esat was one of the few who stated in precise terms what such a union should comprise—a specific area

34 A transparent allusion to separatist Arab aspirations, chiefly in Syria.
35 Probably within the Ottoman Empire.
36 Esat, *İttihad-ı İslâm*.
37 Both terms rendered as *İttihad-ı İslâm*.
38 Esat, *İttihad-ı İslâm*, 7–9.
39 Ibid. 12. Here the term *Jamiʿe-yi İslâm* was used.
40 Ibid. 12 ff.
41 *İslâm devleti*.
42 Esat, *İttihad-ı İslâm*, 21–6.

from the Danube to Istanbul, the entire littoral of the Black Sea, North Africa (including Egypt, Tripolitania, Tunisia, and Algeria), the Red Sea with both its shores, South India, Sumatra, Java and the neighbouring islands inhabited by Muslims, Central Asia (including the Tatar lands, Bukhara, Afghanistan, Iran, Beluchistan, and the Muslims in China)—all obeying the Caliph in Istanbul.[43]

Celal Nuri [İleri] (1877–1939), one of the most widely read and appreciated journalists of his day,[44] as well as a Member of Parliament in the Young Turk decade, was, like Ziya Gökalp, of two minds regarding the relative merits of Pan-Islam and Pan-Turkism. Although he seems to have ultimately inclined in favour of the latter, the former attracted him for a while and he devoted a good deal of attention to it. In addition to works on other topics and several pamphlets in Turkish touching on Pan-Islam, such as the 64-page *Pan-Islam and Germany*,[45] he wrote in 1913 a lengthy book on Pan-Islam (possibly the most extensive treatment of the subject up to that time). Entitled *Pan-Islam: The Past, Present and Future of Islam*,[46] it was published in Turkish and translated into Arabic eight years later, with the further subtitle 'Views about World Civilization and its Political and Social Doctrines'.[47]

This book was evidently intended to be a comprehensive work. After surveying the nature of Islam as a religion, the author discussed the main groups of Muslims: Arabs, Turks, Indians, Chinese, Malaysians, smaller groups, and Persians—distinguishing between Sunnites and Shiites. He then

[43] Esat, *İttihad-ı İslâm*, 31–2.

[44] See about him, among other works, Özer Ozankaya, 'Cumhuriyet'i hazırlayan düşünce ortamı', esp. 130ff. J.M. Landau *Tekinalp, Turkish Patriot, 1883–1961*.

[45] Celal Nuri [Ileri], *İttihad-ı İslâm ve Almanya*. This could also be transl. as *The Union of Islam, and Germany*. In a way, it is a sequel to his longer book on Pan-Islam, discussed below.

[46] *İttihad-ı İslâm: İslamın mazısı, hali, istikbali*. A de luxe edn. was printed shortly afterwards, in 1913, acc. to the *Revue du Monde Musulman*, 25 (1913), 357–8. Its title could also be transl. as *The Union of Islam* ... I have a copy of this book in my possession.

[47] Id., *Ittiḥād-al-Muslimīn: al-Islām, māḍīh wa-ḥāḍiruh wa-mustaqbaluh: Naẓariyyāt fī muduniyyat al-ʿālam wa-madhāhibih al-siyāsiyya wa-'l-ijtimāʿiyya*. Its title could also be transl. as *The Union of Muslims*. A copy can be found in the Oriental Library, Durham University, England.

described the various types of colonialism and imperialism from which the Muslims had been suffering—under Great Britain, the United States, Russia, France, Japan, and Germany. Afterwards he dealt at some length with the union of Islam.

The following summarizes the points which are of Pan-Islamic interest. While acknowledging that 300 million Muslims could not be expected to form a single nation,[48] he warned that Islam frowned on their division and condoned it only if this did not harm its own role as a bond between the Muslim nations. Neglect of Islam caused Muslim nations to draw apart from one another and thus facilitated foreign aggression against them separately. The Christian powers were stirring up mutual hatred between Muslim groups, as when they roused the Bosnians, then the Albanians, to demand independence from Turkish 'despotism'; or the French encouragement of a revival of the Berber language, to foster their separate identity; or the British, Italians, and French stirring up the Arabs against the Turks (the author recommended local government for the Arabs and the use of Arabic in education and the courts to preempt this and foster, instead, a special relation between Turks and Arabs within a Muslim union). Without Muslim solidarity, France and Great Britain would occupy more Muslim lands, including Hijaz—thus separating Sultanate and Caliphate and dooming not only the Ottoman Empire but Muslim unity as well. There were large masses of Muslims who could be united, provided Sunnites and Shiites drew closer together and that various groups did away with the disorder in their views and morals, thinking, and behaviour—and opt instead for method and discipline. This should enable Muslims, from the black man in a tent on the shore of the Niger River to the rich bey in his Istanbul palace, to share in the hope for a Muslim union—their conscious reply to current nationalisms. Such a union was not merely a requirement of the faith, but an imperative need as well, against the powers' political, economic, and other interests (which the author examined, one after the other).

[48] This was an argument for the continued existence of the Turks as a separate group.

Muslim unity existed in faith and should be put into practice, just as Pan-Slavism, Pan-Latinism,[49] and Pan-Germanism had been. Pan-Islam would be stronger than other unions, since unity is implied in it and since it would be based on Islamic sentiments of brotherhood and on common morals and customs. The author noted the Muslims living in peace with one another more than at any past time[50] and argued that the feeling of commonality ought to be encouraged, in order to better resist foreign intrigues. Muslims were making the same music, but they needed a conductor for their orchestra.[51] Islam, a great force, past, present, and future, could—and should—be a bond uniting all Muslims, despite foreign rule and differences of place, government, economics, and language. Indeed, the whole of Europe was afraid of Islam and Pan-Islam. Consequently, Muslims ought not to conceal their thoughts about an Islamic union, but rather demonstrate them openly.

Among the elements common to all Muslims—and, presumably, leading to an Islamic union—the author emphasized the following: (i) The Caliphate, which was renewed by the Ottoman ruling family—and both Turks and Arabs should be grateful for this. The Caliphate was, and would be, an immense force uniting all Muslims everywhere, which was why the British were striving to diminish its stature by fostering alternative caliphs. (ii) The Holy Pilgrimage, which, besides constituting a religious duty, had social and political implications, as the pilgrims were delegates of Muslim communities world-wide, meeting in Mecca on an appointed day—thus making Hijaz a hundred times more valuable for the Ottoman state than the straits between Asia and Europe. (iii) General and improved education with a common basis, to spread the idea of the Islamic union, as well as universities in Mecca, in order to turn Hijaz, a religious centre, into a scientific one as well: Mount ʿArafāt, in Mecca, could be of immense unifying

[49] This Turkish term, *Pan-Latinizm*, refers to a policy aiming at linking up the French, Belgians, Italians, Romanians, the peoples of Ibero-America, and Haiti. Several intellectuals toyed with this dream in the years immediately preceding World War I, but never formed a movement to promote it.

[50] Celal Nuri seems to forget here what he himself had written earlier about the tensions between Turks and Arabs.

[51] Obviously the Caliph, or the Ottoman Empire, within the framework of a Pan-Islamic union.

value for the Muslims, just as the Capitol was for the Romans and the Acropolis for the Greeks. (iv) A common religious literature, and non-religious also, to instil into all Muslim hearts, East and West, the idea of the Islamic union; special Friday sermons, for this purpose, ought to be sent to all mosques over the world, since the mosques served as literary–political–social–educational clubs for Muslims everywhere.

In conclusion, Islamic unity would dam Christian hatred of Islam, smash European aggression, and change the double standards of international law (which imposed different criteria on Europeans and others). Far from being reactionary or fanatic, the Islamic union could be a civilizing force greater than that of Europe in the eighteenth and nineteenth centuries, thanks to its humane character. Pan-Islam would not mean extending the Ottoman Empire to comprise all Muslims, but, rather, preserving the sentiments of unity in the hearts of Muslims and working jointly for the future of Islam. It would compel all colonialist powers to come to terms with Islam or risk losing their possessions—for Islam could shake the entire world with its force, renascent in the near future in a young, united Islamic nation.

Celal Nuri's book has been presented here at some length, as it is probably not only the most explicitly detailed work on Pan-Islam in the Young Turk era, but is also characteristic of the way Pan-Islamic ideology was being expressed before 1914. It is a clear exposé of the need for an Islamic union to oppose European aggression and Christian hatred (as these were perceived then) by uniting Muslim forces on moral and intellectual bases, but also with very definite political and some economic implications. Although it disclaimed that the Ottoman Empire had any ambitions for territorial expansion, the reader was not left in doubt that the Caliph would forge the Islamic union and lead it, using the Pilgrimage, education, and mosque propaganda for this end.

The author has also ably put together a fair sample of the views about Pan-Islam which were then being propounded in the Ottoman press. One gets a still clearer idea of some of these from the appendixes to this book: as was then often customary, the author added a few letters and comments, some already published, others written especially for that work. The former included an article on Pan-Islam by Namık Kemal, published forty years previously.[52] The latter comprised an article entitled 'Truth, Not Fantasy',[53] written by a man of letters named Süleyman Nazîf (1869/70–1927).[54] Bewailing the loss of Tripoli (in Libya) and of the Balkan territories by the Ottoman Empire and to Islam, Nazîf considered this another link in the continuing Crusades which could be stopped only by a general Muslim awakening. The Muslims were stirring up in China, India, Africa, and elsewhere, and this ought to give them hope in all territories lost to Christian rulers.

Said Halim Pasha (1863–1921) differed from some other exponents of Pan-Islam in the Young Turk era in at least two respects—he was an Arab and he had a meteoric career in the government of the Ottoman Empire. As we shall see, this was reflected in his writings on Pan-Islam. Born in Cairo, he was the grandson of Muḥammad ʿAlī and the son of Ḥalīm Pasha who, having been frustrated in his hopes of becoming the ruler of Egypt, went to Istanbul. Said Halim Pasha joined Young Turk circles and was active on their behalf abroad, returning to Istanbul after the 1908 Revolution. After a succession of lesser official positions, he became in 1912 the Secretary General of the ruling Committee of Union and Progress, in 1913 Minister for Foreign Affairs, then for three years Grand Vizir, until February 1917. He was assassinated in Rome after the end of the war.[55]

More than Celal Nuri and some other Ottoman intellectuals of the Young Turk era, Said Halim was totally committed to an Islamist point of view. His writings[56] emphasized the central position of Islam in social as well as in political behaviour.[57] He maintained that Islam, as a unity of the eternal

[52] In the periodical *İbret*, in 1872, as already mentioned above, in my introduction.

[53] *Hayal değil, hakikat*.

[54] On whom see Ş. Karakaş, 'Süleyman Nazîf'.

[55] A recent work about him is Ercümend Kuran's 'Türk düşünce tarihinde Arap kültürlü aydın', whose footnotes mention additional materials. See also Kara, i, pp. xxxvi–xl. Abdul Qayyum Malik, 'Prince Saeed Halim Pasha'.

[56] Said Halim generally wrote in French, since he could not do so elegantly enough in Turkish. His articles, transl. into Turkish, were published in such periodicals as *Sebilür-Reşad* or issued as pamphlets.

[57] For examples, cf. Berkes, *The Development of Secularism in Turkey*, index.

truths of equality and solidarity, had no fatherland.[58] No less characteristic was his saying that a Muslim's fatherland was the place in which the *sharīʿa* prevails.[59] In a lengthy article in French, he argued that the *sharīʿa* should determine both the law and government of Muslim populations.[60] As Said Halim also stressed the non-racial, non-territorial character of Islam, he could not but consider the merits of Pan-Islam as well. Being a statesman, he gave some consideration to ways and means and, acknowledging that it was impossible to achieve immediate all-Muslim union, or *İttihad-ı İslam*, argued for an interim federation or *Aile-yi İslam*,[61] furthering a common all-Islamic society, which would lead to a future union of Muslim nations.[62] His origins as an Egyptian, rather than a Turk, could conceivably have influenced his thinking in this direction, which was more easily acceptable to Arabs and other Muslim groups.

For our discussion, his most significant work is *İslamlaşmak* (Islamization),[63] a little book—more accurately, a long pamphlet—published in 1337/1918 in a Turkish version[64] and incorporated, shortly afterwards, in a volume of Said Halim's collected pamphlets and articles entitled *Buhranlarımız* (Our Crises), which has been reprinted several times since.[65] The last chapter of *İslamlaşmak* discusses the 'political concepts of Islam'.[66] In it, Said Halim expresses strong reservations about uncontrolled mass absorption of European concepts and customs over the previous hundred years, stating his own case for preferring the Islamic heritage in education, society, government, and politics. Innovations, sponsored by foreigners, had led to moral anarchy and were noxious for the Ottoman

58 Cf. Saab, *The Arab Federalists of the Ottoman Empire*, 200.

59 B. Lewis, *The Middle East and the West*, 108.

60 This was written as late as 1921—his last paper before he was assassinated. See his 'Dates pour servir à la réforme de la société musulmane', esp. 43–54.

61 Literally, a 'Family of Islam'.

62 See also Tunaya, *Islamcılık cereyanı*, 84.

63 To be distinguished from 'Islamicization', i.e. converting people to Islam. I am using 'Islamization' in the sense of strengthening the 'Islamism' (*İslamcılık*), discussed above.

64 I have been unable to find the French original if, indeed, it was published in French at all.

65 I shall refer to Said Halim's *Buhranlarımiz* in the 1985 reprint.

66 'İslamın siyaseti anlayışı', ibid. 163–83.

Empire and Islam. None the less, Muslim verities would ultimately prevail and the Ottoman Empire, in which the Caliph was situated, would unite all Muslims and head all Muslim nations. Civilized Muslim nationalism was destined to be the alternative to Western-inspired self-centred, arbitrary, and savage nationalisms, which had found such a terrible outlet in the World War. As there were no English mathematics, German astronomy, or French chemistry, so there did not exist a Turkish, Arab, Persian, or Indian Islam. All Muslim nations should be united in Islam.[67] The implication seems to be that such a union would be capable of resisting both the cultural penetration and the political designs of Europe.[68]

3. *The Young Turk Leaders and Pan-Islam*

The failure of Pan-Islam to produce tangible results during Abdülhamid's reign, despite the efforts which he and his assistants had invested, has already been discussed. A contributory reason may well have been the indifference shown by most of the general public both in the Ottoman Empire[69] and abroad. This attitude changed to some extent in the period under discussion. The Young Turk era had begun with a military and political revolution; competitive strife between various factions was ever present. Hence, at least in Istanbul and other cities and towns—and certainly among urban élites—an increased measure of interest in politics could be noticed. This naturally affected views about Pan-Islam, too, and we have noted above some of the discussions about it.

However, what was crucial was the attitude towards Pan-

[67] I have emphasized those works on Pan-Islam which adopted a political viewpoint. Other writings were largely of an apologetic character, such as Refik Azmzadeh's *İttihad-ı İslam ve Avrupa*, i.e. *Islamic Union and Europe*, which was available to me only in a later Arabic translation, entitled *al-Jāmiʿa al-Islāmiyya wa-Ūrubbā*. The author (1865–1925), better known as Rafīq al-ʿAẓm, a well-known Syrian researcher on Islam, wrote his book in Turkish in *c.*1907 and had it published in 1911. In it, he defended the idea of an Islamic union against its detractors, chiefly in the European press, and argued that Islam was as respectable a bond for Muslims as race was for Europeans.

[68] Cf. I.L. Fadyeyeva, *Ofitsial'niye doktrini i idyeologii v politikye Osmanskoy Impyerii*, 198–9.

[69] E. Pears, 'Turkey, Islam, and Turanianism', 378.

Islam of the leading decision-makers themselves in the Committee of Union and Progress. These, although by no means of one mind, were at first interested in what their group's name actually stood for—in the union of the Ottoman Empire and in its progress in military, political, and economic affairs. This implied, on the one hand, an active Ottomanist policy, granting equal rights to all the religious communities and ethnic groups in the empire, to encourage their identification with it and integration; and, on the other hand, welcoming Western science and ways of life.

The leaders of the Young Turks found it difficult to identify fully with Pan-Islam, an ideology which had been fostered by their arch-enemy, Abdülhamid, with meagre results. They could hardly have hoped to emulate his achievements in Pan-Islam (modest though these were). Furthermore, as pointed out by K.H. Karpat,[70] they were not so sure of the value of such a supra-national ideology as Pan-Islam in an age of surging nationalisms. Some Young Turks regarded Islam and Pan-Islam as a reactionary force, damaging Ottomanism and injurious to the advance of modernization. Moreover, their top leaders were not very pious;[71] some were definitely irreligious.[72] None the less, as a leadership group, they continued their support of Pan-Islam,[73] partly because Muslims who had emigrated to Turkey from Russia were quite influential in the Committee of Union and Progress; partly because, having moved their central committee from Salonica to Istanbul in 1912, they needed there the support of the Ottoman corporations (which had strong religious perspectives to their political thinking); and partly because Pan-Islam was basically an expansionist policy, suited to their mood, as they were seeking compensation for their retreat from previous borders. Also, they could always employ it to put pressure on Great Britain

[70] K.H. Karpat, 'The Turkic Nationalities', 124.

[71] E.F. Knight, *The Awakening of Turkey*, 63–4. Cf. Landau, *Pan-Turkism*, 45. Id., 'Pan-Islam and Pan-Turkism in the Final Years of the Ottoman Empire', 43–4, J.O. Voll, *Islam Continuity and Change in the Modern World*.

[72] H. Luke, *The Old Turkey and the New*, 143.

[73] This included contacts with the increasingly powerful Muslim fraternities of North Africa, in particular the Sanūsiyya. See A. Edwards, 'The Menace of Pan Islamism', 645–57.

and France,[74] even though foreign political commentators advised them to refrain from this.[75]

It follows that, for the Committee of Union and Progress, commitment to Islam and Pan-Islam was a matter of political expediency rather than a matter of principle. Well aware of the potent force of religion, they brought about the deposition of Abdülhamid in 1909 in the time-hallowed manner of referring to the Ulema and obtaining a legal decree, *fatwā* (in Turkish, *fetva*), from the Şeyhülislam to this effect.[76] On a Pan-Islamic level, when the Cretans announced their union with Greece in 1908, the Committee of Union and Progress threatened the European powers that, should they support this act, they would have to reckon with the wrath of Muslims everywhere.[77] Shortly afterwards, as early as 1909, the Committee of Union and Progress dispatched a delegation to the annual pilgrimage in Mecca, seeking Islamic support for their new regime. The delegation attempted to persuade the pilgrims that the new regime deserved their sympathy and that all Muslims should forge a united front against European aggression.[78]

A while later, at the 1910 congress of the Committee of Union and Progress, it was decided to continue to employ Pan-Islamic policies;[79] its 1911 congress, in Salonica, elaborated on this further.[80] Indeed, the resolutions of the 1911 congress in this respect reveal a great deal of the Committee's conception of the then present and future uses of Pan-Islam:[81]

A congress of delegates, summoned from all the Moslem countries of the world, ought to meet annually in Constantinople to discuss questions of interest to the Moslem world. Branches of the Committee

[74] H. Stuermer, *Zwei Kriegsjahre in Konstantinopel*, 159–60. Feroz Ahmad, *The Young Turks*, 154–5. E.J. Zürcher, *The Unionist Factor*, 23, 76. Landau, 'Pan-Islam and Pan-Turkism', 44.

[75] See, among others, R. Pinon, *L'Europe et la Jeune Turquie*, 133.

[76] Luke, 143.

[77] Ibid. 144.

[78] Report no. 51 of the Government of India, Foreign Dept., External B, Oct. 1909, referred to by R.L. Shukla, 'The Pan-Islamic Policy of the Young Turks and India', 303, 306.

[79] 'Exploiting the Crescent', 7.

[80] Pinon, *L'Europe et la Jeune Turquie*, 133–5.

[81] *The Times*, 27 Dec. 1911, quoted in *Mohammedan History*, 65–6. See also below, app. Q. Cf. D.S. Margoliouth, *Pan-Islamism*, 11–12. 'Turkey, Russia and Islam', 115.

should be formed in all Moslem countries, especially in Russia and Persia. The Mohammedans of Russia ought to be persuaded to make revolutionary propaganda among Russian soldiers. As many Tatars as possible should be induced to become members of the seven branches of the Committee which already exist in Russia. Efforts should be made to bring about an understanding between Persia and Turkey, with the ultimate object of effecting a political and economical union between the two countries. The Turks in Bulgaria and in Bosnia and Herzegovina, who should be advised not to emigrate, should be organized in such a way that they would be in touch with the Committee of Union and Progress. Large numbers of Turkish boys from Bulgaria ought to be educated in Turkey, and subsequently sent back as masters to the Bulgarian schools. Schools must be opened with the object of pushing the Turkish language among Pomaks [Moslem Bulgarians], in the hope of making them forget the Bulgarian language. Turkish teachers should also be sent to Bosnia and Herzegovina, and attempts made to persuade the Turks in these provinces not to favour the Serbian aspirations and to learn German rather than Serbian in the schools.

However, an important faction in the leading circles of the Committee of Union and Progress, probably inspired and led by Enver Pasha (1881–1922), had earlier opted for Pan-Turkism, as the leading state ideology. Initiated by Tatars and others in the Tsarist Empire in the late nineteenth century and imported into the early twentieth-century Ottoman Empire, this ideology aimed at a cultural and subsequently political union of all elements of Turkic origin.[82] At the 1909 congress of the Committee of Union and Progress in Salonica, it was adopted as a two-pronged policy,[83] reaffirmed at the 1910 congress:[84] (i) To mobilize the support of all Turks in the Ottoman Empire, by encouraging their political, cultural, and economic preponderance, in order to ensure the survival of the empire by basing it on its most trustworthy component. (ii) To work for the solidarity and eventual union of all groups of Turkic origin everywhere, in order to compensate for the recent loss of

[82] Landau, *Pan-Turkism in Turkey*, and sources mentioned in its footnotes and bibliography.

[83] FO 882/15, V. Vivian's memorandum, dated 30 July 1917, in the files of the (British) Arab Bureau, Egypt.

[84] AE, NS, Turquie, vol. 7, esp. fo. 155, Levant no. 67, reporting to Pichon, French Minister for Foreign Affairs, dated in Salonica, 17 Nov. 1910.

territories. This policy found expression, in practical terms, in a rapid Turkification of culture and economy within the Ottoman Empire, and a propaganda drive among Turkic groups outside it.

The years immediately preceding World War I thus witnessed the concurrent pursuit of a four-directional policy, to suit in various degrees the different ideologies then prevalent in the Ottoman Empire. Modernization was advocated for the benefit of the local élites sympathizing with Western values and of European public opinion; Ottomanism, for many minority groups, as the keynote of consensual internal politics; Pan-Islam, for the Turks, Arabs, and other Muslim groups within the empire and abroad; and Pan-Turkism, for the Turks of the empire and Turkic groups in Russia and elswhere.[85] This overall policy reflected not only the differences of opinion within the leadership of the Committee of Union and Progress,[86] which included traditionalists,[87] but also the wish, desperate in its intensity, to recruit support in every possible quarter at a time of great national danger. As is usual in such cases, however, the multisided ideological approach had, besides its obvious assets, certain liabilities. The traditional-minded resented the sympathy shown to the modernizing elements; the Muslims and the Turks criticized the attempts, at the beginning of Young Turk rule, to grant minority groups (Christians and Jews) equality; later, the non-Muslim elements were unhappy about the renewed majorization of Muslims in public life and business; and the non-Turkish elements were incensed at the introduction of the Turkish language in schools (for instance, at the expense of Arabic) and the added economic opportunities offered the Turks in business and commerce (generally at the expense of the Christian groups).[88]

The Committee of Union and Progress soon adopted a more aggressive stance in Pan-Islamic activities and propaganda.[89]

[85] Landau, 'Pan-Islam and Pan-Turkism', 44.

[86] X, 'Les Courants politiques', 176–7.

[87] PA, Türkei no. 198, vol. 6, German Ambassador Marschall to Bethmann Hollweg, dated in Istanbul, 22 Apr. 1911.

[88] Pears, 'Turkey, Islam, and Turanianism' 379. al-Jundī, *al-Yaqẓa al-Islāmiyya* ..., 105–12. J. Pozzi, *Le Khalifat et les revendications arabes*, 14–15.

[89] 'Exploiting the Crescent', 7.

It encouraged the establishment in Istanbul of associations of immigrant Muslims.[90] Among its early Pan-Islamic moves abroad, it encouraged the establishment of an association in Iran, about 1910, entitled *Ittiḥād ol-Islām* and prompted it to act for an all-Islamic union with the Ottoman Empire.[91] Discounting the significance of Sunnite–Shiite differences, the Committee of Union and Progress repeatedly attempted to draw closer to Iran, as a part of its grand design to set up a union with Iran and Afghanistan as a first step to a wider Pan-Islamic one. In 1911, for this purpose, the Committee sent agents to Van, in Eastern Anatolia, to enter Iran.[92] Early in the same year, it was involved in a meeting of Ulema, held at Najaf, an important Shiite centre of Iraq. The meeting decided that the differences of views between Sunnites and Shiites were not important enough to prevent an alliance and cooperation between the Ottoman Empire and Iran.[93]

A German consular report from Baghdad provides us with some additional information about Pan-Islamic activities sponsored by the Committee of Union and Progress in Najaf, early in 1911. The Imam of the Great Mosque and the protector of Caliph ʿAlī's grave in Najaf was preaching Pan-Islam intensively there; in addition, he was distributing quantities of Pan-Islamic pamphlets, provided by officers of the Committee. One of these travelled to Baghdad to organize a mass meeting, in which a Pan-Islamic *fatwā* by this Imam would be read. The British Consul-General in Baghdad, fearing that the meeting might turn into an anti-British demonstration, persuaded the Ottoman governor to ban it and send the officer back to Najaf.[94] Efforts for a Sunnite–Shiite *rapprochement*, implying an Ottoman-Persian one as well, in a Pan-Islamic perspective, continued, finding support in a Persian weekly, *Ḥabl-ol-Matīn*,[95] published in Calcutta.

[90] Marunov, 39, lists these associations and attributes to them a Pan-Islamic character.

[91] M. Larcher, *La Guerre turque dans la guerre mondiale*, 434. Lancher was a French general staff officer who could read Turkish.

[92] Shukla, 304, based on a report by Lowther, British Ambassador in Istanbul.

[93] Reported in a letter published in the Egyptian monthly *al-Manār*, February 1911, summarized by Margoliouth, *Pan-Islamism*, 13.

[94] PA, Orientalia Generalia 9, vol. 4, German Consul Hesse's report to Bethmann Hollweg, dated in Baghdad, 20 Jan. 1911.

[95] See *Revue du Monde Musulman*, 25 (1913), 349–50.

There are also some indications of agents being sent by the Committee of Union and Progress, to publicize Pan-Islam, during the Balkan Wars of 1912–13, and to recruit Kurds, again in the name of Pan-Islam.[96] However, success seems to have been rather limited, mainly because most of the above, perhaps all, were provisional, *ad hoc* measures. The Committee of Union and Progress increasingly felt the need for regular activists and institutionalized bodies. In 1911, Mehmet Cavid (1875–1926), one of the Committee's leaders, proclaimed the need to train scholars and mullahs, in order to spread the message of Pan-Islam in any place inhabited by Muslims.[97]

A Pan-Islamic league, set up in Istanbul in January 1913 with the inspiration and funding of the Committee of Union and Progress, became a more durable organization than its predecessors. Named the Benevolent Islamic Society (*Cemiyet-i Hayriye-yi İslamiye*),[98] it was made up of Turks, Egyptians, Tunisians, Tripolitanians, Arabs from Yemen and Hijaz, and Indians.[99] It was soon obvious that its humanitarian–philanthropic and educational platform was a cover for its Pan-Islamic designs.[100] One of its founder members, Shakīb Arslān (1869–1946), has left us an eyewitness account of this society's establishment in his autobiography.[101] He joined it since he was given to understand that it aimed at improving relations between Turks and Arabs. However, it was meant also to undertake political and military operations in countries where the Ottoman army could not openly intervene,[102] such as Tunisia.[103]

Thanks to a report by Sierra, the Consul-General of Italy in Cairo, more details of this association are known.[104] It was con-

[96] Fadyeyeva, *Ofitsia'lniye doktrini*, 200.

[97] Marunov, 40, based on the Russian state archives.

[98] Erroneously called, in one of the sources, *Cemiyet-i Hayriye Hamiye*.

[99] *Mohammedan History*, 65, 68. See also G. Samné, *Le Khalifat et le Panislamisme*, 15. Zeine, *Arab–Turkish Relations and the Emergence of Arab Nationalism*, 76.

[100] Samné, 15–16.

[101] Shakīb Arslān, *Sīra dhātiyya*, 100–2. See my transl. below, app. K.

[102] R. Simon, *Libya between Ottomanism and Nationalism*, 122–4.

[103] For which see a report in the Tunisians archives, repr. by B. Tlili, 'Au seuil du nationalisme tunisien', 227 ff.

[104] Sierra's original report, undated, is enclosed in MBZ, dossier 453–4 (A. 190) A. 190, 14704), Netherlands Ambassador's dispatch no. 1029/219, to the Minister for Foreign Affairs, dated in Ramleh (Cairo), 8 July 1914.

sidered important enough by the Committee of Union and Progress for the Ottoman Crown Prince, Yusuf İzzeddin (1857–1915), to give his patronage and Said Halim (who was soon to become Grand Vizir) to be its first secretary-general. The association was managed by distinguished personalities connected to Pan-Islam, such as Sharīf ʿAlī Ḥaydar Pasha (Senator from Mecca, 1865/6–1935), the above Druse Shakīb Arslān (Member of Parliament from Hawran, of whom more will be said presently), ʿAbd al-ʿAzīz Shāwīsh (?1872–1929) (whose important Pan-Islamic activity will be dealt with below), Enver Pasha's uncle, Halil Bey [Kut] (1881–1957), the Tunisians Ṣāliḥ al-Sharīf (1866/7–1921) and ʿAlī Bāş Ḥamba (1879–1919),[105] and Yūsuf Shatwān Bey (?1870–1952),[106] the Society's second secretary-general (formerly a delegate from Benghazi to the Ottoman Parliament, whose involvement must have been especially worrying for Sierra, as Italy was still having some trouble in that newly conquered area). The association was also putting out, in Istanbul, a fortnightly (with Yūsuf Shatwān as the publisher responsible), entitled *Jihān-i Islām* (The World of Islam), in Turkish, Arabic, Persian, and Urdu—obviously for both internal and external consumption, in a Pan-Islamic spirit; and it was supporting and inspiring in Egypt a nationalist newspaper, entitled *al-Shaʿb* (The People). According to reports from Istanbul about this association,[107] in 1915 it was still publishing *Jihān-i Islām* (then as a monthly), and was busy with subversive activities in enemy territories.[108] Yūsuf Shatwān was the central figure in both association and monthly. The association's activities seemed to be important enough for Baron Max von Oppenheim (1860–1946), Director of Oriental Affairs in the German Ministry for Foreign Affairs during

[105] On Ṣāliḥ al-Sharīf al-Tūnisī and ʿAli Bāş Ḥamba, see Bardin, 190–6. Cf. *Türk Ansiklopedisi*, xviii. 428, s.v. ʿAlī Bāş Ḥamba and his brother Muḥammad were characteristic of those Tunisians who believed in the feasibility of an anti-French uprising and looked to the Sultan–Caliph for Pan-Islamic guidance in the matter. See Cleveland, *Islam against the West*, 92.

[106] On these, cf. Simon, 357 n. 33.

[107] MBZ, dossier 453–4 (A. 190), A. 190 (22061), Netherlands Ambassador's dispatch no. 6005/20642, to the Minister for Foreign Affairs, dated in Pera, 14 May 1915, apps. IV and V, being reports by the Embassy's dragoman, dated 11 May 1915.

[108] Simon, 124, based on Italian documents.

World War I, to visit its headquarters in Istanbul and congratulate it on its services.[109]

A parallel Pan-Islamic association, entitled *Encümen-i Terekki-yi İslam* (Society for the Progress of Islam) was set up early in 1913 in Geneva. Its bulletin, bearing the same name, but with a French subtitle, *Progrès de l'Islam*, started to appear in Frebruary 1913 and called for a close *rapprochement* between all Muslim nations and co-operation in intellectual and economic activities—against the designs of France and Great Britain.[110]

These activities of the Committee of Union and Progress, albeit significant, were sporadic.

4. *The Ottoman Empire and the World War*

The significance of Pan-Islam appeared to increase with the Ottoman Empire's entry into World War I in 1914. For the empire, one of the important objectives of this war was the liberation of both co-religionists and people of ethnic relations from the foreign yoke.[111] From the proclamation of the Holy War (to which we shall refer presently) to Enver's repeated calls to India's Muslims to rise against the British[112] and to the appellation of 'Army of Islam' for the corps advancing in 1918 towards Baku to liberate the Muslims in Russia,[113] the Pan-Islamic card was played time and again by the Government of the Committee of Union and Progress. This was done at both the official and unofficial levels.

Over the years, several explanations have been offered as to why the Ottoman Empire entered the war—and did so as the ally of the Central Powers. None of the reasons adduced seems fully conclusive, although an answer may lie in their combined impact. We are interested here solely in the Pan-Islamic

[109] Acc. to the Turkish newspaper *Tanin* (Istanbul), 14/27 Apr. 1915.

[110] Zeine, 76–7. ʿAbd al-Raḥmān al-Rāfiʿī, *Muḥammad Farīd ramz al-ikhlāṣ wa-ʾl-taḍḥiya*, 364–6. We shall discuss this periodical below, at the end of ch. II.

[111] Feroz Ahmad, *The Young Turks*, 154–5.

[112] Mohammad Sadiq, *The Turkish Revolution and the Indian Freedom Movement*, 45.

[113] B. Lewis, *Middle East*, 107. U. Trumpener, *Germany and the Ottoman Empire*, 185.

factor. The consideration of liberating the Muslims outside the empire's borders (and thus, incidentally, securing safer frontiers), important enough to be mentioned in the empire's proclamation of war on Russia,[114] was one of the war's major objectives. While the Young Turk leaders understood that they were unable to liberate the Muslims alone, they could still choose their allies between the Central Powers and the Entente. In the former, Austro-Hungary was ruling over Muslims in Bosnia and elsewhere, while Italy had annexed Tripolitania and Cyrenaica only shortly before the outbreak of the war. However, Great Britain, France, and Russia, for their part, each ruled over tens of millions of Muslims. Moreover, the Young Turk and German leaders seemed to have many interests in common, at least in so far as the objectives of the war were concerned. Both sides hoped, for their respective ends, to eliminate Russia from Balkan and Middle Eastern affairs and to shake the colonial empires of Great Britain and France.[115] The Turks wished, also, to reassert Ottoman pre-eminence in the Muslim world and to regain their lost territories.[116] Holding on to Medina against all odds[117] and the repeated attempts to liberate Muslim populations in the Caucasus and Azerbaijan and set up independent states there[118] were indications of Enver's Pan-Islamic and Pan-Turkish tendencies.[119] As for the Germans, they had no serious misgivings about Pan-Islam and could point to the friendly attitude of Kaiser Wilhelm II and other German personalities towards both the Ottoman Empire and the Muslim world.

Indeed, it seems that the German leadership, in basing so much of their global politics, in both peace and war, on Pan-Islamic solidarity, had overrated its practical significance at

[114] Landau, *Pan-Turkism in Turkey*, 52.

[115] E.M. Earle, *Turkey, the Great Powers and the Bagdad Railway*, 276.

[116] Trumpener, 113–14.

[117] PA, Grosses Hauptquartier, Türkei 41, vol. 2, Under-Secretary of the Ministry for Foreign Affairs Zimmermann's cable no. 387, dated in Berlin, 14 Mar. 1916. Cf. unsigned report no. A 28095, dated in Berlin, 14 Oct. 1916.

[118] W. Zürrer, *Kaukasien 1918–1921*, ch. 1, esp. 28 ff.

[119] This was perceived by the Germans, also. See PA, Grosses Hauptquartier, Türkei 41, vol. 6, General von Lossow's cable no. 592, dated in Berlin, 6 July 1918; ibid., vol. 7, aide-mémoire presented by the Ottoman Embassy in Berlin on 10 Sept. 1918.

that time,[120] as had the Austrians (according to their own archives).[121] The grand design of Kaiser Wilhelm II and the German General Staff regarding Pan-Islam seems to have counted on a general uprising of Muslims in India, Afghanistan, Turkestan, and North Africa, thus weakening militarily the extremities of, respectively, Great Britain, Russia, and France, while the German armies were to deal death blows at their vital parts.[122]

It was Baron Max von Oppenheim, who very probably[123] was largely responsible, in the pre-World War I years, for persuading the German top political leadership about the merits of a Pan-Islamic policy. As adviser, or *Legationsrat*, to the German Consulate-General in Cairo from 1885, von Oppenheim spent more than thirty years among Muslims. He published several works, including one on his travels, *Von Mittelmeer zum Persischen Golf durch den Hauran, die Syrische Wüste und Mesopotamien* (1899–1900), and an even larger one, *Die Beduinen* (1939–68). Thanks to his fluency in Arabic, his many contacts in North Africa, Syria, and elsewhere, his travels, and his regular reading of the local press, von Oppenheim became an expert in Islamic matters; his services, indeed, were called upon frequently, as in the crisis over Morocco in 1905. He wrote hundreds of reports,[124] mostly analytical and interpretative, about political, cultural, and religious affairs in Muslim communities, from Morocco to China and from the Caucasus to Yemen. He devoted special attention to the politics of Pan-Islam, which he considered of particular importance, perhaps somewhat exaggeratedly. Von Oppenheim perceived political Pan-Islam as defensive, rather than aggressive, and from 1898,[125] noted an increase in Pan-Islamic sentiment in Egypt[126]

[120] Ahmed Emin, *Turkey in the World War*, 120.

[121] H. Gardos, 'Ballhausplatz und Hohe Pforte im Kriegsjahr 1915', esp. 262 ff.

[122] J.H. Rose, '1815 and 1915', 15.

[123] The only study of von Oppenheim's career, R.L. Melka's 'Max Freiherr von Oppenheim', unfortunately devotes less than a page to his pre-1918 activities.

[124] These are preserved, in 12 vols., in PA, Orientalia Generalia 9, no. 1, as well as in other series of PA.

[125] Ibid. vol. 3, von Oppenheim's report, dated 5 July 1898, entitled 'Die Panislamische Bewegung'.

[126] e.g. ibid., vol. 2, von Oppenheim's no. 33, to Hohenlohe-Schillingfürst,

and elsewhere, which he thought was responsible even for the increased readership and distribution of Muslim newspapers (he listed 344 newspapers in 1908).[127] Von Oppenheim was himself accused of dabbling in Pan-Islamic propaganda,[128] an allegation which he denied categorically, attributing it to an anti-German campaign.[129] However, he may well have stretched truth a little to make a point, as in his reporting that Aga Khan had told him, in an interview in 1909, that Hindus and Muslims would rise against the British, in case of war.[130]

In von Oppenheim's own view it was essential for Germany to persist in its policy of friendship with the Muslims, especially with the Ottoman Sultan—even more so since Great Britain and France feared the power of Pan-Islam,[131] and the former, at least, aimed at abolishing the Caliphate or dominating it.[132] In his own words,

> In a great European war, especially if Turkey participates in it against England, one may certainly expect an overall revolt of the Muslims in the British colonies ... In such a war, these colonies would be, along with Turkey, the most dangerous enemy of an England strong on the seas. British soldiers would be unable to invade Inner Turkey and, in addition, England would need a large part of its navy and almost its entire army [just] in order to keep its colonies.[133]

dated in Cairo, 23 Apr. 1897; vol. 4, his no. 91, to the same, dated in Cairo, 28 May, 1900; vol. 8, his no. 282, dated in Cairo, 18 May 1906 (Afghans and other Muslims wished to enlist and fight for Islam); vol. 10, his no. 387, to von Bülow, dated in Cairo, 16 May 1908.

[127] PA, Orientalia Generalia 9, no. 2, vol. 1, his no. 406, to von Bülow, dated in Berlin, 30 Nov. 1908 and encs.

[128] Chiefly, in the British and French press. See e.g., *Journal des Débats* (Paris), 10 Feb. 1906, repr. in *Les Pyramides* (Cairo), 15 Feb. 1906.

[129] PA, Orientalia Generalia 9, no. 1, vol. 8, von Oppenheim's letter to the German representative in Egypt, dated in Cairo, 16 Feb. 1906.

[130] Ibid., vol. 11, his no. 412, to von Bülow, dated in Cairo, 20 Feb. 1909. It is most unlikely that Aga Khan, pro-British and circumspect as he was, would offer such opinions to a German diplomat. His real views in the matter were published later, as follows, 'No Islamic interest was threatened and our religion was not in peril. Nor was Turkey in peril ... Turkey has so disastrously shown herself a tool in German hands'. See *The Times*, 22 June 1916.

[131] PA, Orientalia Generalia 9, no. 2, vol. 1, his no. 307, to von Bülow, dated in Berlin, 8 Aug. 1906.

[132] PA, Orientalia Generalia 9, no. 1, vol. 10, his no. 398, to von Bülow, dated in San Stefano, 26 May 1908.

[133] Ibid.

Such arguments convinced not merely the German Ambassador in Istanbul, von Wangenheim,[134] but also Kaiser Wilhelm II, who regularly read von Oppenheim's reports and, in the case of the last-quoted, noted happily in the margin 'gut, richtig!' (good, correct!). Indeed, one may view much of the Kaiser's Pan-Islamic and pro-Ottoman policy—his visits to Istanbul, Jerusalem, and Damascus, and later (in March–April 1905) to Tangier,[135] his protection of the Muslims in China and donations to Muslims elsewhere,[136] and, lastly, his overall support for a Pan-Islamic policy before and during World War I—as largely influenced by the understanding of von Oppenheim (who, during the war, headed the German 'Intelligence Office for the East'). This is significant since Kaiser Wilhelm II did not just reign, but ruled, participating actively in major policy decisions, including military ones. Hence his attitude to Pan-Islam had an impact. It is suggestive that, as late as November 1918, the Kaiser was still looking for ways and means to persuade the sheikh of the Sanūsīs (on Pan-Islamic grounds, no doubt) to harass the British in Egypt.[137] The Austro-Hungarian Ministry for Foreign Affairs, incidentally, was no less certain about the power of Pan-Islam.[138]

In addition, an entire battery of German pamphlets, chiefly intended for the German-reading public, discussed in detail German options for benefiting militarily from Pan-Islam and elaborated during the war years such themes as the mutual friendship between Germany and the Ottoman Empire or the Muslim peoples, the Pan-Islamic hold of the Sultan–Caliph over all Muslims, and the expectation that the call for Holy War in Istanbul would arouse these Muslims against the Powers of the Triple Entente.[139] The fact that most appeared in 1914

[134] PA, Orientalia Generalia 9, no. 1, vol. 5, von Wangenheim's no. 325, to Bethmann Hollweg, dated in Istanbul, 30 Sept. 1912.

[135] Stead, 'The Problem of the Near East', 588–90.

[136] Details in 'Est-ce la fin d'un rêve?', *La Dépêche Algérienne*, 12 Dec. 1909.

[137] PA, Grosses Hauptquartier, Türkei 41, vol. 7, von Grünau's cable no. 548, to the German Ministry for Foreign Affairs, dated in the General Headquarters, 15 Nov. 1918.

[138] See, for 1914, O. Hoetzch, ed., *Die internazionale Beziehungen im Zeitalter des Imperialismus*, ser. 2, 6/1, 227, document no. 297.

[139] Among these are the following: in 1914, C.H. Becker, *Deutschland und der Islam*; R. Tschudi, *Der Islam und der Krieg*; H. Grimme, *Islam und*

and 1915, tapering off later, was a sign of German disillusionment[140] with Pan-Islam. There was hardly any response in Iran, India, and elsewhere,[141] while only isolated cases of desertion by Muslims in both the British[142] and French forces occurred, and no uprisings. This contrasted sharply with such German predictions as the following, 'We should not give up hope of the entire Islam being mobilized like an army . . . by the order of its Caliph and the proclamation of the Holy War'.[143] Perhaps the explanation for the absence of any substantial Pan-Islamic response was the one offered by an anonymous Turk, in late 1916, that a Muslim response could be expected to follow Ottoman victories, not defeats.[144] None the less, energetic German propaganda among Muslims continued, indeed, almost to the end of the war, as we shall see later.

A great deal has been written about the *Jihad*, or Holy War in Islam,[145] and some of this has been discussed in the context of modern times.[146] It seems that *Jihad* had become a theme of interest for many Muslims in the years preceding World War I.[147] The Ottoman proclamation of *Jihad* on 11 November 1914—and of war—was introduced by five *fatwās*, or legal opinions, of Şeyhülislam.[148] The *fatwās* enjoined all Muslims

Weltkrieg; H. Grothe, *Deutschland, die Türkei und der Islam*; G. Diercks, *Hie Allah! Das Erwachen des Islam*. In 1915, E. Jäckh, *Die deutsch-türkische Waffenbruderschaft*; R. Schäfer, *Islam und Weltkrieg*; id., *Der deutsche Krieg, die Türkei, Islam und Christentum*. J. Froberger, *Weltkrieg und Islam* (1916). G. Galli, *Wesen, Wandel und Wirken des heiligen Krieges des Islams* (1918).

140 C.S. Hurgronje, *The Holy War 'Made in Germany'*, 34 ff., criticized several of the above German publications. For Hurgronje's own involvement with Pan-Islam, see G.S. van Krieken, *Snouck Hurgronje en het Panislamisme.*

141 Gardos, 277.

142 G.W. Bury's *Pan-Islam* mentions a few cases, from personal experience.

143 Tschudi, 14.

144 PA, Grosses Hauptquartier, Türkei 41, vol. 2, German Ministry for Foreign Affairs' no. A 31697, to von Grünau, dated in Berlin, 7 Dec. 1916, enc., being an anonymous, undated memorandum by a Turk who claimed to be a friend of Germany.

145 Cf. E. Tyan, 'Djihād', *EI*², s.v.

146 The most recent discussion in English seems to be R. Peters's *Islam and Colonialism*.

147 At least this was the conclusion, in 1910, of C.S. Hurgronje; see his 'Over Panislamisme', esp. 96.

148 English transl. of both the *fatwās* and the proclamation by G.L. Lewis, 'The Ottoman Proclamation of Jihād in 1914', 157–63. The footnotes refer to

to hasten to the *Jihad* against Russia, Britain, and France, also encouraging those in lands dominated by the above three powers, as well as Serbia and Montenegro, not to serve in their military forces, nor to take up arms against Germany and Austria, under penalties of sinning and meriting the fires of hell. The proclamation accused Russia, Britain, and France of enslaving millions of Muslims for their own base interests and of persecuting Muslims in India, Asia, and Central Africa. Hence the Caliph was summoning all Muslims to *Jihad*. He has ordered general mobilization and appealed to Muslims in both foreign-dominated and independent countries to join the Ottomans, with life and property, in the *Jihad*. The Crimea, Kazan, Turkestan, Bukhara, Khiva, India, China, Afghanistan, Iran, and Africa were specifically mentioned. The proclamation then reminded the Muslims that unity was the basis of their religion and that all should join the fighting against the mortal enemies of Islam—Russia, Britain, and France. Ulema and the press spread the message; the Ulema of Najaf reproclaimed the *Jihad* for the Shiites of Iran.[149] In a separate circular, which the Committee of Union and Progress sent to its local branches, it specified the war aims as it saw them, and among them: 'Religious considerations drive us towards liberating the Islamic world from the domination of the infidels'.[150] Enver tried to achieve this by military attacks in the Caucasus and Azerbaijan and by dispatching agents to stir up Muslim unrest in Turkestan.[151]

There are several reasons for the failure of the proclamation of the *Jihad* and the *fatwā*s to have much noticeable effect outside the Ottoman Empire, even though they were trans-

the Ottoman text and to contemporary French translations in the Istanbul newspaper *La Turquie*. A German transl. can be found in AE, Guerre 1914–1918, vol. 1650, no. 273, fo. 148, enc. in a report, fo. 147, of the French Minister in the Netherlands to Delcassé, French Minister for Foreign Affairs, dated in the Hague, 18 Nov. 1914; a somewhat different transl. ibid., fos. 178–80.

[149] Hoetzch, ed., ser. 2, 6/2, 487, document no. 570, Russian Ambassador Korostowecz's cable no. 586, to the Russian Minister for Foreign Affairs, dated in Tehran, 16/29 Nov. 1914.

[150] Quoted by T. Swietochowski, *Russian Azerbaijan, 1905–1920*, 76.

[151] Enver boasted about these agents; see PA, Grosses Hauptquartier, Türkei 41, vol. 1, General von Lossow's cable no. 14281 P, to General von Falkenhayn, dated in Istanbul, 22 Aug. 1916.

lated into Arabic, Persian, Urdu, and Tatar,[152] no doubt for mass distribtion in the empire and abroad, among the forces of the Entente.[153] Indeed, its impact was limited even within the empire itself. Among these reasons are the following: (i) Continued absence of any elaborate organization in Pan-Islamic activities which would have allowed for an efficient, systematic distribution of the proclamation and its exploitation. (ii) Administrative efforts of the Entente to prevent the entry of the proclamation into their Muslim territories and the prompting of Muslim Ulema and dignitaries there to exhort their co-religionists to disobey it. (iii) Reservations of many Muslims regarding the Government of the Ottoman Empire and, more particularly, the irreligiousness of the Committee of Union and Progress leaders (who, after all, had dethroned the legal Sultan–Caliph, Abdülhamid). (iv) Acquiescence of some parts of the foreign-dominated Muslim populations in their status (e.g. in India and Morocco). (v) Alternative priorities of Muslims within the Ottoman Empire, such as among nationalist Arab groups, whose aspirations could be served better by the downfall and disintegration of the empire. (vi) The fact—peculiar from an Islamic perspective—that the Sultan–Caliph was warring with certain Christian powers against others, not exactly a characteristic of the *Jihad*;[154] and, truly, the proclamation *had* ordered Muslims to refrain from attacking Germany and Austria. The Ottoman authorities had an equally serious problem with Italy, an ally of Germany and Austro-Hungary at the time; hence a representative of the Sultan in Libya convoked the principal Ulema and Arab notables in Tripoli, in December 1914, and read to them a cable, received from Istanbul, declaring that the *Jihad* did not affect Italy, the friend of the Ottoman Empire (*sic*!);[155] the message was then printed

[152] Peters, 91. See also Hurgronje, *The Holy War*, 32–4. Gardos, 260ff. E. Layer, *Confréries religieuses musulmanes et Marabouts, leur état et leur influence en Algérie*, 10–11.

[153] AE, Guerre 1914–1918, vol. 1651, fo. 5, dated in Bordeaux, 3 Dec. 1914.

[154] Cf. AE, Guerre 1914–1918—Turquie, vol. 867, fos. 217–18, note prepared in the Section de l'Afrique Occidentale et Equatoriale, dated in Paris, 3 Dec. 1914.

[155] NA, RG 59, 865c.00/14, US Consulate's no. 820, to the Secretary of State, dated in Tripoli, 12 Dec. 1914.

in Arabic and posted on the walls of the local mosques.[156]

All this considered, perhaps one ought not to see the proclamation of the *Jihad* as a complete failure. It certainly did not produce any dramatic results, in terms of numbers. As mentioned above, there were only scattered instances of Muslim desertions from the military forces of the Entente and no wide-scale uprising in its Muslim-inhabited territories. There were numerous and emphatic expressions of loyalty towards Great Britain by India's Muslims and towards France by North African ones.[157] However, matters did not develop as smoothly as Entente sources and later historians would have us believe. A secret French war report of 1 June 1916 is enlightening in this respect.[158] The following is a translation of the relevant passage:

> Turkey, Tripolitania, the Libyan desert, and Darfur have risen for the Holy War, without any doubt. Through the press, one gets the impression that India has remained loyal, but we know that the Dutch Indies have known some effervescence, Iran has been gravely troubled, Afghanistan excited and Egypt trembling. We are also aware that almost all the troubles which have burst in Africa, here and there, were coloured by religion; there is no doubt that the words 'Holy War' have been pronounced and exploited. True, North Africa, which comprises twelve to fifteen million souls, has given us more than 120,000 soldiers; but one cannot say that Islam in the Maghrib has entirely declared itself for us, except by word of mouth ... all the religious manifestations have generally been made in the name of Islam: they failed, indeed, but they caused no end of trouble to the Entente Powers.

Again, there were indeed no signs of large (or even small) foreign Muslim contingents arriving to fight for the Ottoman Empire and Pan-Islamic ideals; but no steps had been taken to bring them, and in countries governed by the Entente Powers they were not allowed to leave, naturally enough. However, one does not have any statistics of those Muslims who might have joined the war effort of the Entente, due to the proclama-

[156] NA, RG 59, 865c.00/15, US Consulate's no. 820, to the Secretary of State, dated in Tripoli, 16 Dec. 1914.

[157] A good number of the French ones were assembled and published in *Revue du Monde Musulman*, 29 (Dec. 1915), 1–386.

[158] AAT, Archives de la Guerre, 7N2106, *L'Islam dans la guerre*, mimeographed, 2 (1 June 1916) = 'Théâtres musulmans de la guerre', 169–71.

tion of the *Jihad*, and did not. Moreover, the qualitative gains of the proclamation were not so insignificant as the quantitative ones. Quite a few brilliant and dedicated Muslims hurried to Istanbul to share in what they considered the War of Islam. We shall presently discuss several of them and their service to the Ottoman war effort and the cause of Pan-Islam. Among the first to arrive was ʿAlī Pasha, the son of the Emir ʿAbd al-Qādir (1808–83), who had led a resistance movement to the French Occupation in North Africa. Yet another was Kâmil Pasha—a grandson of Shaykh Ali (1795–1871)—who had been the hero of the Muslim fighting against Russian domination in the Caucasus. While the latter occupied himself with forming new cavalry units of volunteers for the front, the former busied himself with Muslim prisoners of war in German camps, ensuring their good treatment and attempting to enlist them on the side of the Central Powers.[159] He was joined there by Abdürreşid İbrahim—already mentioned as a Pan-Islamic writer—and others.[160] The Muslims serving in the Entente forces were a matter of concern for both Ottomans and Germans; and the latter published (jointly, it seems) a book in Arabic and French on the Muslims in the French forces, contrasting this with their situation in the armies of the Central Powers.[161]

5. *Pan-Islamic Activity during the War*

Following the proclamation of *Jihad*, Pan-Islamic activity during World War I was (as it had been in Abdülhamid's reign) of two kinds—official and unofficial.

An example of the former is a joint Turco-German mission to Afghanistan, which reached Kabul in August 1915, reportedly bearing messages from the Ottoman Sultan and the German

[159] Reported in *La Turquie*, 30 Mar. 1915. See also AE, Guerre 1914–1918, vol. 1652, fo. 67, enc. in the Minister for War's no. 1929–9/11, to the Minister for Foreign Affairs, dated in Paris, 21 Apr. 1915.

[160] Bräker, *Kommunismus*, i. part 1, 60.

[161] The book by El-Hadj Abdallah (an active Pan-Islamist in World War I) was entitled, in French, *L'Islam dans l'armée française (guerre de 1914–1915)*. The French replied with a book of their own, *L'Islam dans l'armée française: Réplique à des mensonges*, of which a copy is available in AAT, Archives de la Guerre, 7N2104, attached to a letter of the Ministère de l'Intérieur—Police Spéciale, dated 2 June 1916.

Kaiser. Its purpose was to persuade the Emir of Afghanistan, Habibullah Khan, to join the cause of Pan-Islam, to abandon his neutrality, and make war against the British in India. The mission failed, due to the reluctance of the Emir to commit himself. However, its impact was considered so potentially dangerous by the British Chief Commissioner of the North West Frontier Province, Rous-Keppel, as to double the financial allocations to the local tribes, lest they be influenced by foreign Muslim agents.[162] In the same year, the Ṣalāḥiyya school for higher Islamic studies was set up in Jerusalem by Cemal Pasha (1872–1922), then Governor of Syria, as a Pan-Islamic centre to compete with al-Azhar in British-ruled Egypt (the school, however, later became a focus of Arab nationalism).

Much—probably, most—of the Pan-Islamic activity during the war years was carried out far from the limelight. It appears that, among other projects, Enver Pasha (then Minister for war), wishing to stir trouble among the Muslims of British India, suggested an expedition into India to Mustafa Kemal (1881–1938) (then a senior officer and later the founder and president of the Turkish Republic); but Mustafa Kemal refused this offer as too fanciful.[163]

A large part of the unofficial Pan-Islamic activity of the Committee of Union and Progress was carried out by the *Teşkilat-ı Mahsusa*, or 'Special Organization', set up by Abdülhamid for clandestine work. Its agents continued to serve the Committee, parallel with the agents of the Benevolent Islamic Society,[164] devoting much of their energy to guerrilla and sabotage activities among the Muslim subjects of the Triple Entente, chiefly in India and Central Asia, but also to Pan-Islamic propaganda there and in other areas (such as China).[165] Indeed, propaganda was the main Pan-Islamic activity of the Committee (and the Germans) during World War I.

[162] Lal Baha, 'Activities of Turkish Agents in Khyber during World War I', 185–92. On the activities of Maulvi Obeidullah Sindhi, a Sikh convert to Islam and an ardent Pan-Islamist, see E.C. Brown, *Har Dayal*, 205.

[163] Lord Kinross, *Ataturk*, 82–3. Kemal Arıburnu, *Atatürk*, 73. Of course, Enver may also have wished to send Mustafa Kemal far away from the Ottoman Empire.

[164] Simon, 123–4.

[165] A. Mango, 'Turkey in the Middle East', 226. Zürcher, 83–4, 129 n. 1. Landau, 'Pan-Islam and Pan-Turkism', 45.

6. *Pan-Islamic Propaganda during the War*

Propaganda, sometimes useful in peace-time, becomes all-important in war-time. Pan-Islamic propaganda was sponsored and fostered by both the Ottoman and German Governments during World War I. As early as 18 August 1914, Baron Max von Oppenheim, in two memoranda to Reichskanzler Bethmann Hollweg, emphasized the importance for the war effort of having an officially co-ordinated plan of translating war news for the Muslim peoples and suggested employing Muslim lecturers at various Orientalist institutes in Germay, both natives and foreigners. Printed in ten to twenty thousand copies, the translations ought to be distributed in the Maghrib and the Orient. He further suggested recruiting for the war effort the Islamic fraternities, the people of Mecca, and prominent Muslim businessmen, as well as exiles from Egypt, India, and Russia living in Istanbul—in co-operation with the Ottoman Government. A special war effort aimed at India's 70 million Muslims, preferably via Afghanistan, was recommended.[166] So a special translation bureau was set up, with von Oppenheim to head it; and the post office was ordered to plan for distributing its products.[167] Considerable amounts of time and money went into publishing and distributing numerous pamphlets among Muslims everywhere—at least seventy-two, by an unofficial count in Switzerland, in mid-1916—in Arabic, Turkish, Persian, Tatar, Urdu, and Chinese.[168]

While much of the material (perhaps most) was drawn up by the Ottomans, the bulk of the agents distributing it were Germans, active from the very start of the war in places far away from one another: the whole of North Africa, from Morocco to Egypt,[169] as well as Syria, Iran, and Kurdistan.[170]

[166] PA, WK no. 11, vol. 1, von Oppenheim's two memoranda to Bethmann Hollweg, both dated in Berlin, 18 Aug. 1914.

[167] Ibid. memorandum signed 'D.S.S.' and dated in Berlin, 23 Sept. 1914.

[168] J. Grande, 'Germany's Press Propaganda'. This was written in Berne on 29 July 1916.

[169] AE, Guerre 1914–1918, vol. 1650, fos. 2–3, the French Minister for Foreign Affairs to Defrance, French Chargé d'Affaires in Cairo, dated 20 Aug. 1914; fo. 31, Havas Agency telegram, dated in Rome, 20 Oct. 1914.

[170] Ibid., fo. 35, the French Commissaire in Morocco, General Lyautey, to the French Minister for War, dated in Rabat, 21 Oct. 1914.

While there is little detailed information about the Ottoman distribution of propaganda, a British report from 1917, based on French Intelligence sources, tells us about the German methods.[171] German Pan-Islamic propaganda was printed in Berlin. Some was transported by submarines to Morocco; that intended for Tunisia, Algeria, and other parts of North Africa was dispatched from Switzerland, disguised as parcels, via Italy; German consular officials were very active too, from those in Latin American countries to those in Switzerland, Spain and Portugal, Baghdad, Batavia, Peking, and Shanghai (all of whom had established contacts with local Muslim communities).

Co-operation between the Ottomans and Germans was based on consultation. There are at least two reports of joint meetings in Berlin. The first, in March 1916, was attended by the Ottoman Ambassador and Military Attaché in Berlin, several Turkish and Arab dignitaries, and four representatives of a newly formed Committee for the Rights of the Turco-Tatar Muslims in Russia.[172] At the second meeting, in September 1916, Enver Pasha led a delegation of Muslim personalities from Turkey, Afghanistan, Iran, Central Asia, Egypt, and North Africa.[173] Both meetings aimed at intensifying Pan-Islamic propaganda. Moreover, joint Ottoman–German offices were set up in Istanbul, Berlin, and Berne.[174] These comprised Turkish and non-Turkish Ulema, such as the Tunisians Ṣāliḥ al-Sharīf, sometimes called Ṣāliḥ al-Sharīf al-Tūnisī (who lectured at the two meetings in Berlin and was instrumental in setting up there a Deutsch–Islamisch Gesellschaft für Islamkunde)[175] and ʿAlī Bāsh Ḥamba,[176] along with other public

[171] FO 395/151, file 183632.

[172] *Tanin*, 28 Mar. 1916. See also AAT, Archives de la Guerre, 7N2104.

[173] The report was written by the Italian Admiralty, on 16 Oct. 1916. A copy can be found in AAT, Archives de la Guerre, 7N2104, in a French Intelligence report, sent to the French Minister for War on 28 Oct. 1916.

[174] Peters, 93. T.R. Sareen, *Indian Revolutionary Movement Abroad*, 163–4.

[175] PA, Orientalia Generalia 9, vol. 5, Ṣāliḥ al-Sharīf's memorandum of Feb. 1916, and Martin Hartmann's comments; vol. 6, M. Hartmann's letter to Zimmermann, Under-Secretary of the German Ministry for Foreign Affairs, dated in Berlin, 17 May 1916.

[176] For whom see ibid., vol. 8, German Ambassador Johann-Heinrich von Bernstorff's cable no. 1246, to the Ministry for Foreign Affairs, dated in Istanbul, 10 Oct. 1917.

figures like Shakīb Arslān[177] and the deposed Egyptian Khedive ʿAbbās Ḥilmī (reigned 1892–1914). The ex-Khedive set up in Lausanne an Office Musulman International, also called Mohammedanische Gesellschaft,[178] whose 8-page statutes were published in 1916. Although allegedly philanthropic, it had political Pan-Islamic goals, as when its secretary general, Lutfi Bey (former Ottoman Consul-General in Paris), wrote to United States' President Woodrow Wilson, in February 1917, requesting him to see to it that the war did not become a crusade against Islam.[179] These groups published numerous manifestos, pamphlets, and some periodicals—all intended to rouse the Muslims governed by Great Britain and France, and more particularly those serving in their military forces, to revolt.[180] Special efforts were made by Ulema in prisoner-of-war camps in Germany, to instil Pan-Islamic views into Muslim prisoners.[181] Indeed, the Germans proclaimed officially that they were not making war on Muslims and that Muslim prisoners of war captured by the Germans would immediately be set free and sent to Istanbul; such an announcement, in German, Spanish, and Arabic, was exhibited, in October 1914, at the German Consulate in Rabat[182] (incidentally, this was not honoured subsequently).

The entire Ottoman press, not unexpectedly, was harnessed for Ottoman and Pan-Islamic propaganda. The languages employed were determined by the readership and so I will consider publications according to the languages used. *Brotherhood*, an English monthly published in Istanbul during 1918,[183] was

177 For whose activities see Cleveland, *Islam against the West*.

178 PA, WK, no. 11, vol. 8, German Ambassador von Romberg's no. 2217, to Bethmann Hollweg, 'Geheim', dated in Berne, 6 Oct. 1916, enc. a report by Adviser Marum on the matter. PA, Orientalia Generalia 9, vol. 6, Marum's no. 1427, to Bethmann Hollweg, dated in Berne, 6 July 1916 and enc.; Marum's no. G. 1047, dated in Berne, 13 Oct. 1916; von Romberg's no. 2695, to Bethmann Hollweg, dated in Berne, 26 Nov. 1916. See also *Gazette de Lausanne*, 9 Oct. 1916, and *Le Matin* (Paris daily), 31 Dec. 1916.

179 PA, Orientalia Generalia 9, vol. 7, von Romberg's no. 513, to Bethmann Hollweg, dated in Berne, 23 Feb. 1917, enc.

180 Bury, 28–30, 80ff. Samné, 16–17.

181 Peters, 93–4.

182 AE, Guerre 1914–1918, vol. 1650, fo. 29, the French Minister for Foreign Affairs to Ambassador Bompard in Istanbul, dated in Bordeaux, 19 Oct. 1914.

183 IO has a copy, 1/6 (June 1918), of 32 pages. Before 1918 an Urdu version was published.

aimed at India, inciting its Muslim population to seek independence. *Jihān-i Islām*, however, was more versatile. It was continued during the war and put under the supervision of the Benevolent Islamic Society, the organization responsible for Pan-Islamic activities.[184] Edited by Yūsuf Shatwān, it was run, after 1914, by the Ottoman Ministry for War, headed by Enver Pasha.[185] In India, at least, it was distributed free of charge in many cities and towns; when its import was banned during the war, it was smuggled in.[186] Its greatest asset was that it was published in four major languages—Turkish, Arabic, Persian, and Urdu. These are only rarely combined in one issue, as when an urgent and grave matter was to be communicated to all Muslims, for instance, the proclamation of *Jihad* and the *fatwā* of the Şeyhülislam supporting it.[187] More usually, parallel editions in the four languages, of eight pages each, were published. A characteristic example is an issue containing an editorial headed, 'The best deed, next to Belief, is the *Jihad*'.[188] This was followed by an address of the Sultan-Caliph Mehmed Reşad to the Egyptians, saying that, although in the past he had chosen to be silent about the British occupying Egypt, he had always been fond of Egypt and would soon send an army to save it from foreign rule and return it to Egyptian administration. He ended by calling on all Egyptians to participate in the *Jihad*. Muslim legal advice followed, then news of the war.

The bulk of the magazines and newspapers, reflecting Ottoman-inspired propaganda, was in Arabic. Those published abroad, for example in North and South America, were generally supportive of the Ottoman war effort without, however, showing any marked interest in Pan-Islam, as their readers were mostly Christians from Syria and Lebanon who had emigrated to the New World. Examples are *al-Hacaik*[189] of Buenos Aires, *al-Faraed*[190] and *Ottomano*[191] of San Paolo[192] A greater

[184] The mast reads: *Taḥt ḥimāyat al-Jam'iyya al-khayriyya al-Islāmiyya.*
[185] Sareen, 165–6. An incomplete set is available at the Beyazıt Library, Istanbul. Several issues are stored in IO.
[186] Sharif al-Mujahid, 'Pan-Islamism', 110.
[187] *Jihān al-Islām*, 21 (undated).
[188] Ibid. 53. The Arabic version is *Afḍal al-a'māl ba'd al-īmān al-jihād.*
[189] Arabic, *al-Ḥaqā'iq.*
[190] Arabic, *al-Farā'id.*
[191] Arabic, *al-'Uthmānī jarīda 'Uthmāniyya jāmi'a.*
[192] A few issues of each, from 1916, are in the IO.

emphasis on Pan-Islam was evident in the Arabic newspapers published within the empire with Ottoman guidance and funding; understandably so, for the loyalty of some of the Arabs to the Sultan-Caliph needed reinforcement. The respected Sharīf Ḥusayn in Mecca, indeed, was one of the prominent Arabs suspected by the Ottomans, as his name had been mentioned in the European press as a potential alternative Caliph; he was duly persuaded to join Pan-Islamic propaganda.[193]

The number of Arabic newspapers and the effort invested indicate that a wide readership was envisaged. One paper was *al-ʿAdl* (Justice), published in Istanbul in 1908–23, in Turkish and Arabic, which attacked the injustice of British rule in Egypt and the Sudan and of French rule in Algeria, in the name of Islam and Pan-Islam.[194] In Damascus, *al-Raʾy al-ʿĀmm* (Public Opinion) followed the Ottoman and Pan-Islamic line;[195] for instance, it sharply attacked 'Sharīf Ḥusayn's betrayal'[196] (in switching his support to the British), using Islamic sources to present it as a Pan-Islamic betrayal. An Arabic newspaper, set up in Damascus on 29 April 1916, *Jarīdat al-Sharq* (The Newspaper of the Orient),[197] had on its editorial staff such renowned journalists of the time as Muḥammad Kurd ʿAlī (1876–1953), ʿAbd al-Qādir al-Maghribī (1867–1956),[198] and Shakīb Arslān.[199]

An Arabic weekly, clearly committed to Pan-Islamic propaganda, entitled *al-ʿĀlam al-Islāmī* (The Islamic World), started to appear in Istanbul in April 1916.[200] It was edited by ʿAbd al-ʿAzīz Shāwīsh. This was joined soon afterwards by a German weekly, appropriately entitled *Die Islamische Welt* and issued in Berlin from November 1916 to August 1918,[201]

193 C.E. Dawn, *From Ottomanism to Arabism*, 23, 51; and cf. pp. 83–4.

194 See e.g. *al-ʿAdl*, 8 June 1916, 1–2, and 24 July 1916, 2. Both issues are available in IO. Other issues are available in Hakkı Tarık Üs Library, Istanbul.

195 e.g. *al-Raʾy al-ʿāmm*, 26 Aug. 1916, available in Hakkı Tarık Üs Library, Istanbul.

196 Ibid. 1.

197 Some issues are available in the Oriental Dept. of the British Library (London), shelfmarked OP 237; others in the Firestone Library, Princeton.

198 Whose book *Jamāl al-Dīn al-Afghānī* ... has already been mentioned.

199 Cleveland, 'The Role of Islam', 87–8.

200 Its complete title was *al-ʿĀlam al-Islāmī jarīda siyāsiyya, dīniyya, ijtimāʿiyya, adabiyya, yawmiyya* (*sic*). An issue of 24 pages 1/5 (17 May 1916) is available in IO, shelfmarked Ar PP 44. The Bibliothèque Nationale, Paris, has all issues from 25 May 1916 to 17 May 1917.

201 A complete set is stored in the Bayerische Staatsbibliothek, Munich.

with German financing.[202] Although *al-ʿĀlam al-Islāmī* and *Die Islamische Welt* were each attuned to a different readership, they both claimed that the Ottoman Empire was speaking for all Muslims everywhere and appealed to them to unite and join the *Jihad*, liberate themselves, and revive the ancient Islamic community in its pristine glory.[203] As Shāwīsh had persuaded Enver, the Minister for War, of the importance of defending the interests of Turkey and Islam at home and abroad, both periodicals enjoyed the moral and financial support of the Ottoman Ministry for War.[204] However, the Arab version was more noticeably inflammatory, the German more matter-of-fact.

The former strived, first and foremost, to enlist Muslim support for the Central Powers and vilify the Entente.[205] However, both in its editorials and in its reporting of the war and printing news about Islamic lands and peoples, it consistently advocated a *rapprochement* between the Muslims themselves; this was the panacea for all the diseases of the Muslim world.[206] A 'common initiative of co-operation'[207] was strongly recommended. That Pan-Islam was the best and only way to achieve this was spelt out in an article by Abdul Malik Hamsa[208] (most other articles were unsigned, but were probably written by ʿAbd al-ʿAzīz Shāwīsh), as follows, 'Pan-Islam is an expression of the strong ties among Muslims, east and west [i.e. everywhere] and of their overall readiness to co-operate with one another and demonstrate solidarity in resisting together any harm that may befall any of them, in the manner and according to the regulations prescribed by the teachings of

[202] Acc. to the memoirs of an Egyptian nationalist involved in the matter, Muḥammad Farīd, *Awrāq Muḥammad Farīd*, 1. *Mudhakkirātī baʿd al-hijra*, 331.

[203] *al-ʿĀlam al-Islāmī* (Istanbul), 1/1 (29 Apr. 1916), quoted by Cleveland, 'The Role of Islam', 88.

[204] PA, Orientalia Generalia 9, no. 2, vol. 5, German Adviser Wilhelm von Radowitz's no. 619, to Bethmann Hollweg, dated in Istanbul, 7 Oct. 1916

[205] For a characteristic example, see 'al-ʿĀlam al-Islāmī wa-duwal al-iʾtilāf al-muthallath', *al-ʿĀlam al-Islāmī*, 22 June 1916, 1–4; 29 June 1916, 1–4.

[206] 'al-ʿĀlam al-Islāmī, amrāḍuh wa-ʿilājuh', ibid., 6 July 1916, 1–6; 13 July 1916, 1–6; 20 July 1916, 1–6; 27 July 1916, 1–6; 30 July 1916, 1–6; 10 Aug. 1916, 1–6; 17 Aug. 1916, 1–6; 31 Aug. 1916, 1–4; 14 Sept. 1916, 1–7; 28 Sept. 1916, 1–4; 5 Oct. 1916, 1–5.

[207] 'al-Tashabbus al-jamʿī al-taʿāwunī', ibid., 6 July 1916, 5–6.

[208] This is how he wrote his name in the German way.

Islam'.[209] The practical conclusion enjoined on all Muslims was to go to war with the infidels,[210] for the final victory of 'the [Islamic] religion and state'.[211] Praising the Ottoman Sultan as the leader of Islam, it appealed to all Muslims for allegiance and support for him.[212] A logical follow-up to this appeal was the violent attack on Sharīf Ḥusayn of Mecca for rebelling against the Sultan and the interests of Islam.[213]

In *Die Islamische Welt's* sixteen issues (some were double ones), covering a total of 1,008 pages, emphasis was laid on Islam, the Ottoman Empire, the relations of Germans with Islam and the Turks; the policies of Great Britain and France in Muslim lands were condemned, and the right of Egypt and other Muslim countries to freedom—with German help—was championed. Pan-Islam was a recurring topic. For instance, Abdul Malik Hamsa, one of the two editors (the other was the ubiquitous ʿAbd al-ʿAzīz Shāwīsh)[214] explained in some detail his views on Pan-Islam in the very first issue.[215] This was the concept of a brotherly union of the Muslims dispersed throughout the world, striving together for progress and the crushing of hostile elements. It also implied mutual assistance and joint defence against all attacks, regardless of nationality, race, or colour. He further maintained that Pan-Islam did not conflict with patriotism, since both wanted the Muslims to shake off foreign domination. In another article, Hamsa called for the channelling of Pan-Islam into a Turkish–Arab alliance.[216] And an anonymous article on the Caliphate as an Islamic bond[217] argued for the need of maintaining the Caliphate, due to its religious and *political* functions of uniting all Muslims.

[209] 'al-Jāmiʿa al-Islāmiyya', *al-ʿĀlam al-Islāmī*, 9 Nov. 1916, 14–18.

[210] 'Wa-aḥabb al-Muslimīn naḥw dawlat al-Khilāfa al-ʿUẓmā', ibid., 28 Sept. 1916, 15–19.

[211] Ibid. 19: *al-Dīn wa-'l-dawla.*

[212] 'al-Khalīfa al-aʿẓam wa-'l-ittiḥād al-Islāmī', *al-ʿĀlam al-Islāmī*, 30 Nov. 1916, 1–4.

[213] 'Abū Jahl al-qarn al-tāsiʿ ʿashar', ibid., 30 July 1916, 9–24; 28 Sept. 1916, 4–12; 5 Oct. 1916, 5–15.

[214] I. Rothmann, 'Ägypten im Exil (1914–1918)', 11.

[215] Abdul Malik Hamsa, 'Der Panislamismus: Seine Bedeutung und seine Grenzen', 18–20.

[216] Id., 'Der Panislamismus: Seine praktische Ziele', 384–6.

[217] 'Das Kalifat als geistiges Band', ibid. 361.

In addition to several Pan-Islamic periodicals of World War I, quite a few manifestos and propaganda tracts with Pan-Islamic proclivities have survived, mainly in some public libraries and the archives of the warring states. We shall examine a sample of the more characteristic ones, starting with several publications in Arabic. Most of these date from 1915 (as their preparation, after the Ottoman Empire's entry into the war in November 1914, took some time) to 1918.

Many are anonymous, like a 30-page pamphlet smuggled into Egypt in 1915[218] which said, 'Every nation honoured by the faith of Islam should have the same feelings towards other Islamic nations as towards herself. If any of these nations is placed in the hands of infidels, all other Muslim nations are religiously bound to help that nation to rid herself of the government of infidels'. A description of the sorry state of the Muslim world followed, country by country, along with a call to *Jihad*.

Another anonymous Arabic publication, issued in Istanbul but undated, was entitled *The Muslim Revival in the Seat of the Supreme Caliphate*,[219] rubber-stamped with an insignia (which leads one to suppose that this was also the name of an association). It started by asserting that 'The Muslim revival aims at regaining the robbed rights of all Muslims in the rest of the inhabited world. It is headed by His Majesty the mightiest Caliph ... Oh, Muslims, you are aware of the hostility of France, England, Russia, and Italy, the four Christian states which wish to see all Muslims under their domination'. After having described the iniquities of the Christian states toward Muslims, the broadsheet ended as follows,

> These four states have agreed between them to obliterate the Muslim state, to enter the seat of the Caliphate [i.e. Istanbul], to depose the Caliph from his throne; then to enter the Holy Kaʿba and to transfer

[218] FO 371/2495, file 133531, a report by Sir H. McMahon to Sir Edward Grey, dated in Ramleh (Egypt), 6 Sept. 1915, transmitting extracts (in transl.) of this pamphlet, printed at the Khayriyya Press in Istanbul and published by the Religious Society of Defence at the Seat of the Caliphate. Acc. to British Intelligence, ʿAbd al-ʿAzīz Shāwīsh might have authored it (but no proof is adduced).

[219] *al Nahḍa al-Islāmiyya bi-Dār al-Khilāfa al-ʿUlyā*. This broadsheet is available in IO, shelfmarked Arab PP 73.

the Prophet's tomb and the Caliphate to their countries, transforming the Holy Kaʿba into a church; and to appoint a ruler for the Muslims, subject to them, whom they would name Caliph ...

The conception that the war was one of all Christians against all Muslims was drummed in, also, in several other anonymous leaflets, calling on Indians, Egyptians, and Afghans to desert the Allied forces and join the army of Islam.[220] A characteristic example follows.[221] The British attack was directed against the entire Muslim community (*al-Umma al-Islāmiyya*). These enemies of Islam spread their venom and caused dissension between Arabs and Turks, Indians and Afghans, thus weakening the overall support for Islam and the Caliph's throne. Thus were Muslims deceived and induced to fight the sole surviving Muslim kingdom representing the Islamic Caliphate—the Ottoman state, the sole defender of the Muslim flag and the Islamic religion. Muslims should not fight their brethren-in-faith, or else the pillars of Islam might crumble.

Some tracts were signed by their authors. Shaykh Ṣāḥib Zādeh Asʿad (1855–1928), who claimed to be a leader[222] of the Naqshbandiyya Fraternity in Damacus, published in 1915 a 15-page pamphlet in Arabic (based on his article in the Damascus periodical *al-Ra'y al-ʿĀmm*) entitled *An Important Proclamation to the World of Islam*.[223] In it, he praised the Sultan–Caliph for his decision to proclaim the *Jihad* and called on all Muslims to participate in it, for the defence of religion and fatherland, against Britain and France, the enemies of Islam

[220] Examples are: *Ayyuhā al-ikhwān al-Muslimūn*, 1 p. (IO, Arab PP 90). *Ayyuhā al-ikhwān fī al-dīn ayy khayr tantaẓirūn min al-Inglīz*, 1 p. (IO, Arab PP 84). *Ayyuhā al-Hindiyyūn wa-'l-Miṣriyyūn al-bu'asā' alladhīn atʿasahum al-ḥaẓẓ wa-'l-qadr* (IO, Arab PP 81): Muslims who care for their religion ought to join the Ottoman army, whose only concern is to guard the honour of Islam.

[221] *Ilā man fī juyūsh al-Inglīz min ikhwāninā Muslimī al-Hind wa-Miṣr wa-'l-Afghān*, 2 pp. (IO, Arab PP 76). This one is signed by its author, Jīlānīzādeh al-Sayyid Muṣṭafā Baghdādī.

[222] Since he signed, at the end of the article, *Ḥāfiẓ sajjādat al-ṭarīqa* (Guardian of the rug [of the leader] of the Fraternity), this appears to denote that he was an important sheikh in the Fraternity. Actually, he was the nephew of the founder of the Naqshbandiyya–Khālidiyya, Shaykh Khālid Shahrazūrī.

[223] *Bayān hāmm li-ʿālam al-Islām*. A copy can be found in the Bayerische Staatsbibliothek, Munich.

and the persecutors of Muslims;[224] and against Russia, which had treated Muslims even worse. Germany and Austro-Hungary were said to be real friends of Muslims.[225]

The same arguments were used by another sheikh, but he laid even greater emphasis on the principle that all Muslims should participate in a *Jihad* whenever one of them was attacked. This was Ṣāliḥ al-Sharīf al-Tūnisī, already mentioned as an active leader of the Benevolent Islamic Society and a dedicated propagandist of Pan-Islam.[226] At the end of 1914, the sheikh wrote a pamphlet in Arabic, entitled *The Truth of the Jihad*,[227] which was published in Berlin in the following year (and translated into French in 1916, in Berne,[228] probably for distribution in North Africa). He had had a chequered career: born in 1866/7 in Tunis and educated in the Zaytūna Mosque, he emigrated in 1906/7 to Damascus, then (after the Young Turk Revolution) to Istanbul, thence to Derna, to fight in the Tripoli war against the Italians—where he collaborated with Enver Pasha, into whose service he entered.[229] Thereafter, he applied himself to Pan-Islamic propaganda in Istanbul and Berlin.[230] He composed several further tracts. The first, a 20-page pamphlet, was published at the end of 1915 in Arabic and entitled *A Commentary on French Intrigues against Islam and the Caliph*.[231] Conceived as a reply to an article in *Le Monde* (Paris) of 10 November 1915, it defended the Caliph's role

[224] Among the more bizarre accusations of discrimination, one reads, ibid. 7, that the penalty for a tramway treading over a Muslim is 6,000 fr., for a Jew 12,000, and for a Frenchman 18,000.

[225] As'ad mentions, ibid. 12, a Turkish–German Friendship Society.

[226] On his activities during World War I, see Bardin, 190–5, and P. Heine, 'Ṣaliḥ ash-Sharif al-Tunisi'.

[227] al-Shaykh Ṣāliḥ, *Haqīqat al-Jihād* (Berlin, 1915). I have been unable to find this work and have read it only in its French transl. (see n. 228). Acc. to Heine, 89, 92, 94, it was also transl. into German and entitled *Die Wahrheit über den Glaubenskrieg*.

[228] Schaich Salih Aschscharif Attunisi, *La Vérité au sujet de la Guerre Sainte*, 14 pp. The transliteration of the author's name is suggestive of German influence.

[229] Ibid. 13–14.

[230] *Tanin*, 28 Mar. 1916, reported that the sheikh had lectured in Berlin before Pan-Islamists from various countries. See also AAT, Archives de la Guerre, 7N2104.

[231] Ṣāliḥ al-Sharīf, *Sharḥ dasā'is al-Fransīs ḍidd al-Islām wa-'l-Khalīfa*. A copy can be found in AAT, Archives de la Guerre, 7N2104.

as leader and protector of Islam, then explained that he was responsible for the unity (or union) of Islam (*Ittiḥād al-Islām*).[232] This was an elaborate discussion of the principles of Pan-Islam, as they were very probably perceived in those Istanbul circles in which Ṣāliḥ al-Sharīf moved at the time: a union of hearts in believing in Allah, agreeing on the same morals, respecting the Prophet, obeying the Caliph, fighting together any would-be aggressor, being ready for martyrdom in repelling the would-be foreign rulers, and joining all other Muslims in following the Caliph's command. He then ended by declaring that the entire Muslim world was united in its effort to end foreign domination. Sharīf's writing was no less anti-British than anti-French[233] and, in a shorter pamphlet, undated (probably from 1915), he attacked the former, in Arabic, for *The Last British Perfidy against the East and its People.* [234] He accused them of having agreed to partition the Ottoman Empire among themselves, the French, and the Russians, dealing a death-blow to the Caliph and, therefore, to all Muslims—who were called upon to unite under the leadership of the Ottomans and fight the British. For good measure, he incorporated many of those arguments in another anti-British manifesto,[235] in which he accused Great Britain of aiming at the obliteration of the Caliphate.[236]

The topic of *The Islamic Caliphate* was the central subject of a 20-page pamphlet, published in 1916, in Arabic,[237] by

[232] Ibid. 10–16. The same author, signing in full Ṣāliḥ al-Sharīf al-Tūnisī, produced yet another anti-French pamphlet, entitled *Bayān tawaḥḥush Fransā fī al-quṭr al-Tūnisī al-Jazāʾirī*, 21 pp. This attacked France for its exploitation of Muslims in North Africa. A copy is available in IO.

[233] Although, having been born in Tunisia, he had a personal score to settle with France.

[234] Ṣāliḥ al-Sharīf al-Tūnisī, *Ākhir ghadr li-'l-Inglīz li-'l-sharq wa-ahlih.* A copy can be found in AAT, 7N2103.

[235] Id., *Brīṭāniyā al-ʿuẓmā wa-hadhayānuhā al-fiʿlī wa-'l-qawlī fī ḥaqq al-Khilāfa al-Islāmiyya*, 1 p., undated. A copy can be found in IO (Arab PP 69).

[236] Ṣāliḥ al-Sharīf al-Tūnisī survived the war and, after the defeat of the Ottoman Empire and the Central Powers, fled to Switzerland, where he was reportedly active in an association, under the patronage of Enver Pasha, carrying on Pan-Islamic propaganda in all Muslim lands. See FO 371/5220, file 2233, memorandum by Politis, Greek Minister for Foreign Affairs, dated in Athens, 1–14 Mar. 1920. Acc. to Tlili, 219 n. 1, al-Tūnisī died in 1921.

[237] ʿAbd al-ʿAzīz Shāwīsh, *al-Khilāfa al-Islāmiyya.* A copy can be found in the Pamphlet Collection on Islam at the Middle East Centre, St Antony's College, Oxford.

ʿAbd al-ʿAzīz Shāwīsh (1872/6–1929)[238] who has already been mentioned and will be discussed more fully below, in the section on Pan-Islam in Egypt. His pamphlet on the Islamic Caliphate was written in reply to certain articles in the British and French press of the time. It was also a violent attack on the colonial policies of Great Britain and France in Muslim lands, both of which he accused of wilfully ignoring the significant force of 300 million Muslims. He then reproached them with writing against the Caliphate, imputing this to a design of theirs to divide the Muslims united behind the Caliph. This plan he attributed to their awareness, along with Russia, of the Caliphate's renewed role in political affairs and to their fear of the combined force of all the world's Muslims, allied with Germany's might.[239] Shāwīsh assured the Entente Powers that those Muslims who had not yet volunteered for the *Jihad* would do so at the first opportunity; and he appealed to all Muslims in the conquered lands, as well as to the Sultan of Morocco and to Muslim leaders in Central Africa, to join the Caliph in the *Jihad*.

Shāwīsh's Pan-Islamic propaganda, as that of others which we have mentioned, is of a general character, addressed to all Muslims everywhere. However, near the end of this pamphlet, he addressed it to certain specific groups. A wide range of other materials, indeed, were aimed more particularly at individual groups. One such instance is a manifesto, written by Shāwīsh, in Istanbul, in 1915 and later seized by British Military Intelligence.[240] This was directed at Egypt, when the Ottomans were setting out on their campaign to conquer it. The manifesto affirmed that the sole motive for the entry of the Ottoman Empire into the war was to liberate Muslim and Ottoman Egypt from foreign domination.

Another example is a 12-page booklet in Arabic entitled *The Russian Atrocities*,[241] with suitable photographs, which seems

[238] The best biography of Shāwīsh, viz., Anwar al-Jundī's *ʿAbd al-ʿAzīz Shāwīsh*, 62, expressed some doubt about his birthdate.

[239] Shāwīsh, *al-Khilāfa al-Islāmiyya*, 11–12.

[240] FO 371/5170, file 10708, Weekly Summary of Intelligence Reports, issued by Military Intelligence, Istanbul Branch, for the week ending 5 Aug. 1920, 4–5.

[241] *al-Faẓāʾiʿ al-Rūsiyya*. A copy can be found in AAT, Archives de la Guerre, 7N2104.

to have been addressed in the main to Muslims in the Tsarist Empire, and probably elsewhere, too. While most of the savage acts related were perpetrated against Germans, the booklet was distributed among readers of Arabic, with obvious intentions.

Other leaflets were avowedly intended for North Africa, such as one by Lieutenant El-Hadj Abdullah, entitled *Les Musulmans de l'Afrique du Nord et le 'Djéhad'*, distributed by German agents in late 1915.[242] A one-page manifesto in Arabic, distributed in 1916–17, was addressed to 'Our brethren in Tunisia, Algeria, and Morocco'.[243] It called on all Muslims to demonstrate Muslim solidarity and refuse to serve in the French armies, since that would mean fighting their brethren-in-faith. Other materials were earmarked for individual countries in the same area. A 4-page *Open Letter to His Majesty Yūsuf, Sultan of Morocco*, in Arabic,[244] was distributed in 1916–17. It informed Yūsuf of the Ottoman and German military successes and called on him to drive out the French and join the Ottoman Empire in a Pan-Islamic effort to win the war and obtain independence for all Muslims. At about the same time, two one-page leaflets were printed in Arabic, for distribution among Muslims in Algeria. Entitled, respectively, *Disparaging Islam in the French Forces*,[245] and *The Opposition of Algerian Notables to the Civil Code*,[246] they accused the French of grossly insulting their 20,000 Muslim soldiers in their religious requirements, from prayer to burial; and of unlawfully mobilizing and sacrificing additional Muslim soldiers. Again, the implications are obvious.

A particularly interesting 4-page manifesto, in Arabic, was

[242] AAT, Archives de la Guerre, 7N2104, enclosed with a letter from the French Minister for the Navy to the French Minister for War, dated in Paris, 15 Dec. 1915. See also below, app. N.

[243] Unsigned and without any title, place, or date. Seems to have been printed in a North African press. Starts, after offering praise to Allah, with: 'Ikhwānanā ahl Tūnis wa-'l-Jazā'ir wa-Marrākish'. A copy can be found in AAT, Archives de la Guerre, 7N2104.

[244] *Kitāb maftūḥ li-Mawlāya Yūsuf Sulṭān Marrākish*. A copy can be found ibid.

[245] *Iḥtiqār al-dīn al-Islāmī fī al-ṣufūf al-Fransiyya*. Seems to have been printed in a North African press. A copy can be found ibid.

[246] *Mu'āraḍat wujahā' al-Jazā'ir min al-Muslimīn ḍidd al-qānūn al-ahlī*. Seems to have been printed in a North African press. A copy can be found ibid.

addressed, very probably in 1916 or 1917, to the Sanūsiyya Fraternity.[247] It roundly accused the Entente Powers of planning to destroy all Muslim states, openly aimed at the Sanūsiyya Fraternity's apprehensions of having France to their west, Egypt to their east, and Italy right in their midst. The fraternity was praised for its determined resistance to Italian aggression and for military prowess impregnated with religious sentiment. Its prestige was very high in Tunisia and Algeria, too, and it was enjoined to declare the *Jihad* in the whole of Africa: 'The Muslim World is expecting you to declare the *Jihad*, so that the rays of your guidance would illuminate all parts of Africa, and, with the deliverance of Africa, the Islamic State would rise in importance and all Muslims would get back the rights stolen from them by these [above-mentioned] despotic states'.[248]

Pan-Islamic propaganda, directed during World War I at readers of Arabic, has been given particular attention, since it was doubtlessly considered critical for the Ottoman Empire. Some of it was directed, naturally enough, at the Arabs living within the empire, to foster their identification with it; but other methods, some of them less gentle, were also used by the Ottomans for these Arabs (particularly in Syria). However, the bulk of the Muslim Arab population lived in the lost territories of Egypt, Libya, Tunisia, and Algeria, while in Morocco and some parts of Africa there lived additional Arab groups in various degrees of dependence on the Entente Powers. The Ottoman leaders and their Pan-Islamic propagandists were greatly tempted to address all these, in order to obtain support and throw a spanner into the military machine of the Entente. Moreover, these Arab concentrations were relatively close and easily accessible. Except for Muslims inhabiting the shores of the Black Sea, others in the Tsarist Empire, India, and the Dutch Indies were further away, with often limited means of access—and carefully watched over, respectively, by the Russians, the British, and the Dutch. Successful efforts by Pan-Islamic agents were made, early, in the war, to propagate the cause via Muslim pilgrims coming from those countries to

[247] H.Kh., *Ayyuhā al-Sāda al-Sanūsiyyūn*. Seems to have been printed in a North African press. A copy can be found ibid.

[248] Ibid. 4.

Mecca and Medina.[249] However, the Entente Powers limited strictly the Pilgrimage of their subjects during the war years, so that the Ottomans had to rely on a few emissaries, operating clandestinely and probably disheartened by the meagre results achieved in Central Asia, India,[250] and South-east Asia; or to attempt other ways, with German assistance, for distributing their written propaganda. Even then, this was carried out mostly in the Middle East and North Africa, for instance among Muslims from India, serving in the British forces in Egypt, during 1915—that is, before the Turco-German attack on the Suez Canal.[251]

Perhaps because of these limitations to their operational possibilities, the Ottoman authorities responsible for Pan-Islamic propaganda adapted their writings to a varied public. Thus a special leaflet was prepared in Istanbul soon after the Ottomans entered the war, reportedly in 50,000 copies.[252] This was a considerably enlarged version of the proclamation of the *Jihad*, emphasizing the need for Muslim solidarity and the urgency of delivering other Muslims from foreign rule: Every Muslim ought henceforth to consider himself as a soldier. The entire Muslim world should unite in action for the Holy War. The Muslim populations (each addressed by its own name) ought to rise and fight. Another Arabic pamphlet, 30 pages long, was being circulated in Aleppo, in April 1915, entitled *A Universal Proclamation to All the People of Islam*. A copy was obtained by the American Consul in that city, J.B. Jackson, who had the

[249] See AAT, Archives de la Guerre, 7N2144, Hamelin's report to the French Minister for War, no. 5174–8/II, dated 21 Nov. 1914.

[250] The results of Pan-Islamic propaganda in India will be discussed below, in ch. IV.

[251] A newspaper in Urdu was distributed then, urging India's Muslims to desert and unite with the Muslims of Iran, Egypt, China, and Morocco, against the British. See FO 371/2495, file 133531, enc. in Sir H. McMahon's dispatch to Sir Edward Grey, dated in Ramleh, 6 Sept. 1915; and ibid., file 125269, for a copy of this newspaper, *Ghadar* or *Ghadr* (Revolt). This was very probably the organ of a radical movement in the Punjab, bearing the same name, with a Pan-Islamic orientation; see Sharif al-Mujahid, 'Pan-Islamism', 111.

[252] I have seen only a French version and I have no way of ascertaining in which other languages it may have been written. See AAT, Archives de la Guerre, 7N2103, enc. in the letter of the French Minister for Foreign Affairs to the Minister for War, no. 1217, dated in Bordeaux, 14 Nov. 1914. Cf. AE, Guerre de 1914–1918, vol. 1650, fo. 15, Bompard, French Ambassador in Istanbul, to the French Minister for Foreign Affairs, dated in Pera, 14 Oct. 1914.

most relevant passages translated and sent to Henry Morgenthau, United States' Ambassador in Istanbul.[253] This pamphlet elaborated the one published in 1914, urging solidarity and unity and calling for an out-and-out attack on non-Muslims everywhere. However, perhaps the most interesting item was its appeal for an economic war by all Muslims, including the non-payment of taxes to foreign non-Muslim rulers. Here may lie the seeds of economic Pan-Islam, about which we shall have more to say later.

It is difficult to assess the success of the Pan-Islamic propaganda activities of the Ottomans and the Germans during World War I; on the whole, they seem to have had more impact in Africa than in Asia. One way to evaluate the results is by considering the reactions of the Entente Powers. Great Britain and France were very worried about the possible results of this propaganda and invested considerable efforts and funds in intercepting it and countering its arguments; the archives of both are replete with such information,[254] as are the German ones. The British and the French, no less than the Germans, had their 'own' Muslims, well-versed in Islamic and Pan-Islamic argumentation. One example is Shaykh ʿAlī al-Ghayātī, an Egyptian graduate of al-Azhar, who was living in Switzerland during World War I and writing in the French and pro-Entente Arabic press.[255] He argued that the Istanbul Caliphate had no authenticity and that Sharīf Ḥusayn, a direct descendant of the Prophet Muḥammad and a member of the Prophet's Quraysh tribe, was the appropriate person to assume these functions.[256] They also had their own Orientalists. Edouard Montet, a Pro-

[253] For the Arabic version, see NA, RG 59, 867.00/762, enc. in US Consul J.B. Jackson's no. 258, to the US Ambassador in Istanbul, Henry Morgenthau, dated in Aleppo, 8 Apr. 1915, appending a selective English transl. Copies of the dispatch and transl. can also be consulted in MBZ, dossier 453–454 (A. 190), A. 190 (213). For selected passages see below, app. M.

[254] See e.g. FO 395/64, file 192843; FO 395/151, files 176962 and 183632; AAT, Archives de la Guerre, esp. 7N2104.

[255] PA, Orientalia Generalia 9, no. 2, vol. 4, German Ambassador von Romberg's no. 1036, to Bethmann Hollweg, 'Geheim', dated in Berne, 21 May 1916, enc.

[256] *Gazette de Lausanne*, 9 June 1917. A year earlier, a German Orientalist had already noted that the press of the Entente was increasingly propagating among the Arab population of the Ottoman Empire the idea of an Arab Caliphate. See report, dated 11 July 1916, in PA, Orientalia Generalia 9, vol. 6.

fessor of Arabic at the University of Geneva, wrote in 1916 *L'Islam et la France,* a leaflet answering El-Hadj Abdullah's *L'Islam dans l'armée française.*[257] The French even set up their own Arabic weekly, entitled *al-Mustaqbal* (The Future), starting on 1 March 1916. This rejected the claims of the Ottoman Empire and Germany to lead all Muslims, while stressing France's service to Islam; a characteristic article reproached the Germans with having created Pan-Islam in the image of Pan-Germanism and with using it for their own ends.[258] Among the contributors was ʿAlī al-Ghayātī.[259] The English press, at home and abroad, ridiculed the Kaiser's claim to be the patron and protector of Islam.[260] No less, they tried to prove that Germany was against Islam, in a 1916 War Office pamphlet.[261] Their press attacked alleged German transgressions against the *sharīʿa,* in the East German possessions,[262] and accused Turkey of overt repudiation of Islam and desecration of Prophet Muḥammad's grave.[263] Some of the arguments were repeated in leaflets, which British airplanes dropped over Khan Yunis and al-ʿArish on 18 May 1916[264] (a move which, in hindsight, was a part of the preparations for the military advance into Palestine).

7. *Pan-Islam in Egypt*

It is hoped that, in examining Pan-Islam in several specific regions during the Young Turk era, a better perspective of the Pan-Islamic relations of these areas with Istanbul may be obtained. Here we will look closer at the response—or should

[257] *Journal de Genève,* 22 July 1917. See also PA, Orientalia Generalia 9, vol. 8, von Romberg's no. 2279, to the Reichskanzler, dated in Berne, 26 July 1917.

[258] *al-Mustaqbal,* 65 (25 May 1917).

[259] e.g. on France and Islam, *al-Mustaqbal,* 11 (12 May 1916).

[260] e.g. *South China Morning Post* (Hong Kong), 10 June 1916.

[261] War Office, *Anti-Moslem Germany.*

[262] e.g. 'Germany as Oppressor of Islam: Testimony from East Africa', *The Times,* 13 Mar. 1917; repr. in *The Egyptian Gazette* (Cairo), 8 May 1917.

[263] *The Near East* (London weekly), 30 Mar. 1917. The accusation of desecration was taken up, also, by *La Dépêche Marocaine* (Tangier), 18 Oct. 1917.

[264] PA, Orientalia Generalia 9, vol. 6, German Consul Braude's no. 67, to Bethmann Hollweg, dated in Jerusalem, 3 June 1916.

one say, the feedback—to the centre's exertions to spread this ideology in the periphery, concentrating on three cases: Egypt, Libya, and non-Ottoman Africa; India will be dealt with separately, later.[265]

Egypt, as the most populated Arab country and a significant cultural centre, had had a lengthy and influential role in Muslim history—and still was an important centre at the time under discussion. The ancient al-Azhar Academy in Cairo was still a focus for local and foreign students seeking to acquire the religious scholarship of Islam. Distinguished thinkers, like al-Afghānī, resided in Egypt in the 1870s (before he started his Pan-Islamic activity). All this, combined with the fact that, from 1882, Egypt was under British Occupation—with the Ottomans insisting throughout on their special relationship with it—made it a perfect target for Pan-Islamic propaganda. None the less, the impression given by some European newspapers that a trend towards Pan-Islam pervaded the Egyptian masses[266] was grossly exaggerated.

Sentiments of solidarity with their co-religionists undoubtedly became stronger among Egypt's Muslims, during the 1897 Turco-Greek war, when they seemed to identify with the Sultan–Caliph.[267] Muslim solidarity was even more obvious during the 1906 Akaba conflict, in which the British administration in Egypt was poised against the Ottoman Sultan in a frontier disagreement;[268] and, again, following the Danishway incident, in Egypt itself in the same year, between British soldiers and Muslim *fallāḥīn*.[269] Not coincidentally, more than a thousand Muslims in Aligarh called on India's Viceroy to use his influence in preventing an armed conflict between the Sultan–Caliph and Great Britain over the frontier at Akaba.[270]

265 See below, ch. IV.

266 For instance, in a 4-column letter from Cairo to *Le Temps* (Paris daily) of 22 Aug. 1906, 1. See also PA, Orientalia Generalia 9, no. 1, vol. 10, von Oppenheim's no. 387, to von Bülow, dated in Cairo, 16 May 1908.

267 PA, Orientalia Generalia 9, no. 1, vol. 2, von Oppenheim's no. 33, to Hohenlohe-Schillingfürst, dated in Cairo, 23 Apr. 1897.

268 Ibid., vol. 8, von Oppenheim's no. 286, to von Bülow, dated in Cairo, 21 May 1906. See also *The Tribune* (London), 12 July 1906; *Le Gaulois* (Paris), 19 July 1906; *Le Rappel* (Paris), 20 July 1906.

269 *The Tribune*, 23 Aug. 1906.

270 PA, Orientalia Generalia 9, vol. 2, A. Quadt's reports, dated in Simla, 21 and 28 May 1906, respectively.

The Egyptian press, however, although Pan-Islamically involved at the time, was careful to reject the label of fanaticism.[271] It goes without saying that Pan-Islamic sentiment rose, once again, in 1911, during the war in Libya, and was expressed in a boycott of Italian goods.[272] During World War I, the press—even the moderate *al-Ahrām*—focused on German injustice towards Muslims, calling on them everywhere to support Great Britain and not what was considered anti-Muslim Germany.[273]

As in Turkey itself, Pan-Islamic discussions were largely (but not solely) part of the wider debate within Islamist circles. Some of this was strictly on religious premisses, such as the 1910 printing in Cairo of a medieval Treatise of Unity (*al-Risāla al-Ittiḥādiyya*) by Muḥammad Ibn ʿAlī ʿAbd al-ʿĀlim. Arguments were expressed in the local press too, mostly in Arabic, which—contrary to that in the provinces under the direct rule of the Ottoman authorities—enjoyed relative freedom from censorship. Thus one could read articles committed to political Pan-Islam in certain Egyptian newspapers, in the later years of Abdülhamid's reign and during the Young Turk era. It is quite likely that Pan-Islamic propaganda in some newspapers was financed from Istanbul.[274] *al-Umma* (The Islamic Community) published some articles,[275] as did the Cairene *al-Dustūr* (The Constitution), too: for example, it argued that after the experience of Muslims everywhere with foreign rule and exploitation, the only remedy was for them to unite religiously and politically, hoping for the leadership of the Caliph;[276] and a short while later, it called for a political union between Turks, Arabs, Egyptians, and Afghans, to form one continuous bloc, which would be joined subsequently by

[271] Examples from *al-Mu'ayyad* and others in PA, Orientalia Generalia 9, vol. 2, German Consul-General von Grünau's report to von Bülow, dated in Cairo, 18 July 1906.

[272] Ibid., vol. 4, German Consul-General and Diplomatic Agent von Hatzfeldt-Wildenburg's report to Bethmann Hollweg, dated in Cairo, 10 Nov. 1911.

[273] e.g. unsigned article in *al-Ahrām* (Cairo), 9 May 1916.

[274] Cf. 'Le Panislamisme turc', 59–61.

[275] Hubert, 15.

[276] *al-Dustūr*, 28 May 1909. See also MBZ, dossier 452 (A. 190), A. 190 (11579), enc. in the dispatch of Netherland's diplomatic representative in Egypt to the Dutch Minister for Foreign Affairs, dated in Cairo, 29 May 1909.

Turkestan, the Muslim areas in Africa, and the Muslims in India.[277]

However, it was the larger newspapers, with an Islamist readership, which could decide the shaping of public opinion in Egypt versus Pan-Islam. Of these, the Cairene daily *al-Mu'ayyad* (The Supported, i.e. by Allah) was perhaps the most respected, in pre-World War I days. Established in December 1889,[278] it soon reached a circulation of 8,000 to 9,000,[279] a high figure in Egypt of the Young Turk era. Its editor, ʿAlī Yūsuf (1863–1913), while keeping on good terms with Egypt's Khedive ʿAbbās Ḥilmī,[280] was mostly appreciated in Islamist circles and his photograph could be seen in many Egyptian homes.[281] He was an energetic defender of Islam and a frank critic of any suggestion for a *modus vivendi* between European civilization and Islam.[282] He frequently favoured Pan-Islam in *al-Mu'ayyad*, for example in 1908, when the newspaper wrote 'The Caliphate of the Sultan (*Khilāfat al-Sulṭān*) remains the strongest bond uniting Muslim peoples, not merely the Turks, to the Ottoman Empire, whose power consists of them. Take away this title of the Sultan and the empire would decline and become a second-rate state'.[283] This approach hardly changed after the Young Turk Revolution and the deposition of Abdülhamid, following which ʿAlī Yūsuf published, in Arabic, a 39-page *Announcement Concerning the Attitude of al-Mu'ayyad towards the Ottoman State*.[284] In it, he reiterated that the editorial policy of *al-Mu'ayyad* had consistently been to back the Ottoman and Islamic state, Ottomanism, and Pan-Islam ('al-Jāmiʿatayn al-ʿUthmāniyya wa-'l-Islāmiyya'),[285] even

[277] *al-Dustūr*, transl. in *La Turquie* (Istanbul), 3 Aug. 1909.

[278] Abbas Kelidar, 'Shaykh ʿAli Yusuf', 10. An incomplete set of *al-Mu'ayyad* is available at Istanbul University's Central Library.

[279] PA, Orientalia Generalia 9, no. 1, von Oppenheim's no. 91, to Hohenlohe-Schillingfürst, dated in Cairo, 28 May 1900. Cf. Hubert, 14–15.

[280] See ʿAbbās Ḥilmī's glowing tribute to him, in his memoirs, published in *al-Miṣrī*, 13 May 1951, 6.

[281] I am indebted to the late H.A.R. Gibb for this information.

[282] Kelidar, 'Shaykh ʿAli Yusuf', 14–15. For the newspaper's politics in the Hamidian era, see PA, Orientalia Generalia 9, no. 1, vol. 8, von Oppenheim's no. 246, to von Bülow, dated in Cairo, 21 Feb. 1906.

[283] Quoted in Naṣr, 97. English transl. mine.

[284] ʿAlī Yūsuf, *Bayān fī khuṭṭat al-Mu'ayyad tijāh al-Dawla al-ʿUthmāniyya al-ʿAliyya.*

[285] Ibid. 33.

though he considered the Committee of Union and Progress a secret, dangerous association.[286]

ʿAlī Yūsuf's support of Pan-Islam became even more evident from the beginning of the Ottoman–Italian war in 1911. The newspaper's issue of 26 November 1911, for instance, not only published a letter concerning the Pan-Islamic merits of a Sunnite–Shiite *rapprochement*, but another soliciting contributions for the war in Libya, as follows, 'Are ye waiting to see your countries withdrawn from you one by one till the turn comes to the most revered and sacred, and ye will find yourself humiliated whereas ye once were strong, and divided whereas ye once were united?'[287] In interviews with the European press, ʿAlī Yūsuf characteristically reiterated his suspicions of all the European powers and the support of Muslims for their Caliph, the military débâcle in the Balkan Wars notwithstanding.[288] Finally, a small political party he set up and headed, *Ḥizb al-iṣlāḥ* (The Party of Reform), supported the Sultan on both Pan-Islamic and Ottoman premisses.[289]

Also in 1911, one of the leading Islamists in Egypt, Muḥammad Rashīd Riḍā (1865–1935),[290] a leading disciple of Muḥammad ʿAbduh and editor-publisher of the respected monthly *al-Manār* (The Lighthouse) in Cairo, expressed his views on Pan-Islam. Although he was not involved at the time in any Pan-Islamic activity, he submitted to the Ottoman authorities in Istanbul a memorandum recommending the establishment of a 'College of Propaganda and Instruction' (*Madrasat al-daʿwā wa-'l-irshād*). However, he obtained no encouragement there, possibly because Riḍā was known in Istanbul not only as a noted Islamic scholar and writer, but also as an activist for Syrian Arab nationalism.[291] Riḍā published the

286 Ibid. 21.

287 Transl. by Margoliouth, *Pan-Islamism*, 13.

288 André Duboscq, 'Egypte—les projets du Panislamisme'. This refers to 1913. See also MBZ, dossier 452 (A. 190), A. 190 (6914), enc. in a report by Netherlands' Embassy in Paris, to the Dutch Ministry for Foreign Affairs, dated 28 Mar. 1913.

289 Naṣr, 68 ff.

290 On whom see C.C. Adams; M.H. Kerr; and Aḥmad al-Sharabāṣī, *Rashīd Riḍā ṣāḥib al-Manār ʿaṣruh wa-ḥayātuh wa-maṣādir thaqāfatih.*

291 It appears that Riḍā was suspected of intending his college to undermine the Ottoman Empire in favour of an Arab one. See Mohammed Farid, *Les Intrigues anglaises contre l'Islam*, 24–6.

essentials of his memorandum in *al-Manār*, first the general programme,[292] then a detailed plan of studies and administration.[293] The former stipulated that the students would be selected from promising candidates in the Muslim countries, especially those in need of Islamic knowledge, such as Java, China, and North Africa. The college, which would provide the students with all their material needs, would emphasize morals, manners, and a spirit of unity, eschewing both racial nationalism and politics. Studies would cover a broad knowledge of Islamic and general studies, as well as languages. The latter stressed that the orientation of the curriculum would be the training of teachers and preachers for foreign countries. Riḍā, indeed, emphasized that this college would train students for work outside the Ottoman Empire, since the empire was already served by another college in Istanbul itself.[294] This proviso, as well as the shunning of politics, was very probably intended to dispel in Istanbul any suspicion of competition.

The Young Turks no doubt perceived, before World War I, the possibilities of Pan-Islamic and Ottoman propaganda in Egypt, among Islamist and other circles, and forged some ties with them.[295] These contacts were governed by prudence,[296] as in the case of Riḍā's proposal, since they did not wish to antagonize the British rulers of Egypt unnecessarily. But Egyptian support for the Ottoman side, on Islamic and Pan-Islamic grounds during the 1911–12 Libyan War, although not unanimous,[297] was not forgotten. When the Ottoman Empire entered World War I, on the side of the Central Powers, the need to manage British susceptibilities disappeared. On the contrary, Egypt became the object of an approaching Ottoman–German military attack, launched from Syria and represented as part of the pursuit of Pan-Islam; in preparation, Pan-Islamic propaganda was duly intensified as early as December 1914, less than a month after the Ottoman Empire's proclamation of

292 *al-Manār*, 14/1 (Jan. 1911), 52–3.
293 Ibid. 14/10 (Oct. 1911), 785–800; 14/11 (Nov. 1911), 801–21.
294 Ibid. 14/1 (Jan. 1911), 53.
295 Pinon, *L'Europe et la Jeune Turquie*, 133–5.
296 Knight, 64–5.
297 On which see Jamāl Zakariyā Qāsim, 'Mawqif Miṣr min al-ḥarb al-Ṭarāblusiyya'.

war. A strongly worded Pan-Islamic manifesto was sent to various notables in Egypt, appealing to their loyalty to the Caliph and the Khedive ʿAbbās Ḥilmī (then in Istanbul) and urging them, in the name of Islam, to revolt against the Occupying Power. The manifesto maintained further that the war ought to free all Muslim peoples from the foreign yoke. The Ottoman Empire intended, through its mighty army, to unify the entire world of Islam.[298] These arguments were repeated during the war, time and again. The British naturally responded in kind, for instance in getting a group of Ulema from al-Azhar, in January 1916, to publish a proclamation to Muslim soldiers from Arabia, Hijaz, Syria, and Iraq, who were fighting with the Ottoman forces. It roundly accused the Turks of deceiving the Arabs and robbing them of Koran and Caliphate; by contrast, the British and French were presented as befriending the Arabs and encouraging the use of their language.[299]

The limited impact of Pan-Islamic propaganda in Egypt, in the Young Turk era, can be explained not only by British countermeasures, but more so by the competition between two ideologies, nationalism versus Islam (and Pan-Islam). We shall attempt to clarify this by analysing the politics of the Nationalist Party in Egypt, the only one entirely committed at that time to 'Egypt for the Egyptians' as its central ideal.

During the last years of Abdülhamid's reign, there seem to have been some tenuous connections between his Pan-Islamic propaganda and Muṣṭafā Kāmil (1874–1908),[300] the founder and first Chairman of the Nationalist Party.[301] These have

[298] AE, Guerre 1914–1918, vol. 1651, fo. 25, report no. 446 by the French Minister in Egypt, Defrance, to Delcassé, Minister for Foreign Affairs, dated in Cairo, 6 Dec. 1914, and enclosures.

[299] I have not seen the Arabic original, but a French transl. can be found in AAT, Archives de la Guerre, 7N2104, enc. with a letter from the French Minister for Foreign Affairs to the Minister for War, no. 257, dated in Paris, 21 Jan. 1916.

[300] V. Chirol, 'Islam and the War', 492–3, exaggerated the importance of these connections. A German report from 1900 asserted that Abdülhamid subsidized Kāmil's newspapers even at that time, but adduced no proof. See PA, Orientalia Generalia 9, no. 1, von Oppenheim's no. 91, to Hohenlohe-Schillingfürst, dated in Cairo, 28 May 1900.

[301] The most useful studies on these relations are: D. Walker, 'Muṣṭafā Kāmil's Party', esp. part 3, 79–113; and F. Steppat, 'Nationalismus und Islam bei Muṣṭafā Kāmil'.

not been established beyond doubt and are mostly based on several quotes from Kāmil's speeches[302] and writings.[303] However, a closer examination of his views[304] establishes that he was a keen Egyptian patriot, attached to Islam[305] and esteeming it as a potent factor in education and public life, but primarily committed to Egypt's independence from the British.[306] In this light, his professions of loyalty to the Ottoman Sultan are less signs of support for a supra-national Pan-Islamic community than for the interests of Egypt, as Kāmil wished to mobilize Ottoman action against British rule in Egypt. Thus, in what reads very much like a Pan-Islamic article, which he wrote in 1895 and printed in the Parisian *La Nouvelle Revue*,[307] Kāmil warned Great Britain that its anti-Islamic policies (as he perceived them) would cause all the world's Muslims to rise against it, under the Caliph's leadership.

Kāmil's pragmatic political (rather than religious) approach to Pan-Islam[308] resurfaced in another three-column article which he published six years later, again in Paris, in the daily *Le Figaro*.[309] In it, he expressed his conviction that Europe was determined to destroy Islam both as a religious and as a political force; thus threatened, the Muslims were rising to protect Islam and the Ottoman Empire. Pan-Islam, according to Kāmil, was not fanatic, as European propaganda implied, but

[302] Examples in Muḥammad ʿAmāra, *al-Jāmiʿa al-Islāmiyya wa-'l-fikra al-qawmiyya ʿind Muṣṭafā Kāmil*, esp. 56–58. See also Muṣṭafā Kāmil, *Miṣr wa-'l-iḥtilāl al-Injlīzī, passim*. ʿAlī Kāmil, *Muṣṭafā Kāmil Bāshā fī 34 rabīʿan*, ii. 121–34.

[303] Such as *Difāʿ al-Miṣrī ʿan bilādih*, 25–31.

[304] See e.g. his *Kitāb al-mas'ala al-sharqiyya*.

[305] Steppat, 'Nationalismus und Islam', 281 ff. makes a persuasive case for this. See also *Vossische Zeitung*, 8 Sept. 1898.

[306] In addition to the above refs., see: C.C. Adams, 220. ʿAbd al-Raḥmān al-Rāfiʿī, *Muṣṭafā Kāmil bāʿith al-ḥaraka al-qawmiyya*, 83–4, 346–54. A. Goldschmidt, 'The Egyptian Nationalist Party', 320, 323, 328, 331–2. Id., 'The National Party from Spotlight to Shadow', 12–13, 21. E.I.J. Rosenthal, *Islam in the Modern National State*, 118–19. Fatḥī Raḍwān, *Muṣṭafā Kāmil*, 233–46. J. Heyworth-Dunne, *Religious and Political Trends in Modern Egypt*, 8–9. Ankush B. Sawant, 'Nationalism and National Interest in Egypt', 136–7. Amiri Tamura, 'Ethnic Consciousness and its Transformation in the Course of Nation-building', 107. P. Smoor, 'Krachtlijnen tussen de polen Islām en nationalisme in Egypte', 1114–15.

[307] Moustafa Kamel, 'L'Angleterre et l'Islam', 835–7.

[308] See also Steppat, 'Nationalismus und Islam', 287.

[309] Moustafa Kamel, 'L'Europe et l'Islam'.

a principle of Islam which made the unity of Muslims an essential condition of their power. However, the recent 'disguised Crusade' by Europe against Islam brought the Muslims closer to their Sultan–Caliph, for he personified for them Islam and its power in their struggle with Europe. Later, after 1904, with the waning of French support for Egyptian nationalism following the Franco-British Agreement in that year, Kāmil seemed willing to make some further concessions to Pan-Islam. This was intertwined with his seeking alternative support in Germany for the cause of Egyptian nationalism. In an article on Wilhelm II and Islam, which Kāmil wrote for a Berlin daily in late 1905,[310] he asserted that all 300 million Muslims, from Tangier to Peking, had been following closely the German Kaiser's activities in the cause of Islam and considered them as important moves in their favour; they all respected and admired him as their friend and patron.[311] In yet another article published in France,[312] Kāmil maintained that, although no enlightened Muslim believed in an anti-European league of all Muslim peoples, Turkey could not survive isolated; the spirit of Pan-Islam could exist and would enable those peoples to co-operate closely in their own interests. The expression of such views may have moved Lord Cromer, the powerful Agent and Consul-General of Great Britain in Egypt, to opine at the time that the Egyptian nationalist movement was 'deeply tinged with Pan-Islamism'[313] although, at times, it was 'not easy to recognize the Pan-Islamic figure under the nationalist cloak'.[314]

Following Muṣṭafā Kāmil's death, there gradually developed a more obvious opening in the Nationalist Party's attitude towards Pan-Islam. The new Chairman of the Party, Muḥammad

[310] Id., 'Kaiser Wilhelm und der Islam'.

[311] A reply to Kāmil's article, criticizing it, was published in *The Egyptian Gazette*, 25 Oct. 1905.

[312] 'Patriotisme et Panislamisme', *Le Temps* (Paris), 8 Sept. 1906. al-Jundī, *'Abd al-'Azīz Jāwīsh*, 15–16, referred to another(?) article, *Le Temps*, 8 Nov. 1906.

[313] *Egypt. No. 1 (1907): Reports of His Majesty's Agent and Consul-General on the Finances, Administration, and Conditions of Egypt and the Soudan in 1906*, 4. The report was submitted to Sir Edward Grey on 3 Mar. 1907.

[314] Ibid. 6. Cromer defined Pan-Islam, ibid. 5, as follows: 'A combination of all the Moslems throughout the world to defy and to resist the Christian Powers'.

Farīd (1868–1919), although a Muslim, was not a particularly pious one. However, throughout his leadership of the Party, he was as vehemently anti-British as his predecessor had been, orating and writing against them. It was natural for him to visit Istanbul in 1909 and establish relations with several of the Young Turk leaders.[315] When he left Egypt for ever in 1912, it was again to Istanbul that he went and he seems to have been won over to Pan-Islam. After all, of the three ideologies predominating in the Ottoman capital at the time, he had little to choose from: Ottomanism was reprehensible to him, for it implied the return of Egypt to the Ottoman Empire, not its independence; Pan-Turkism was irrelevant for an Egyptian Arab. So Farīd increasingly involved himself in Pan-Islamic politics and propaganda, drawing around him in Istanbul, then in Berlin and Geneva, other exiled leaders of the Nationalist Party.[316] He published, in 1912, a book in Arabic on the history of the Ottoman state, ending with the accession of Mehmet Reşad and the reforms of the Committee of Union and Progress.[317] Then he started contributing articles to the Committee's newspapers, calling for Ottoman and Pan-Islamic support for Egypt's independence.[318] Moving briefly to Geneva, he founded a Pan-Islamic society, *Encümeni-i Terekki-yi İslam*, already mentioned, and edited its bulletin during 1913–16.[319] Returning to Istanbul in 1914, he gave a lecture in Arabic (with a Turkish translation), entitled *Fī ittiḥād al-Islām* (About Pan-Islam), lauding Pan-Islam,[320] as he did in several of his subsequent press articles. Back in Switzerland in 1915, Farīd published an 84-page collection of articles entitled *Études sur la crise ottomane actuelle 1911–1912—1914–1915*. After having condemned Christian Europe's intentions towards Islam, he appealed to all the Muslims of the world to cast aside any illusion about Europe's goodwill:

> O nations musulmanes! abandonnez toute idée de race et unissez-vous au sein reconfortant de l'Islam ... Ne formez qu'une grande

[315] Muḥammad ʿAlī Gharīb, *Muḥammad Farīd al-fidāʾī al-awwal*, 25–6, 56 ff. Naṣr, 65 ff. Fatḥī Raḍwān, *Mashhūrūn mansiyyūn*, 11–24.

[316] For these exiles, see Rothmann, 3 ff.

[317] Muḥammad Farīd, *Taʾrīkh al-dawla al-ʿaliyya al-ʿUthmāniyya*.

[318] Id., *Awrāq* ..., i. introduction, 37.

[319] al-Rāfiʿī, *Muḥammad Farīd*, 364–6.

[320] Farīd, *Awrāq* ..., i. 190.

famille, une et indivisible, comme un seul bloc, pour mieux resister à cette invasion qui cherche à traverser le Bosphore pour vous submerger! ... Nations musulmanes! tendez-vous la main les uns les autres; relevez-vous par votre propre effort ...[321]

In 1916, Farīd reorganized Muslims living in Geneva in his Muslim association (with Pan-Islamic characteristics) and attempted to do the same in Lausanne.[322] A year later, he went—together with Shāwīsh[323]—to attend a Pan-Islamic convention in Stockholm, organized by its mayor (despite the Swedish Government ban). Other participants were Muslims originating from Turkey, Tunisia, Algeria, Morocco, Libya, the Tatar areas in Russia, Turkestan, and India. The convention got considerable publicity.[324] In the same year he published his last work, a booklet in French entitled *Les Intrigues anglaises contre l'Islam*.[325] In this anti-British tract, Farīd posed as the champion of Islam and Pan-Islam. In his interpretation, since its conquest of India, Great Britain had had to take the force of Islam into account. It had decided to enslave Islam and use it to further its policy of world domination. It was striving to weaken the Caliphate—as an act against the entire Muslim world. Farīd saw Pan-Islam as an antidote to this, and esteemed it a success in gaining overall Muslim sympathy for, and support of, the Caliph.[326]

The Muslim organization which Muḥammad Farīd set up and headed in Switzerland in the war years merits a more detailed mention. The Société Progrès de l'Islam listed in its statutes (para. 2) the following goals: 'To strengthen the union among the diverse Muslim nations and to assist in their intellectual and economic advance.' The organization's organ, entitled *Bulletin de la Société Endjouman Terekki-Islam (Progrès de l'Islam)*, reflected these aims. Published in French in Geneva, as a monthly, from February 1913 to November 1916,[327] it had

321 Mohamed Farid, *Étude sur la crise ottomane actuelle 1911–12—1914–1915*, 52–4.

322 Farīd, *Awrāq* ..., i. 315–18.

323 For Shāwīsh's Pan-Islamic propaganda in Copenhagen and Stockholm, see FO 371/3057, files 234753, 234766, and 241852, all from Dec. 1917.

324 FO, ibid. Farīd, *Awrāq* ..., i. 379.

325 For a summary, see FO 395/152, file 227557.

326 Farid, *Les Intrigues*, esp. 16 ff.

327 An almost complete set is available in the Centre de Documentation Internationale Contemporaine, Paris.

to be cautious in advocating Pan-Islam too openly, as the Swiss authorities were very insistent, during the war years, on their neutrality. So the bulletin championed the cause of Muslims in Turkey, Egypt, the Balkans, and elsewhere, defending their interests and emphasizing their attachment to the Caliph in Istanbul.[328] It supported, of course, the Muslim struggle against Italy in Libya.[329] Farīd himself wrote in it frequently, for instance appealing for all-Muslim solidarity and the recruitment of Muslim military units everywhere, in support of the Ottoman Empire.[330] Another contributor was Halil Halid, formerly Ottoman Consul-General in Bombay, who argued that India's Muslims were compelled, against their will, to remain loyal to British rule.[331]

There were, during World War I, three main groups of Egyptian exiles connected in one way or another with Ottoman Pan-Islamic activities and propaganda. One was led by ex-Khedive ʿAbbās Ḥilmī, who had moved from Istanbul to Vienna. The other two were led, respectively, by two activists of the Egyptian Nationalist Party—Muḥammad Farīd and ʿAbd al-ʿAzīz Shāwīsh. The latter has already been mentioned as the editor of *al-ʿĀlam al-Islāmī* in Istanbul and co-editor of *Die Islamische Welt* in Berlin. He had been encouraged in his career by the Grand Vizir Said Halim, a convinced Pan-Islamist. After Said Halim's dismissal, the groups of Farīd and Shāwīsh drew closer to one another. However, it was the latter who remained the leading figure among the Egyptian propagandists of Pan-Islam in the Young Turk era. Shāwīsh[332] had already been involved in Islamic and Pan-Islamic matters before the war. As editor of *al-Liwāʾ* (The Standard), the Arabic daily of the

[328] e.g. 'L' Islam en Chine', *Bulletin de la Société Endjouman Terekki-Islam (Progrès de l'Islam)*, 1/5–6 (Aug.–Sept. 1913), 323.

[329] Ibid. 1/7–9 (Oct.–Dec. 1913), 434.

[330] Mohamed Farid, 'La Guerre mondiale, la Turquie et l'Islam', *Bulletin de la Sociéte Endjouman Terekki-Islam*, 3/5–6 (Oct.–Nov. 1915), 195–219. See also his 'Intrigues contre le Califat ottoman, Les Lieux Saints et l'Islam', ibid. 4/3 (Oct.–Nov. 1916), 159–69.

[331] Halil Halid, 'La Loyauté forcée des Indiens', ibid. 3/7 (Dec. 1915), 302–13. This was transl. from his articles in *Berliner Tageblatt*, 21 Oct. 1915.

[332] For his career, see al-Jundī, *ʿAbd al-ʿAzīz Jāwīsh*.... Raḍwān, *Mashhūrūn mansiyyūn*, 25–54. Salīm ʿAbd al-Nabī Qunaybir, *al-Ittijāhāt al-siyāsiyya wa-'l-fikriyya wa'-l-ijtimāʿiyya fī al-adab al-ʿArabī al-muʿāṣir*, 307–14, 358–65.

Nationalist Party, from 1908, Shāwīsh attacked simultaneously the enemies of Egypt and the foes of the Islamic world. Shāwīsh believed fervently in the right of Egypt—and the entire Islamic world—to independence and constitutionalism, and in the unity of the Islamic world, headed by the Ottoman Government. These views grew even stronger after his visit to Istanbul in 1909 and were at least partly responsible for his anti-Coptic views, which caused a rift in the Nationalist Party between the two communities.[333] Accused of incitement and sentenced to prison, Shāwīsh took refuge in Istanbul for a while, and in 1912 he edited 164 issues of *el-Hilal el-Osmani* (The Ottoman Crescent), a Turkish daily of the Committee of Unity and Progress,[334] in which he preached the interests of Egypt and the Muslim world in an anti-British and anti-French spirit and attacked Italy for its invasion of Libya. In 1913, Shāwīsh edited an Arabic newspaper, *al-Haqq yaʿlū* (Truth Will Prevail), calling on all Muslims to unite against the Christians; thousands of copies were reportedly distributed in Beirut and elsewhere, free of charge.[335] The offices of these newspapers soon became meeting places for Muslims from Egypt, Turkey, India, Java, Algeria, and elsewhere.[336] His war activities have been described above, in connection with Pan-Islamic propaganda. October 1918 found him in Istanbul, prevented by the Germans from going to Germany, along with other Pan-Islamic personalities, whom they no longer wanted as their hour of military defeat approached.[337]

With the end of the war, Pan-Islam was in retreat in Egypt, giving way to a wave of ardent popular nationalism. True, in 1920 Celal Nuri [İleri]'s book on Pan-Islam was published in an Arabic translation and there were some other signs of renewed

[333] Tamura, 107.

[334] Not an Arabic monthly, as in Goldschmidt, 'The National Party', 18. See also Ḥasan al-Shaykha, *Aqlām thā'ira*, 101–20. An issue of this newspaper (no. 133) is preserved in Atatürk Kitaplığı, Istanbul.

[335] 'Turks and Arabs', *The Egyptian Gazette*, 26 Apr. 1913. This article was contributed from Beirut. A copy is available in NA, RG 59, 867.00/539, with US Vice-Consul F.S. Smith's no. 443, to the Secretary of State, dated in Beirut, 6 May 1912.

[336] al-Jundī, *ʿAbd al-ʿAzīz Jāwīsh* . . . , 62, 83–9, 121–2

[337] Rothmann, 14, based on German documents. For Shāwīsh's last years in Turkey and Egypt, cf. al-Jundī, *ʿAbd al-ʿAzīz Jāwīsh* . . . , 138–43.

interest in Pan-Islam. However, its return to an important public role took some time (as we shall explain later),[338] for the political leaders who rose to pre-eminence, in the anti-British movement of 1919 and subsequently, were Egyptian nationalists who distanced themselves from Pan-Islam.[339]

8. *The Libyan War and Pan-Islam*

The 1911 Italian invasion of Libya (more strictly: Tripolitania and Cyrenaica) shocked not only Muslims, but European diplomats, apprehensive of Pan-Islam, too.[340] The Ottoman–Italian war which ensued was, for the Ottoman Empire, a sort of 'general rehearsal' for World War I, in certain respects, including Pan-Islam;[341] especially since the war did not end with a scrupulously adhered to peace treaty, but continued, with the involvement of many other Muslims.

Indeed, one of the most interesting aspects, foreshadowing the development of political Pan-Islam, was the response of Muslims far and near to the Italian invasion of Muslim territory. Right from the inception of this war, Pan-Islam served as a bond for disparate tribes in Libya, as well as between them and the Ottomans, and between both of these and other Muslims within and without the empire. The flow of Pan-Islamic propaganda via books, newspapers, and manifestos, as well as some sermons sponsored by the Young Turks in the mosques and the *zāwiya*s of the fraternities, demonstrated that Pan-Islam was a dominant factor in uniting these diverse

338 See below, chs. V and VI.

339 For details, see I. Gershoni, *Mitsrayim beyn yiḥūd lĕ-aḥdūt*, esp. 35–9. Some Egyptian political leaders had been critical of Pan-Islam even earlier, cf. Nasr, ch. 5.

340 Even such an experienced diplomat as Edward Goschen, British Ambassador in Vienna and Berlin before World War I, noted in his diaries, on 29 Sept. 1911, that Italy's declaration of war would embarrass Great Britain, with its millions of Muslim subjects. See C.H.D. Howard, ed., *The Diary of Edward Goschen 1900–1914*, 245.

341 The most recent study of the Ottoman involvement in Libya and its war with Italy is Simon's *Libya between Ottomanism and Nationalism*. Osman Okyar brings out the interesting point that this war was a rehearsal for Turkey's War of Independence. See his 'Un témoignage français sur Union et Progrès et la défense de la Tripolitaine (1911–1912)', 173.

elements.[342] One may agree, in this context, with the statement expressed in a recent doctoral thesis on this war, that it was the first resistance movement inspired by Pan-Islam against Western Occupation.[343]

The first reaction to the Italian invasion occurred in the Ottoman Parliament, during October and November 1911.[344] A speech by the Sultan, certainly written by the Grand Vizir, reminded Parliament of the warm reception the Sultan had experienced during a visit to the Balkans only four months earlier (due to the excitement of local Muslims at the Libyan War), and called for Ottoman and Islamic solidarity. This appeal was supported warmly by parliamentarians of the Committee of Union and Progress as well as the Opposition. Numerous cables of identification with the Government arrived from Muslim dignitaries and communities both within and without the empire. The war was widely considered as a *Jihad*. Enver Pasha issued a proclamation to the warriors, urging them to fight the enemies of Islam and assuring them of the support of the world's Muslims. The entire Muslim press in the Ottoman Empire and many Muslim newspapers abroad (including Shiite ones in Iran) supported the Ottoman Government and its military forces, on Pan-Islamic grounds, emphasizing the need for unity and union.[345] Although reactions varied, with written and verbal expressions unequally paralleled by donations of money, they were unanimously supportive. On the whole, manifestations of Islamic solidarity were so impressive that they created a backlash of accusations of Pan-Islamic fanaticism, not merely in the Italian press, but in that of other European states as well.

At the start of the Ottoman–Italian war, Shakīb Arslān and ʿAbd al-ʿAzīz Shāwīsh were apparently planning a Maghribi alliance, but they soon gave up and joined a Pan-Islamic one

[342] A.M. Barbar, 'The Ṭarābulus (Libyan) Resistance to the Italian Invasion: 1911–1920', 374–6.

[343] Ibid. 1–2.

[344] This has been very adequately summarized in a forthcoming book by Orhan Koloğlu, *Islamic Public Opinion during the Libyan War 1911–1912*, ch. 2, A (refs. are to the TS of the book). I am grateful to Mr Koloğlu for allowing me to read his work before its publication.

[345] Many examples ibid., ch. 2, B.

instead.[346] Ottoman statesmen and journalists emphasized the Islamic bond with Libya's Muslims, as the only common factor between the Libyans and outside supporters. According to a report in a leading Berlin daily of the time,[347] a group of prominent Pan-Islamists met in that city and drew up an appeal to all the Muslims of the world, whose gist was as follows. The Ottoman Empire, as the Caliphate, could not give up Tripolitania. Renouncing it would both create great difficulties in the Balkans and pave the way for an Arab Caliphate; thus it would mean suicide both for the Caliphate and the state. The Pan-Islamists called on all Muslims world-wide to boycott Italy; moreover, Italians should be taken hostage, wherever encountered. As for the Ottoman Empire, it ought to appeal to all Muslims for assistance in money and manpower.

Many Muslims, far and near, contributed to the war effort against Italy, responding to the emphasis on the religious ties uniting the whole of the Muslim world, with the obligation to participate or help in what was regarded as a Holy War.[348] The consideration displayed, in this instance, by the Committee of Union and Progress for the bond of Pan-Islam, was an act of practical politics and demonstrated its understanding of the realities of the situation.[349] A German observer noted that, due to this war, leading members of the Committee of Union and Progress in Salonica had been affirming their interest in Pan-Islam as a political movement.[350] True, not many volunteers arrived from outside the empire; however, moral and financial support was forthcoming from Muslims as far away as Britain, India,[351] and South Africa.[352]

A British journalist, G.F. Abbott, who joined the Turkish Arab forces in the desert near Tripoli a few weeks after the war started, in order to write a book, has left us a very vivid account[353] of his meeting with one of those voluntary warriors

[346] Cleveland, *Islam against the West*, 92.
[347] *Vossische Zeitung*, 29 Sept. 1911.
[348] Simon, 86–7, 330.
[349] Luke, 146.
[350] PA, Orientalia Generalia 9, vol. 4, Kanzlerdragoman Schwörbel's no. 51, dated in Salonica, 11 Apr. 1912.
[351] Ameer Ali, 'Moslem Feeling', 101 ff. For a more detailed discussion of the response of India's Muslims, see below, ch. IV.
[352] Simon, 87.
[353] G.F. Abbott, *The Holy War in Tripoli*, 261–9.

in the desert. A dervish named Muḥammad al-Kiyānī belonging to the Fraternity of the Qādiriyya, he had been fighting the Italians, haranguing local warriors to excite their Islamic sentiment, and mobilizing fighters in the hinterland. He had travelled on several missions to other Muslim lands, on behalf of the Young Turks, and demonstrated great confidence that the Muslims would unite to drive out the Italians—and, eventually, the French and other non-Muslims—from their newly conquered territories. The dervish spoke some French, English, and German, in addition to Arabic, Turkish, and modern Greek, so that he had had some secular education. What is impressive about al-Kiyānī, as reported by Abbott, is not only the coherent, well-thought-out manner in which he presented his cogent conception of Muslim unity, but his accurate prediction of World War I[354] in Europe (and of its participants), which he considered a godsent opportunity for a Pan-Islamic union to re-establish Muslim rights everywhere.

The precise Pan-Islamic dimension of the Ottoman involvement in the Ottoman–Italian war has never been established[355] and there are obvious difficulties in attempting this at present. There are, however, indications that the Benevolent Islamic Society—the main channel for the Pan-Islamic activities of the Committee of Union and Progress—organized a sizeable share of the non-state Muslim assistance to Libya. In this manner, the war contributed to the institutionalization of Pan-Islam as a force to be employed, in certain patterns and with due organization, in order to assist Muslims militarily, politically, and economically.

The Italian Government was aware of all this and, in order to end the war, made various suggestions, including one which, while not acknowledging the Sultan's suzerainty over Tripolitania, would recognize and uphold there the religious supremacy of the Caliph.[356] This was so stipulated in the 1912 peace treaty, which specifically stated that the Caliph's name would

[354] Abbott's account was published in 1912.

[355] Despite the assertion of a local officer in Tripolitania, a French war correspondent with the Turco-Arab fighters, there obtained a definite impression that the religious element of Pan-Islam was the decisive factor in the political unity achieved among Turks, local Arabs, and others in fighting the Italians. See G. Rémond, *Aux camps turco-arabes*, 59–64.

[356] Cf. E.J. Dillon, 'Tripoli and Constantinople', 253.

be mentioned in the Friday sermons in Tripoli and that Libya's Supreme Religious Judge, or *qadi*, would be appointed from Istanbul.[357] In practice, however, peace treaties little affected the adepts of Pan-Islam—and hostilities against the Italians continued, with Ottoman connivance[358] and Muslim external assistance.[359] True, particularly from the start of World War I, Pan-Islamic ideology was differently interpreted by the Ottomans and the inhabitants of Libya. The former saw it as a means of recruiting Islamic support everywhere for their war effort, while the latter considered it more as a unifying bond to fight the Italian invasion of their country.[360] None the less, the lesson of the Ottoman–Italian war had been learnt: Pan-Islam could be used to win local sympathies and bring closer together elements previously indifferent to one another (Arabs and Turks, Turks and Sanūsīs),[361] and to obtain political support and money abroad.[362] The lesson was applied in World War I.

9. *The Case of Non-Ottoman Africa*

Morocco provides an interesting case for the spread of Pan-Islam into the only North African country which had never been a part of the Ottoman Empire.[363] Both because of religious prestige (Morocco's rulers averred that *they*, not the Otto-

[357] Kohn, *A History of Nationalism in the East*, 46.

[358] Acc. to the diary of the then Grand Vizir and Minister for War, Mahmud Şevket Pasha, several thousand gold coins were dispatched in the Sultan's name (as Caliph) to the sheikh of the Sanūsīs, to assist in continuing the war; see Simon, 105. In March 1912, the Sultan conferred high honours on the sheikh of the Sanūsīs for fighting the Italians; see NA, RG 59, 867.00/352, US Embassy's no. 156, to the Secretary of State, dated in Istanbul, 15 Mar. 1912.

[359] J. Roumani, 'From Republic to Jamahiriya', 158–9.

[360] Barbar, 378.

[361] 'Turkey, Russia and Islam', 114. The Sanūsīs were among the leaders of local resistance to the Italians; see A. Wirth, 'Panislamismus', 437 ff.; Schmitz, *All-Islam!*, 60. They continued to fight against the Entente Powers during World War I; see G. Gubaydullin, 'Krakh Panislamizma vo vtoroy impyerialistichyeskoy voyni', 112–18.

[362] *Mohammedan History*, 68.

[363] I am indebted for some of what follows on Morocco to two recent articles, based on archival research, both by Edmund Burke: 'Pan-Islam and Moroccan Resistance to French Colonial Penetration, 1900–1912' and 'Moroccan Resistance, Pan-Islam and German War Strategy, 1914–1918'.

man sultans, were the descendants of the Prophet Muḥammad and hence deserved the Caliphate) and *raisons d'état* (unwillingness to become bogged down in the diplomacy of the 'Eastern Question'), the sultans of Morocco consistently rejected not only diplomatic relations with the Ottoman Empire, but other official ties as well. Whatever connections existed were unofficial and secret, and generally carried out via Egypt. This held true in the case of Pan-Islam also: Abdülhamid's emissaries who visited Morocco were not encouraged to remain for long.[364]

One of the demands of the Moroccan revolutionaries who deposed the Sultan ʿAbd al-ʿAzīz, in favour of his more militant brother, ʿAbd al-Ḥāfiẓ, was that the latter would seek closer ties with the rest of the Muslim world against increasing French penetration into Morocco's affairs. In 1909–10, a military mission of Turks and Egyptians stayed in Morocco, to train its forces. It also founded a short-lived Pan-Islamic youth organization, 'The Young Maghrib', joined by youngsters from Algeria, Tunisia, and Egypt as well.[365] This was followed by an allegedly philanthropic society, *Jamāʿa Khayriyya*, another Pan-Islamic association based in Tangier, which in reality comprised not only high officials but officers as well.[366] In 1911–12 a Pan-Islamic conspiracy, acting through an association entitled *al-Ittiḥād al-Maghribī* (Union of the Maghrib), planned Moroccan riots against the French, as a step towards more North African uprisings. Turkish and Egyptian officers took part, as did Pan-Islamic groups in Egypt. Even ʿAlī Yūsuf, the editor of *al-Muʾayyad* in Cairo, was involved; his newspaper printed, in 1912, a series of articles calling for the unity of all North Africa's Muslims against the French.[367] Pan-Islamic leaflets were distributed in Moroccan towns. The fact that the French, with superior forces, crushed the rebellion should not discount this renewed Pan-Islamic effort to dislodge them from North Africa. Methods were employed which had been used

[364] 'Le Panislamisme turc', 60. For Abdülhamid's relations with Morocco, on the Pan-Islamic level, see also Çetinsaya, 'II. Abdülhamid döneminin ilk yıllarında "islâm birliği" hareketi', 52–4.

[365] Burke, 'Pan-Islam', 109.

[366] E. Michaux-Bellaire, 'Notes sur le Gharb', 11.

[367] Burke, 'Pan-Islam', 110ff.

previously in the Ottoman–Italian war—attempts by experienced Turkish and Egyptian officers to organize secret Pan-Islamic associations and tribal coalitions; and appeals to Muslim sources in the empire and abroad for moral and financial support.

During World War I, Enver's agents tried to recruit Moroccans for Pan-Islamic activity.[368] In addition, there was a continuous flow of Pan-Islamic propaganda into French-ruled Morocco, some of which has been referred to above. Further information[369] establishes that most of it was sent via a German mission in neutral Spain, operating under the banner of Pan-Islam—for which purpose a group of Arab officers in the Ottoman forces were seconded to it. Written in Istanbul, Berlin, and Madrid, Pan-Islamic tracts calling for anti-French Moroccan co-operation with the Islamic war efforts were distributed. The Pan-Islamic works of Ṣāliḥ al-Sharīf al-Tūnisī[370] seem to have been especially popular in Morocco. The above Ottoman component in the German mission added credibility to the German war propaganda in Morocco and signalled, indeed, the Ottoman Government's readiness to use Pan-Islamic motifs in the war in order to stir up problems for the French rulers. In addition, it sent special agents to Morocco to cause more trouble and afford Pan-Islamic legitimacy to local Moroccan warriors against the French.[371] Even though all these activities failed to change the fortunes of war in Morocco, due to French vigilance and to the victories of the Entente Powers on other fronts, it laid the basis for future Moroccan identification, lacking previously, with Islamic factors in the Middle East.

Considerably less information is available about Pan-Islam in other parts of Africa during that period. Shortly before World War I, members of the fraternities were known to increase the awareness of their Muslim identity in sub-Saharan Africa, which may have included a clearer perception of all-Muslim unity.[372] During the war itself, there was some difference in the penetration of Pan-Islamic propaganda and activity be-

[368] Cleveland, *Islam against the West*, 91.
[369] Collected by Burke, 'Moroccan Resistance', esp. 439 ff.
[370] Burke, ibid., erroneously calls him 'Salah al-Sharif al-Tunisi'.
[371] Ibid. 458–60.
[372] Some evidence in R.M., 'L'Islam et les missions', esp. 103.

tween West and East Africa. In the former case, although the Germans circulated the proclamation of the *Jihad* and similar propaganda, they made little headway in British-ruled areas, such as Nigeria (with its large Muslim population).[373] By contrast, Pan-Islamic propaganda introduced by German and Ottoman agents from southern Tunisia into the French Sahara and other French-ruled territories was more effective, in 1915, provoking rebellions; trouble increased for the French with the Sanūsīs invading the same areas, in 1916, and bringing their own version of the Islamic and Pan-Islamic message.[374]

In East Africa, Pan-Islamic propaganda was no less intensive, due to the proximity of German colonial bases. It assumed a special character, however, of which we are informed by two secret memoranda, written by Captain J.E. Phillips for British Intelligence. Their author, formerly the Chief Political Officer in Uganda and, during the war, a Military Intelligence Officer, wrote a 6-page memorandum on *'Africa for the Africans' and 'Pan-Islam'*, on 15 July 1917,[375] and a 4-page one on *Greek Traders Assisting Germanic–Islamic Propaganda*, on 13 August 1917.[376] In the first, Phillips said that the Germans had been encouraging, in their own areas of Africa, the formation of a black army. In conjunction with the Pan-Islamic propaganda of the Germans, with Islam serving as the cementing factor, the cry of 'Africa for the Africans' could cause real danger—all the more so because Islam in East Africa tended to see itself as a political no less than a spiritual force. Hence, he opined, East Africa could become common ground for Pan-Islam and Pan-Africanism. An African *Jihad* would be welcomed there enthusiastically. During World War I, Phillips concluded, 'printed and written Islamic propaganda with green flags[377] has frequently been intercepted by Uganda Intelligence Agents en route from German East Africa to the Sudan, Congo

[373] Arabic letters, calling for *Jihad* in the Caliph's name, were found there, acc. to reports in the Colonial Office (London), mentioned by Jide Ọsuntokun, 'The Response of the British Colonial Government in Nigeria to the Islamic Insurgency in the French Sudan and the Sahara during the First World War', 99, 106.

[374] Ibid. 101–2.

[375] FO 395/64, file 192843.

[376] FO 395/151, file 176962.

[377] The emblem of warring Islam.

and Darfur. Ex-Sultan Saʿīd, Khalīf of Zanzibar,[378] recently captured, was a leading spirit'.

The second memorandum stated that, since the beginning of the German East Africa campaign, on 8 August 1914, there had been a steady flow of German Pan-Islamic propaganda, frequently intercepted by British Military Intelligence in Uganda, Congo, and Sudan. This generally took the form of printed proclamations in Arabic and Swahili, signed by Governor Schnee and the ex-Sultan Sayyid Khalīfa of Zanzibar.[379] Pan-Islamic agents were liberally supplied with funds. Prominent positions in the future Germano-Islamic Empire and Eastern Africa were promised, in return for spying services or for influencing Sudanese troops to desert. In 1915–16, however, a new phase started. Printed proclamations, with green flags, were dispatched to French Sudan, Darfur, and Eritrea. They set forth that the *Jihad* required of all true Believers not merely a passive, but an active resistance to the Entente; and promised, as a reward, the establishment of an Islamic Empire under benign German protection.

[378] Zanzibar seems to have been an important centre of Pan-Islamic sentiment, consequent on Abdülhamid's efforts there, as pointed out above, in ch. 1.

[379] See n. 378.

III. *Pan-Islam Clashes with the Russian and Soviet Authorities*

1. *The Muslims in Tsarist Russia and Pan-Islam*

In contrast to the uninterrupted land mass of Muslim populations from Morocco to Iran and Afghanistan, nineteenth-century Tsarist Russia's Muslims, living in the huge area between the Crimea and the boundaries of China, were separated by non-Muslim groups. Although the bond of Islam remained a potent source of commonality, ethnic sentiments (some of them of tribal origin), linguistic differences between Turkic dialects (in a majority of cases) and Iranian ones (in a minority) divided them. Large uninhabited areas and others, populated by people of Russian origin and Christian beliefs, separated the Muslim communities from one another. While language barriers were relevant to other Muslim concentrations too, they had more impact in the case of those in Tsarist Russia, when combined with the other factors.

While precise, reliable figures are not available for the individual Muslim groups in Tsarist Russia, the first official census, in 1897, reported their overall number as 13,600,000 out of a total population of 125,600,000, that is, almost 11 per cent. Since most of the areas inhabited by Muslims had been annexed by Russia during the late eighteenth and nineteenth centuries and economic penetration was only just beginning, most Muslims continued their daily life undisturbed. However, in the second half of the nineteenth century, a strong official policy of Christianization and Russification brought about an equally powerful ideological response from segments of the local Muslim intellectual élites. As in the Ottoman Empire, the response of these élites, in a renewed quest for a Muslim identity and its assertion in Pan-Islamic terms, was essentially a reaction to external challenge. However, while the Ottomans were responding to political and military aggression, the Muslims in Russia were faced by the danger of imposed assimilation, on the one hand, and discriminatory treatment for those resisting it, on the other. Even so, because of the conditions mentioned, an uprising against the Russians, coordinated by Pan-Islam, was a practical impossibility.[1] Other means had to be considered.

The late nineteenth century witnessed the rise of small but dynamic élites among the middle class of several Muslim groups in Russia, more among the Turkic (chiefly the Tatars) than the Iranian ones. Some of their spokesmen were acutely aware of the perils of assimilation facing their communities and of the need to find a common denominator between them, preferably with some support from abroad, in order to better resist Christianization and Russification. Probably taking their cue from the spokesmen of Pan-Slavism, they searched for a pan-ideology to assist them. This occurred in the 1870s and 1880s, when, in the Ottoman Empire too, intellectuals were examining and arguing about such general ideologies.

For the Muslims in Russia in search of a distinctive identity, the choice seemed to lie between two main ideologies, Pan-Islam and Pan-Turkism (for modernization, as an ideology, was being espoused by the assimilationists, in the main). The decision was imperative in Russia one or two decades earlier than in the Ottoman Empire. However, as has been pointed out by Alexandre Bennigsen,[2] while in the Ottoman Empire Pan-Islam and Pan-Turkism were rival ideologies, among Russia's Muslims they were complementary. Islam was the main cementing bond among the various Turkic and Iranian groups and was, obviously, better understood and appreciated. Moreover, customs and established structures were heavily impregnated by Islam; alienating their functionaries and mullahs

[1] A British MP, Henry Norman, who visited Central Asia, early in the 20th century, remarked on this, referring chiefly to Bukhara, where mullahs dispatched by Abdülhamid II had been active in Pan-Islamic propaganda. See his *All the Russias*, esp. 288–91.

[2] A. Bennigsen, 'Panturkism and Panislamism in History and Today', esp. 40–1. See also T. Swietochowski, *Russian Azerbaijan, 1905–1920*, 33–4. C.W. Hostler, *Turkism and the Soviets*, 130ff.

meant the provoking of unnecessary opposition. Pan-Turkists in Russia spoke and wrote in Pan-Islamic terms when seeking to forge a new union of forces. The fact that new railroads were being constructed and maritime connections made more easily available were undoubtedly important factors for those Muslims from Russia who were performing the Holy Pilgrimage and stopping in Istanbul on their way.[3] After the end of the Crimean War in 1856, about 140,000 Tatars (more than half of those living in the Crimean Peninsula) left for Turkey; this large-scale migration established ties which were later relevant for the progress of Pan-Islam in southern Russia.[4]

No less significant for the spread of Pan-Islamic ideas among Russia's Muslims was the overall increase among them, in the late nineteenth and early twentieth centuries, in education, literacy, and publishing—much of this with an emphasis on Islam. To select one instance, Kazan, main city of the Tatar population at the bend of the Volga, underwent a striking Islamic renaissance, becoming a centre not merely for Russia's Muslims, but for many abroad as well. In 1902, the Islamic university there had some 7,000 students, and its press published about 250 volumes. In 1907, the Islamic library had 18,700 readers. According to fairly reliable Russian figures, there were in 1912, in the whole of Tsarist Russia, 24,321 Muslim communities, served by 45,339 religious officials, 48 philanthropic associations, 34 societies for Islamic studies, and 87 organizations of experts in Islamic law. Between them, these owned 23 printing-presses, 194 libraries, and were publishing 18 newspapers and periodicals.[5]

This Muslim revival, which was to provide a fertile ground for Pan-Islamic propaganda and activity, was first and foremost a natural response to the pressure of Russification and Christianization; it was also, perhaps, influenced by the ideological ferment in the neighbouring non-Muslim society in the Tsarist Empire. Contacts with the Ottoman Empire have already been mentioned. However, one other special event should be noted:

[3] J.M. Landau, *Pan-Turkism in Turkey*, 7 ff.

[4] S.A. Zenkovsky, *Pan-Turkism and Islam in Russia*, 27. On these Tatars, see A. Fisher, *The Crimean Tatars*.

[5] See H. Bräker, *Kommunismus and Weltreligionen Asiens*, i, part 1, 51, with footnotes.

the Russian defeat in the Russo-Japanese war of 1904–5. To Muslims everywhere, this event proved that an Oriental state can achieve victory over a major Christian European power.[6] To the Muslims in Russia, in addition, it indicated the weakness of the empire; it was then that some of them started considering seriously their own liberation, and they were further encouraged by the 1905 Revolution.[7]

2. *Towards a Pan-Islamic Identity*

As in the Ottoman Empire, a cultural approach to Pan-Islam preceded the political one. In the younger generation of intellectuals in Russia who were grappling with broader solutions to their particular problems as Muslims, the most original personality—and the most active one, perhaps—was Ismail Gasprinskiy (in Tatar, Gaspralı).[8] Gasprinskiy (1851–1914) was well educated and familiar not only with his own Tatar culture, but also with that of the Russians, French, and Turks. He had been considerably influenced by Young Ottoman writers, including their views on Pan-Islam. He served as the mayor of his town in the Crimea, Bahçisaray, between 1877 and 1881, and remained public-minded throughout his life. It is in this spirit that he strived to serve his people and his co-religionists as an educator,[9] journalist, and writer, who was constantly involved in politics. The year 1881 was a crucial one in Gasprinskiy's career. It was then that he resigned his mayoralty to dedicate his whole life to writing and teaching—and, later, to political activity—in the cause of Pan-Turkism and Pan-Islam. It was in that year, also, that he published what may well have been his most important book.

[6] K. Kreiser, 'Der japanische Sieg über Russland (1905) und sein Echo unter den Muslimen', 209 ff.

[7] H.C. d'Encausse, *Réforme et révolution chez les Musulmans de l'Empire russe*, 117.

[8] Various aspects of Gasprinskiy's writing and activity have been dealt with. See e.g. A. Arsharuni and Kh. Gabidullin, *Ochyerki Panislamizma i Pantyurkizma v Rossii*, 12 ff. Fisher, 100 ff. Zenkovsky, ch. 3. Landau, *Pan-Turkism in Turkey*, 9–13. Cafer Seydahmet, *Gaspıralı Ismail Bey* (Istanbul, 1934). Gerhard von Mende, *Der nationale Kampf der Russlandtürken*, 44–61.

[9] For a new book on Gasprinskiy and his educational activity, see Mehmet Saray, *Türk dünyasında eğitim reformu ve Gaspıralı İsmail Bey (1851–1914)*.

Written in Russian, and entitled *Russkoye Musul'manstvo*[10] (Russian Islam),[11] this 50-page work, expanded from a previously published five-article series,[12] foreshadowed Gasprinskiy's future views and activities. Its modest subtitle, 'Thoughts, Notes, and Observations of a Muslim', concealed an eager, cultivated mind. He may have chosen to write in Russian in order to reach a wider readership among the intelligentsia of the disparate Muslim groups, as well as among Russians. According to Gasprinskiy's perceptions, Russia was soon to be the state with the world's largest Muslim community. Due to discrimination and ignorance, he argued, this community did not yet have the affection towards its fatherland which Russia deserved. A better defined governmental policy was necessary. New schools using the Tatar language, including a university, would correct the situation. Imposed assimilation, via Russification, would only antagonize the Muslims. On the other hand, sweeping away ignorance and encouraging education among Muslims, in their own culture as well as in Russian, would bring about a *rapprochement* between Muslims and Russians and a more meaningful involvement in their fatherland's life, as barriers would be done away with. Such accommodation was the only feasible way for inter-communal relations, for Muslims possessed, in their religion and social life, the necessary force to resist any imposed assimilation. Every Muslim community had its own miniature government, binding the individual to the body and following common laws, customs, and traditions—all rooted in Islam. Every such community had its own spiritual leadership, to which any Muslim might aspire, its own mosque and school (*mekteb*), funded by pious foundations (*waqf*) money, while many communities had a more advanced school (*medrese*). Muslim society presented, therefore, a compact mass, living its own particular life. Russification could not really have much effect, for Islamic life would continue within the family, which could not be controlled. On the contrary, pro-

[10] I. Gasprinskiy, *Russkoye Musul'manstvo*.

[11] The term *Musul'manstvo* is broader than *Islam* (which is also used in Russian) and may imply 'the entire Muslim community'. In Gasprinskiy's case, his use of the broader term is very probably intentional.

[12] In the newspaper *Tavrida*, nos. 43–7 (1881).

viding Muslims with education and publishing facilities in their own languages, combined with Russian, would make them better patriots.

Two main themes run through Gasprinskiy's book. The first is an opening towards Russification, emphasizing the particular need of Muslims in Russia to acquire the language of the government. He was attacked for this by his fellow Tatars and other Muslims who, however, did not acknowledge Gasprinskiy's support for the Tatar language and for Muslim civilization and society. The few pages he devoted to the compactness and self-sufficiency of the Muslim society were indicative of his early concepts of Muslim unity and presaged his further writing and activity in the cause of Pan-Islam. After all, considering the severe Tsarist censorship at the time, it would have been nothing short of suicidal for him to argue explicitly for an immediate union of all Russia's Muslims. What he wanted was for the Muslims to preserve their own identity and to acquire enough foreign—in this case, Russian—knowledge to provide better conditions for a subsequent campaign to arouse them.

That this was so may be deduced, also, from Gasprinskiy's further activities. As director of and teacher at a school in Bahçisaray he introduced a 'new style' (in Tatar, *uṣūl-u jedīd*) which earned him and his partisans the appellation of 'Jedidists'. Jedidism[13] consisted not only in introducing better teachers and more suitable textbooks but, more importantly, overhauling the curriculum and changing the principles of instruction. Secular subjects, previously excluded, were introduced into the Islamic schools: arithmetic, history, geography, and the Russian language; the study of Arabic was left untouched—as the basis for Islamic studies and contacts between Muslim Ulema—but Turkish was introduced instead of Persian, to provide a link to Crimea's neighbour and the most powerful independent Muslim state of the day, the Ottoman Empire.[14] Learning by rote was gradually discarded and more active methods of instruction substituted. These schools were

[13] A useful article on Jedidism and its sources is I.S. Braginskiy, 'O prirodye Sryedniaziatskogo Dzhadidizma v svyetye lityeraturnoy dyeyatyel'nosti Dzhadidov'.

[14] For the syllabus, cf. S. Bobrovnikoff, 'Moslems in Russia', 16–18.

soon widely imitated among Muslims in Russia and elsewhere, even as far away as Samarkand and its vicinity.[15] Most of the supporters of Jedidism, which comprised strong Pan-Turkish and Pan-Islamic features, came from the middle class and probably the middle-aged (with the poorer and younger Muslims more attracted towards socialist or revolutionary doctrines).[16]

Gasprinskiy's journalistic activity was no less impressive. In 1883, he started a new semi-monthly, *Tercüman* (The Interpreter), which fostered interest in social affairs, educational reform, and modern science. It advocated nationalism with very noticeable Pan-Turkic nuances and, since Gasprinskiy was careful not to antagonize the mullahs, Pan-Islamic overtones. A union of all Turkic groups (more rarely, Muslim groups) in Russia, with the Ottoman Empire offering its spiritual guidance, was consistently supported, as it was in other publications of his.[17] In order to further this cause, Gasprinskiy invented for *Tercüman* a new lingua franca for all Turkic groups,[18] largely based on the Turkish of Turkey, but freed from many Arabic and Persian accretions and minimalizing phonological differences, so that it could hopefully be understood by Turkic groups practically everywhere; or, as Gasprinskiy put it, 'from the boatman on the Bosphorus to the camel-driver in Kashgar'. That this was no idle dream can be deduced from the fact that, in the 1880s, *Tercüman* had a circulation of about 5,000, with many subscribers abroad, keeping up that figure until the eve of World War I (by which time it had become a daily newspaper).[19]

The slogan of *Tercüman* was 'unity in language, thought and action'.[20] This was a leitmotif of several of Gasprinskiy's other writings, such as a tract on *The Principles of Civilizing the Russian Muslims.*[21] That this unity did not refer solely to the

[15] K.K. Pahlen, *Mission to Turkestan*, 44–6, 74.

[16] Cf. E.D. Sokol, *The Revolt of 1916 in Russian Central Asia*, 65–7.

[17] In 1906, Gasprinskiy started publishing a weekly, *Millet* (Nation), which also reflected Pan-Turkism and Pan-Islam.

[18] *Tercüman* was issued in this new language and in Russian, but from 1905 Russian almost disappeared.

[19] *Tercüman* lasted until 1918.

[20] *Dilde, fikirde, işte birlik̡.*

[21] *Mebādi-yi temaddün-i Islamiyān-ı Rūs* (1901). English transl. by E.J. Lazzerini, 'Ğadīdism at the Turn of the Twentieth Century', 250–8.

cultural field, and that it would be eventually invoked, to draw Russia's Muslims to concerted action, by Gasprinskiy and his associates and disciples,[22] was seen in the early twentieth century, when dramatic change in Russia offered new opportunities for political activity.

In the late nineteenth and early twentieth centuries, Gasprinskiy was only the most prominent of several very capable Muslims in Russia, all active in promoting Pan-Turkism and Pan-Islam. Yusuf Akçura[23] (in Russian, Akchurin), also a Tatar, has already been mentioned as one of the foremost ideologues of Pan-Turkism, as has Abdürreşid İbrahim (Ibragimov), an industrious propagandist of Pan-Islam in Orenburg, St Petersburg, and Kazan. In 1900–9 İbrahim edited a periodical, *Mir'āt* (Mirror), of which twenty-two issues were published, in St Petersburg and then in Kazan, mostly in Tatar, but partly in Arabic. This began as a literary and cultural review, then became increasingly political, with a definite Pan-Islamic appeal.[24]

Such activities were not limited to the Tatars, although they had a major share in them. In the 1880s, Pan-Islam found some support in the press of Azerbaijan as well.[25] A leading Azeri Pan-Islamist in that generation was yet another journalist, Ahmed Ağaoğlu (in Russian, Agayev) (1870–1938).[26] Influenced by al-Afghānī's writings,[27] Ağaoğlu wrote in French and other languages, supporting a renaissance of Islam and the establishment of a Pan-Islamic union, led by Iran; he subsequently changed his mind and campaigned for the Ottoman Empire to head such a union. Later, he emigrated to Istanbul, and became actively involved in the circles of the Committee of Union and Progress after the Young Turk Revolution of 1908.

Pan-Islamic sentiments, fostered by leading personalities such as Gasprinskiy, İbrahim, and Ağaoğlu, undoubtedly had

[22] Cf. d'Encausse, *Réforme et revolution*, 102 ff. referring to Bukhara and Turkestan.

[23] On Yusuf Akçura, see F. Georgeon, *Aux origines du nationalisme turc*. Also *Ölümünün ellinci yılında Yusuf Akçura sempozyumu tebliğleri*.

[24] Lazzerini, 269. Mahmud Tahir, 136. A set can be consulted at the New York Public Library.

[25] Zarevand, *United and Independent Turania*, 9.

[26] On Ahmed Ağaoğlu, see Swietochowski, 34–5.

[27] Ibid. 34.

some impact on Russia's Muslims, but it is difficult to estimate how effective this propaganda was. Some indication is supplied by the correspondence of the Young Turks in Europe, early in the twentieth century, with their co-religionists in Russia. A letter, signed by Dr Nazım and Dr Bahaeddin on behalf of the Committee of Union and Progress, dated in Paris on 22 September 1906 and addressed to the Muslims in the Caucasus, was uncovered by the late Turkish historian Ahmed Bedevi Kuran and reprinted in a shortened version.[28] In it, the mostly secular, politically minded Committee addressed their far-away co-religionists in the name of Pan-Islam, stating that the European powers' fear of Pan-Islam was a good indication of its force. The writers used this letter to establish relations with Russia's Muslims.[29]

Naturally enough, the feedback attests to internal and external Pan-Islamic propaganda among Russia's Muslims. Some of it is expressed in anonymous manuscript letters, written by Muslims in the Caucasus and found in the registers of the Committee of Union and Progress by the late Turkish historian Yusuf Hikmet Bayur.[30] The first letter, dated 22 September 1906, complained about the pitiful situation of the Muslims in the Tsarist Empire. Its writer maintained that, in order to save the Muslims from Russian tyranny, they should seek power, obtainable solely through union (*ittihad*). Since the Christian governments of Europe feared *İttihad-ı İslam* so much, their apprehensions ought to be exploited by uniting all Muslims from Africa to India and from the Adriatic to China. The writer himself considered a Muslim union axiomatic for the safety and well-being (*selamet*) of all Muslims, under the leadership of the Ottoman Government, pursuing together a Pan-Islamic policy (*İttihad-ı İslam politikası*). Another letter employed very much the same arguments, calling on 'Our Brethren-in-Faith' (*İhvan-ı dinimiz*) to unite. A third one urged, more specifically, all Muslims and all Turks to unite, while a fourth, dated 23 November 1906, reported the establishment in every corner (of the Caucasus?) of Islamic associations (*Cemiyet-i İslamiye*).

[28] Ahmed Bedevi Kuran, *İnkılâp tarihimiz ve İttihad ve Terakki*, 209–10.
[29] For selected extracts, see below, app. F.
[30] Bayur, ii, part 4, 89–93.

3. *Pan-Muslim Conventions and Political Parties*

The preparatory work in Pan-Islam (and Pan-Turkism) among Russia's Muslims, which was essentially religious and cultural—with politics only obliquely hinted at, for obvious reasons—bore political fruit from 1905, after Russia's defeat in the Russo-Japanese war and upon the success of its first revolution. Such was the general mood and so great were the hopes for a more liberal, egalitarian society, that several of the more prominent Muslims took active steps for concerted action. Their conventions have as yet received only brief mention, although they were the first concrete signs of Muslim readiness to work for unity, in Pan-Islamic terms, via public deliberation and political decision,[31] leading the way to numerous subsequent Muslim meetings.

In 1904, various Muslim personalities in Russia consulted with each other about enlarging and institutionalizing their contacts on behalf of their co-religionists.[32] This resulted in their first general convention, convoked semi-clandestinely at the fair of Nijni Novgorod, in August 1905, and held on a ship on the Volga, to avoid official intervention. The delegates numbered about 150 and Gasprinskiy was one of the chairmen. The resolutions were remarkable, if not totally unexpected: (i) a union (*ittifak*) of all the Muslims of Russia in one movement; (ii) equality of rights between the Russian and Muslim peoples; (iii) establishment of a constitutional monarchy; (iv) freedom of Muslim education, press, and publications; (v) respect for private property. It is significant that the very first resolution was couched unequivocally in Pan-Islamic terms. Yet another decision—and a more practical one—was to elect a *Majlis*, or parliament, to take care of joint Muslim affairs.

These were momentous decisions, but they lacked an effect-

[31] For the first three conventions, in 1905–6, see Arsharuni and Gabidullin, 23 ff.; A. Le Chatelier, 'Les Musulmans russes', 146–51; 'Les Aspirations politiques des Musulmans russes', 136–41; Swietochowski, 48–50; Fisher, 104; Bräker, *Kommunismus*, part 1, 60–3; d'Encausse, *Réforme et révolution*, 119–20; Landau, *Pan-Turkism in Turkey*, 11–12; A. Bennigsen and C. Quelquejay, *Les Mouvements nationaux chez les Musulmans de Russie*, 57 ff.; R.A. Pierce, *Russian Central Asia 1867–1917*, 255–8.

[32] The results of these deliberations were communicated abroad in a 10-page memorandum in French, dated 5 Apr. 1904 and signed Hadji Ahmad. See PA, Orientalia Generalia 9, vol. 1.

ive organizational framework to carry them out. In the months immediately following this first convention, local meetings deliberated such issues too, and when the Tsar granted, on 17 October 1905, the long expected liberties in Russia, the leaders of the Muslims acted. About a hundred delegates arrived in St Petersburg and managed to hold a second convention (characteristically, in the home of the local Benevolent Islamic Society), despite its being banned by the police. Again, Gasprinskiy presided. The most important decision taken was to set up a Union of Muslims of Russia (*Russiya Müsülmanlarının İttifakı*), to serve as a central body for collective action by all of Russia's Muslims. Even if in practice this body was at first composed almost solely of Tatars and Azeris, it was an important step towards Pan-Islamic organization, at least within the confines of the Tsarist Empire.

A series of local conventions followed, throughout many of the Muslim centres (again, chiefly in the Tatar areas and in Azerbaijan).[33] The heated debates, indicating increasing political awareness, concerned not only the forthcoming, third all-Muslim convention, but the elections to the Duma, the all-Russian parliament, as well. The third convention of Russia's Muslims met, in August 1906, again in Nijni Novgorod. This was the first open meeting allowed by the authorities and was the most important of the three. It was the largest, too, comprising nearly 800 participants. The speeches were bolder and one of the most committed to Pan-Islam came from Abdürreşid İbrahim, who said:

> kindred relations and solidarity among Muslims exist not only in talk or on paper, but also in their souls and blood. As these relations and solidarity are with the 20 million Muslims in Russia, so are they with the ideological relationship of the more than 300 million Muslims in the whole world, on the entire globe. The relationship of Muslims does not depend on a party—consequently, union is necessary not solely among the Muslims in Russia, but among the Muslims on the entire globe.[34]

[33] For a list of these local conventions and the subjects debated, see Le Chatelier, 'Les Musulmans russes', 149–52. Cf. 'Les Aspirations politiques', 138–40.

[34] Text in Arsharuni and Gabidullin, 29–30. English transl. mine.

The major resolution of the third convention[35] was to transform the Union of the Muslims of Russia into a political party also named the Union of Muslims (*İttifak-ul-Muslimin*), in Russian, *Musul'manskaya Partiya*). While the party acted cautiously at first, focusing on educational and benevolent activities, it served as a framework for a section of the Muslim members in the Duma as well. For the first time, a political party had not only adopted a Pan-Islamic name, but was actively concerned, if not with the affairs of all Muslims worldwide, at least with those of all the Muslims in Tsarist Russia. The party's programme[36] emphasized that its main objective was to 'unite the political and social activities of all Muslims in Russia'.[37]

The theory, however, was more impressive than the performance. The conventions had aroused rivalries between the Tatars in the Crimea, led by Gasprinskiy, and those of the Volga, whose spokesman was Abdürreşid İbrahim; and between those groups and the Azeris, led by Topchibashyev of Baku, the chairman of the third convention. Yet another meeting, at Baku in October 1906, was beset by quarrels between the religious and the secular.[38] Worse still, the negotiations and alliances with non-Muslim members of the first Duma, then of the second, third, and fourth,[39] eroded the consensus needed for concerted political moves in the cause of Russia's Muslims and Pan-Islam.[40] Gasprinskiy bewailed missed opportunities in his *Tercüman*;[41] but the *İttifak-ul-Muslimin* party was already declining, due to the resumption of authoritarian rule in the Tsarist Empire from 1908, until the

35 The resolutions of the third convention were reprinted ibid. 149–52.

36 Entitled in Russian, *Proyekt': Ustav' obshchyestva 'Musul'manskoy partii'*, and in Turkish, *'Muslimān ittifākı' cemiyetinin niẓāṃnāmesi*. Both versions, comprising 8 pages each, were printed at Abdürreşid Ibrahim's press in St Petersburg.

37 Ibid. § 2.

38 L. Bouvat, 'Les Derniers Congrès des Musulmans russes', 264–5. 'Les Aspirations politiques', 140–2.

39 Details in Arsharuni and Gabidullin, 35–49.

40 For the points agreed upon in 1907 by a large number of Muslim members in the Duma, see the 16-page *Programma Musul'manskoy parlamyentskoy fraktsii v' gosudarstvyennoy dumi*.

41 Of 6–19 April 1907, quoted by L.B. (= Bouvat), 'Le Parti musulman russe', 388.

party ceased to exist in 1914.[42] This was despite interest in it and support by Russia's Muslims, as evidenced, for instance, in a letter from Astrakhan to the *Revue du Monde Musulman* in Paris,[43] written in late 1909 or early 1910, which said, *inter alia*, 'The Russian Muslim party *İttifak-ul-Muslimin*, a political and national party and its directors ... have played a principal role in this movement. At present all Russian Muslims are united in the common idea that a single compact mass might be organized ...'.[44]

The next step by Gasprinskiy in *Tercüman*, early in November 1907, was a proposal for a general convention of the world's Muslims. As he knew that this would never be permitted in the Tsarist Empire and since he was reluctant to suggest some capital city in Western Europe, lest the Young Turks take over the convention and alter its objectives, he wavered between Istanbul and Cairo; when it became clear that official Ottoman circles (inspired by Abdülhamid's reluctance towards international Muslim gatherings) opposed the entire idea, he decided on Cairo. Gasprinskiy himself went to Cairo to promote his idea,[45] even publishing an Arabic newspaper, *al-Nahḍa* (Revival, or Renaissance), of which three issues appeared in February–March 1908 before it closed down. In *al-Nahḍa*[46] he explained (authoring himself seventeen out of its twenty articles) the need for such a convention, as he had in a festive speech in Cairo before 300 prominent Egyptians, on 1 November 1907. That Turkish speech was translated into Arabic and then reprinted in the Cairene press.[47] Essentially, it appealed for an international gathering to improve the cultural and economic situation of Muslims everywhere, skirting political issues or, as Gasprinskiy himself

[42] Cf. Bräker, *Kommunismus*, part 1, 62.

[43] J. Réby, 'Un cri d'alarme russe', 106–7.

[44] Ibid. 107. English transl. mine.

[45] The entire matter of this proposal for a general Muslim convention in Cairo has been dealt with by Kramer, *Islam Assembled*, ch. 4, and T. Kuttner, 'Russian Jadīdism and the Islamic World: Ismail Gasprinskii in Cairo', 383–424. See also PA, Orientalia Generalia 9, vol. 3, von Oppenheim's no. 392, to von Bülow, dated in Cairo, 20 May 1908. Saray, *Türk dünyasında ...*, 62–71.

[46] For details, Kuttner, ibid. 413–18.

[47] A French transl. of part of this is available, by A.L.C. (= Le Chatelier), 'Le Congrès musulman universel', 498–502.

phrased it, avoiding Pan-Islam, of which the Europeans were afraid.[48] Had the convention been held (it was still being mentioned in the press in 1911), it might well have acquired the dimensions of a cultural, economic, or political meeting. The Egyptian press reacted favourably, in the main, to the idea of an all-Muslim convention,[49] publishing Gasprinskiy's statutes for it.[50] Although the convention never met, it did have an impact, conceptually and organizationally, and it possibly influenced similar conventions which met, in Egypt and elsewhere, during the inter-war years.

When World War I broke out in 1914, Pan-Islamists and Pan-Turkists in the Tsarist Empire found themselves in a dilemma, since their country was fighting their co-religionists, from their own ethnic groups, in the Ottoman Empire. In this clash of loyalties between patriotism and Pan-Islam/Pan-Turkism, the former won. The appeal for *Jihad* from Istanbul had little effect in Russia and, in general, the Muslim masses and their leaders supported the Russian war effort.[51] Gasprinskiy himself had died in 1914. Other prominent Pan-Islamists and Pan-Turkists, like Akçura, Ağaoğlu, and Abdürreşid İbrahim, had emigrated to the Ottoman Empire before the war and were active on behalf of Pan-Islam there, as has already been mentioned. These three leaders were among the signatories, in 1916, of an appeal to US President Woodrow Wilson, on behalf of Russia's minorities, presented to and approved by the Congress of Nationalities in Lausanne.[52]

As far as Muslims within the Tsarist Empire were concerned, it appears that some Pan-Islamic activity, possibly abetted by Ottoman agents, continued in Central Asia during the war. Most of the information available is sketchy and not

[48] A. Le Chatelier, 'Le Congrès musulman universel', 500. Whether his reservations about a politically inspired Pan-Islamic convention were genuine or not, he had to voice them in Egypt, considering British suspicions of Pan-Islam.

[49] Examples in Kuttner and in Ismaël Hamet, 'Le Congrès musulman universel', 100–7.

[50] Published in Arabic in *al-Mu'ayyad* (Cairo), 23 Jan. 1908. A French transl. is available in 'Statuts du Congrès musulman universel', 399–403, where the date at the top of p. 399 should read 23 Jan. 1908 (not 1907).

[51] Arsharuni and Gabidullin, 49 ff.

[52] B. Nikitine, 'Le Bolchevisme et l'Islam', 6–7. Landau, *Pan-Turkism in Turkey*, 15. For the text of the appeal, see 'Les Aspirations politiques ', 146–7.

always reliable. The Tsarist police discovered, in 1915, at least two different hectographed appeals calling on all Muslims—in the name of Islam and Pan-Islam—to free themselves from Russian domination by exhibiting their loyalty to the Caliph and assisting the cause of Islam by either enlisting in the Ottoman army or donating money.[53] The Tsarist secret police, the Okhrana, wrote numerous reports about little-organized Pan-Islamic activity in Turkestan and elsewhere, based on local Muslim networks which met at prayer and collected funds for the Ottoman war effort.[54] Most of the reports were based on hearsay. What is certain, however, is that Pan-Islam had little to do with the uprising of Turkestani Muslims in 1916.[55] This was caused by their natural reluctance to serve in the Tsarist armies (although, of course, it is not impossible that they were refusing also to fight against other Muslims).

Most of Russia's Muslims devoted themselves, during the war, in addition to the problem of survival, to religious and cultural issues. Some voices were raised in support of a religion-oriented Pan-Islam. A Tatar lawyer, Giray Alkın, writing in the *Kazanskiy Tyelyegraf* of 22 October 1914, maintained that Pan-Islam was a spiritual rather than a political concept; while Cemal Validi, a journalist writing in 1916, argued that a Muslim union would be feasible, at some future date, but only on religious—not on political, economic, or even cultural—grounds, since in his view Muslims had never been united except by religion.[56] An interesting report by the Russian Orientalist Gordlyevskiy, published in 1917,[57] informs us that, on the eve of the war, benevolent Muslim societies were being transformed into strictly cultural ones with a focus on libraries, reading-rooms, schools, literacy, and musical *soirées*. In Orenburg, in late 1916, a Muslim association for music and drama was inaugurated. Even before that, a Muslim society for the study of the history, literature, character, and customs of

[53] Sokol, 76, for the texts of these appeals.

[54] Ibid. 74–7.

[55] Ibid. 72–182.

[56] Gubaydullin, 'Krakh Panislamizma vo vtoroy impyerialistichyeskoy voyni', 104–5.

[57] Vl. Gordlyevskiy, 'Musul'manskoye uchyonoye obshchyestvo v' Moskvi', 69 ff.

the Muslims of Russia was founded in Moscow. Initiated by Tatar students there, the society thrived during the war years, despite the difficulties of the time. Sixty members were on its administrative committees—Tatars, Azeris, Sarts, Kirghizes—and other members throughout the Tsarist Empire. Their main objective was to research and spread knowledge about the Muslims, with an emphasis on the Turkic groups. The society prepared reports and provided lectures and lessons (with the financial support of several rich Muslims in Moscow). From all this information, it can be deduced that cultural Pan-Islamic activity took the place of political involvement, due to war conditions and the emigration of several of the more prominent Muslim leaders. The fateful year of 1917 witnessed a renewal of political activity, on a Pan-Islamic scale, involving Russia's Muslims.

In February and March 1917, Muslim leaders held several private meetings as to the ways of using for the benefit of all Russia's Muslims the new situation created after the deposition of the Tsar. From 15 to 17 March, the Muslim members of the Duma met with several of their prominent co-religionists, to set up a Provisional Central Bureau of Russian Muslims, in order to prepare a general convention of the Muslims in Russia.[58] It was noticeable that Muslim leaders and organizations were gradually drawing away from their war-time co-operation with Russian political parties and groups in order to focus on their own Muslim interests. In April, the Muslim members of the Duma requested Muslim groups throughout Russia to hold local and professional gatherings in preparation for an all-Muslim one.[59] The most important of these was the convention of the Caucasus's Muslims in Baku, in April, at which Pan-Islamic arguments were heard.[60]

The most important gathering, however, was the all-Russian Muslim convention, which opened in Moscow on 1 May 1917, attended by between eight and nine hundred participants.[61]

[58] Von Mende, 121 ff. Bennigsen and Quelquejay, *Les Mouvements nationaux*, 64.

[59] R. Pipes, *The Formation of the Soviet Union*, 76.

[60] 'Turkey, Russia and Islam', 128–9. Swietochowski, 89–90.

[61] The proceedings were published in Tatar and subsequently transl. into Russian. See Bennigsen and Quelquejay, *Les Mouvements nationaux*, 65 n. 1.

Representatives of various associations and groups, as well as of all political groupings (the Bolsheviks excepted) were there. The Pan-Islamists in the convention, few but vocal, could initially congratulate themselves that all participants, including the Marxists, agreed that there existed only one Muslim nation. This granted, a violent clash of opinions soon developed, between those advocating one Muslim nation in an undivided centralized Russian state (the motion supported mainly by the Volga Tatars) and those recommending one Muslim nation divided into several states (supported by the Bashkirs, the Crimean Tatars, and the Azeris). The latter won the vote and, although a central federal administration was suggested as a sop to minority voters,[62] the opportunity for setting up a widely supported and politically powerful Muslim union was lost.[63] The National Central Council, elected at the May 1917 convention, did attempt in subsequent months to assume responsibility for the religion, education, and finances of all Russia's Muslims. Some Muslim intellectuals were seriously hoping for a revival of Islam within Russia and without.[64]

The matter was raised again at a second all-Russian Muslim convention in Kazan, in July 1917.[65] In certain respects, the Kazan convention proved to be more Pan-Islamically oriented than the one in Moscow. Not only was the style of the debates both more revolutionary and more Pan-Islamic; the Kazan convention reversed the Moscow decision about a federation. Instead, it passed a resolution affirming the cultural and political unity of all Russia's Muslims.[66] In order to reinforce

For details, cf. *Utro Rossii*, 109 (2/15 May 1917), and *Ryech*, 102 (3/16 May 1917). Shafiga Daulet, 'The First All Muslim Congress of Russia', 21 ff. Cf. *Der Neue Orient* (Berlin), 2/10, Supplement (1917), 525–8.

[62] For the resolutions of this convention, see *Programmniye dokumyenti Musul'manskikh politichyeskikh partiy 1917–1920 gg.*, 9–33. For their analysis, see Tamurbek Davletshin, *Sovyetskiy Tatarstan*, 59–111. See also below, app. O.

[63] Davletshin, 64–9. Bennigsen, 'Panturkism and Panislamism', 42–3. R. Pipes, 76–8. Swietochowski, 91 ff. Landau, *Pan-Turkism in Turkey*, 15–16.

[64] R. Pipes, 78–9.

[65] For the resolutions in Kazan, see *Programmniye dokumyenti*, 34–46.

[66] For an analysis of the resolution, cf. Davletshin, 112–28. The resolutions themselves, transl. from Tatar into Russian, are reprinted there, 337 ff. See also below, app. P.

this resolution, a military council, located in Kazan, was added to the civil administration set up formerly; this started immediately to recruit a Muslim army.[67] The course of the Revolution, however, interfered with such plans and hopes, the meeting of several conventions serving only to emphasize the fragmentation of Russia's Muslims. Any Muslim solidarity demonstrated at the all-Russian conventions crumbled after the October Revolution and during the Civil War.[68]

The fragmentation of Muslim groups and organizations, so obvious in 1917, had tragic consequences for the fortunes of Pan-Islam in the Russian Empire and subsequently in the Soviet Union. An appeal entitled 'Long Live the Free Muslim World!'[69] contrasted with a statement about 'The Myth of One Muslim World'.[70] Numerous groups competed with one another, in 1917 and in the immediately following years, carrying ideological banners of various shades of socialism and communism. Even those who wished to form a large, unified Muslim army within Russia were thinking less of a separate Muslim entity and more of bringing socialism to the whole of the Muslim world.[71] Other groups exhibited nationalist,[72] federalist, or autonomous tendencies.[73] Among the many political parties which were being formed and re-formed by Muslims in those years,[74] only the programme of the Young Bukharans, formulated in 1917, referred to definite Pan-Islamic elements.[75] This was at least partly due to the fact that this party was a continuation of a secret Pan-Islamic association set up by Young Turk agents in Bukhara, in *c.*1910, along with several similar ones in Tashkent and Kokand.[76]

[67] A. Bennigsen and C. Lemercier-Quelquejay, *Les Musulmans oubliés*, 43–4.

[68] A. Bennigsen and S.E. Wimbush, *Muslim National Communism in the Soviet Union*, 20–1.

[69] 'Da zdrastvuyet svobodniy Musul'manskiy mir!'; see Arsharuni and Gabidullin, 59.

[70] 'Mif yedinnom Musul'manskom mirye', ibid. 66.

[71] A. Bennigsen and C. Lemercier-Quelquejay, *Islam in the Soviet Union*, 76 ff, 90 ff.

[72] Id., *Les Mouvements nationaux*, 64–81.

[73] Nikitine, 6–10.

[74] *Programmniye dokumyenti.*

[75] Ibid. 54–9.

[76] R. Vaidyanath, *The Formation of the Soviet Central Asian Republics*, 60–2. It seems that these associations had been carrying out Pan-Islamic and Ottoman propaganda during the Balkan Wars too.

4. *From Conflict with the Russians to Clashes with the Soviet Regime*

Diametrically opposed opinions are offered by analysts of the Soviet impact on the Muslims living under its aegis. Western ones generally see the Soviet regime as continuing the Tsarist traditions of discriminating against Muslim minorities and limiting their individual activities and religious life. Soviet ones argue that the Communist regime has liberated the Muslims from the indignities heaped upon them by the Tsarist Government and its agencies and offers them untrammelled equal opportunities for the pursuit of their religion and occupations. What both camps would very probably agree on is that the Communist authorities have been doing everything in their power to liquidate Pan-Islam as an international ideology competing with their own and, perhaps worse from their point of view, tied up with Islam, one of the religions most abhorrent to the champions of atheism. The fact that large masses of Muslims lie right across the Soviet borders, in one continuous land mass with the Muslims in Soviet Central Asia (and, for a while, the Crimea) has added yet another impetus for the Soviet decision-makers to do away with the vestiges of Pan-Islam.

There were two main stages affecting Pan-Islam in the mutual relations between the Soviet rulers and the Muslims in their areas: the first decade of civil war and consolidation of the new regime; and the subsequent era of centralized rule in the entire Soviet Union. It is not intended to present here a detailed exposé of these relations but, rather, their general background.

In the first stage, the Soviets used their military forces selectively in the Muslim-inhabited regions, as they needed an untroubled situation there in order to bolster their own rule elsewhere. The policy of meeting the minorities half-way, instituted by Lenin, expressed itself in using the Muslim and other groups on a basis of co-operation and promises allowing them, for the time being, complete freedom of religion and education.[77] Some of this propaganda was couched in evident

[77] Cf. B.L. Larson, 'The Moslems of Soviet Central Asia', 68 ff. Sokol, 69–70. Muḥammad Sāmī ʿĀshūr, *al-Muslimūn taḥt al-ḥukm al-shuyūʿī*, ch. 3.

Pan-Islamic phraseology, for instance in Bukhara and Russian Turkestan, as noted at the time by a keen local observer, Lieut.-Col. P.T. Etherton, British Consul-General in Kashgar.[78] For a while, these promises raised hopes of equality, autonomy, even secession, among various Muslim groups, which threw in their lot with the Bolsheviks. Where such promises and hopes were crushed by brutal reality, as in Turkestan, popular movements of Muslims against Russians—and against Bolsheviks—were sometimes led by the mullahs,[79] but more frequently they started from a new ideology, national Muslim Communism.[80] Such was the movement of the Volga Tatars, led between 1918 and 1928 by Mir Said Sultan-Galiyev (1880–1939?); 'Sultangaliyevism' became a derogatory designation, in Communist terminology, for non-Russian Communists trying to adapt Marxism-Leninism to the specific national situation of their group.[81] Such was the movement initiated among the Kazakhs by Turar Riskulov (1894–1938), during the 1920s.[82] So were several others.[83] Perhaps the most violent of these movements was that of the Basmachis, in Bukhara, who fought tooth and nail the Soviet onslaught during the early 1920s, until defeated by political intrigue and the Red Army.[84]

Soviet writers have usually labelled these movements 'bourgeois', 'nationalist', 'Pan-Islamic', and 'Pan-Turkish'—all in a pejorative sense. There was more than a grain of truth in these appellations, but they did not pinpoint accurately the characteristics of national Muslim Communism in the Soviet Union in that period.[85] These characteristics were the emphasis on a common Islamic heritage, resistance to the impact

[78] FO 371/8033, files 2073 and 2544, Etherton's report on Kashgar, dated 3 Apr. 1922, the last chapter of which discusses Pan-Islam. Other examples in R. Pipes, ch. 4.

[79] G.J. Massell, *The Surrogate Proletariat*, 28.

[80] Bennigsen and Lemercier-Quelquejay, *Les Musulmans oubliés*, 58 ff.

[81] On Sultangaliyevism, cf. M.S. Sultan-Galiyev, *Stat'i*. See also R. Pipes, 168 ff.

[82] Bennigsen and Lemercier-Qulquejay, *Les Musulmans oubliés*, 60–1. Vaidyanath, *Formation*, 106–9.

[83] Bennigsen and Lemercier-Quelquejay, *Les Musulmans oublíes*, 61–3. Vaidyanath, 105 ff. Davletshin, 129 ff. Swietochowski, 173–84.

[84] For the Basmachi movement, see J. Kunitz, *Dawn over Samarkand*, ch. 8; and the more recent A.I. Zyevyelyev *et al.*, *Basmachyestvo*.

[85] Bennigsen and Wimbush, *Muslim National Communism*, esp. 21–30.

of European civilization (including the Russian), preservation of Muslim landed property, the need for a Muslim autonomous Communist Party, and the conception that all Muslim peoples constituted one nation which, if it had to be divided, ought to be broken into as few units as possible—with the hope of being again reunited into one Muslim state.[86] Indeed, during the first decade of Soviet rule, national Muslim Communists busily conducted a campaign of propaganda in all the Muslim Republics for their Pan-Islamic and Pan-Turkic ideas, and quite a few works on history, novels, and textbooks bore the mark of that propaganda.[87]

Strong Islamic and Pan-Islamic sentiments were an important contributory impetus in the second stage, chiefly from 1928, for the Soviets to do all in their power to destroy national Muslim Communism, by successively liquidating the Muslim leaders and its cadres and then taking a series of administrative and ideological steps. The destruction of Muslim unity does not seem to have been an end in itself, but rather a means towards preventing separatist trends and Pan-Islamic inclinations by integrating the Muslims into the Soviet state and society.[88]

The Soviet regime proved to be suspicious of both subnational loyalties (tribes) and supra-national ones (Pan-Islam, Pan-Turkism) and proceeded to challenge these by breaking up Muslim units, especially the larger ones, and remoulding them into new, national ones. Each was to have new demarcation lines (frequently artificial), with its own language (encouraged to draw apart from kindred ones), cultural and historical background (carefully supervised), economic interests (stimulated to compete with other units), and proletarian culture (promoting the atomization of entities). A frontal attack on Islam (and

[86] Sultan-Galiyev even had a name for this future state—Turan. For these characteristics, cf. Bennigsen and Lemercier-Quelquejay, *Les Musulmans oubliés*, 63–5. Parallel Pan-Islamic thinking was exhibited by another Tatar, Mullah-Nur Vahitov (1885–1918), who envisioned a Tatar–Bashkir Muslim state and called for the liberation of all Muslims suffering under European imperialism. See A. Bennigsen, 'Marxism or Pan-Islamism', 55–66. Davletshin, 167–79.

[87] A. Bennigsen, *Self Determination in Soviet Central Asia*, 7–8. Id., 'Panturkism and Panislamism', 43–4.

[88] A. Bennigsen and M. Broxup, *The Islamic Threat to the Soviet State*, 35.

the denigration of Pan-Islam) have been an integral part of the Soviet Union's atheistic policy and its desire to fragment Islam within its frontiers.[89] Inducements were offered for *rapprochement* with the Russians and sanctions applied to achieve all this[90]—with force applied when necessary. Thus, even before World War II—and certainly since—totally new conditions were in evidence in the Muslim Republics of the Soviet Union.[91]

These new conditions, however, were not without serious problems for the Soviets. By encouraging national entities, they risked the creation of another demon, from their point of view. The Pan-Islamic identity had not disappeared and in some cases, such as Azerbaijan, had bridged over Sunnite–Shiite differences.[92] Islam has survived, resisting in various ways.[93] A resilient national–religious symbiosis has emerged,[94] resulting in Islam remaining a major force in the daily life of the Muslim Republics. As for Pan-Islam, it seems to have beaten a retreat since the late 1920s. Hardly anyone remembers nowadays the message of Sultan-Galiyev and his likes, *a fortiori* that of his predecessors. Its main potential of revival appears to be within the framework of 'Parallel Islam', the Muslim fraternities which have survived and are active in the Soviet Union as the only authentic non-Communist mass organization in the Muslim Republics.[95] These fraternities keep alive a sustained interest in Islam among the masses and provide the supra-national awareness which is a key element of Pan-Islam.[96]

[89] Ibid. 38. Bennigsen, 'Panturkism and Panislamism', 45–6. T. Rakowska-Harmstone, *Russia and Nationalism in Central Asia*, 232. Baymirza Hayit, *Islam and Turkestan under Russian Rule*, 73–219. Davletshin, 217–310.

[90] D. Nissman, 'Iran and Soviet Islam', 53. A. Bennigsen, 'Islamic or Local Consciousness among Soviet Nationalities?', 178–80. J. Castagné, 'Le Bolchévisme et l'Islam', 36–8.

[91] Two recent handbooks relating to this are Shirin Akiner, *Islamic Peoples of the Soviet Union*, and A. Bennigsen and S.E. Wimbush, *Muslims of the Soviet Empire*.

[92] C. Lemercier-Quelquejay, 'Islam and Identity in Azerbaijan', 51.

[93] For which see Bennigsen and Broxup, 62–87.

[94] M. Rywkin, *Moscow's Muslim Challenge*, 84–92.

[95] A. Bennigsen, 'Muslim Conservative Opposition to the Soviet Regime', 334–8. Id. and Chantal Lemercier-Quelquejay, *Le Soufi et le commissaire*, 50–5. Bennigsen and Broxup, 74–7.

[96] See e.g. the case of the Kazakhs: Bennigsen and Wimbush, *Muslims of the Soviet Empire*, 72; and for an opposite case among the Turkmens, ibid. 105.

5. *Russian and Soviet Attitudes to Pan-Islam*

Attitudes to Pan-Islam by the governmental authorities in Tsarist Russia and in the Soviet Union are more relevant to its development than those in Great Britain, France, the Netherlands, or Italy. The British, French, Dutch, and Italian views, while not uniform, tended to see Pan-Islam as a negligible (or even non-existent) quantity or, conversely and more frequently, to exaggerate its importance and its potential appeal to Muslims everywhere. However, while these four Western powers were mainly concerned with the effects Pan-Islam might have on the Muslims in their remote colonies, for both Tsarist Russia and the Soviet Union it was in their very midst; furthermore, their past wars with those Muslims were of relatively recent vintage. Here lies, perhaps, the partial explanation for their continuous concern with Pan-Islamic activity among their 'own' Muslims, which at times expressed itself in violent antagonism.

As early as the late nineteenth century, the Russian authorities—at several levels—worried about what seemed to them the growing impact of Pan-Islamic and Pan-Turkish ideas.[97] The government attempted to counteract the spread of Pan-Islam by influencing Russia's Muslims against it via their religious officials and their schools.[98] Apprehension grew, at the Court in St Petersburg and in the highest government circles, after the Young Turk Revolution. The Russians, by that time, had moved somewhat closer to the British, out of their distrust of German policies. Syergyey Dimitrovich Sazonov (1860–1927), the newly appointed Minister for Foreign Affairs, confided in the British, in November 1910, that he was worried by German support for Pan-Islam.[99] In the following month, the Tsar told Sir G. Buchanan, British Ambassador to St Petersburg, that 'Russia was likely soon to be confronted with similar difficulties[100] in Turkestan. An active propaganda was being carried by the Mullahs, and His Majesty feared that we were

[97] H.C. d'Encausse, 'The Stirring of National Feeling', 174–5.
[98] Gubaydullin, 104 ff, 108 ff.
[99] G.P. Gooch and H. Temperley, eds., *British Documents on the Origins of the War 1898–1914*, 567, Sir F. Bertie, British Ambassador in Paris, to Sir Edward Grey, FO 371/1016, no. 422, 'Confidential', dated in Paris, 23 Nov. 1910.
[100] To those of the British with the Muslims in India.

both threatened with a serious Pan-Islamic movement'.[101] A fortnight later, Buchanan reported that Pyotr Arkad'yevich Stolipin (1862–1911), the Russian Minister for the Interior, had shared with him his concern about Pan-Islamic activity in Russia.[102] And in the reply from the Foreign Office, Sir A. Nicolson told Buchanan that he himself had learnt from the Russians, during a visit to Russia shortly beforehand, that the Young Turks were intensifying their Pan-Islamic activity in South Russia and Turkestan.[103]

It seems that the suspicions and apprehensions of the Russian authorities were, at least partly, due to their confusing the religious universality of Islam with the political message of Pan-Islam,[104] which became something of an ever-present spectre for the Tsarist secret police.[105] On 18 December 1910, the Ministry for the Interior ordered the Okhrana to watch carefully persons suspected of Pan-Islam, which included supporters of Jedidism as well.[106] A number of reports of the Okhrana have survived;[107] frequently they combined what Goethe has called (in quite a different context) *Wahrheit und Dichtung*.[108] Thus, in 1913, the files of the Turkestan office of the Okhrana stated that the main objective of Pan-Islamists was political struggle with the Russian Empire, whose regime they considered as the main obstacle in their efforts to obtain self-determination for all Muslims.[109] In the years immediately preceding the outbreak of World War I, it was suspected in Tsarist Russia, time and again, that the Ottoman Empire, governed by the anti-Russian Committee of Union and Progress, was sending agents for Pan-Islamic and Pan-Turkish activities into Russian areas inhabited by dense Muslim populations. These suspicions were expressed not solely in the

[101] Gooch and Temperley, 583, Sir G. Buchanan to Sir Edward Grey, FO 371/1016, no. 488, 'Secret', dated in St Petersburg, 15 Dec. 1910.

[102] Ibid. 598, Sir G. Buchanan to Sir A. Nicolson at the Foreign Office, no. 635, 'Private', dated in St Petersburg, 29 Dec. 1910.

[103] Ibid. 600–1, Nicolson to Buchanan, no. 637, 'Private', dated in London, 3 Jan. 1911.

[104] Bräker, i. 72–3.

[105] E. Kirimal, *Der nationale Kampf der Krimtürken*, 29.

[106] Baymirza Hayit, *Turkestan zwischen Russland und China*, 190 n. 41.

[107] Examples in Arsharuni and Gabidullin, *passim*, and in their appendices.

[108] An example from 1911 is transl., in part, below, app. J.

[109] The document is quoted in Hostler, 117–18.

police reports, but in the Russian press as well. For example, in 1911, the *Novoye Vryemya* (New Times) raised the spectre of a Pan-Islamic peril and accused the Young Turks of strongly supporting a Pan-Islamic campaign.[110] In 1913, a long article on 'The Question of Pan-Islam', in the Orientalist journal *Mir' Islama* (The World of Islam) of St Petersburg, expressed concern about the impact of this movement on Russia's Muslims;[111] while another, in the same journal, maintained at the time that, even if Pan-Islam was utopian, it might become, under certain circumstances, a political threat.[112]

In the Soviet Union, official suspicion of Pan-Islam continued and even increased; some Bolsheviks, too, were suspected at the time (wrongly, as it turned out) of supporting Pan-Islam.[113] Much of this antagonism was due to the international and supra-national character of Pan-Islam; indeed, the Soviet Union's foreign enemies were accused of fomenting it.[114] There was, however, an added ideological dimension, when compared with attitudes to Pan-Islam in Tsarist Russia. If the three pillars of Communist doctrine are proletarian internationalism vanquishing nationalism, scientific atheism replacing religion, and a combination of both principles producing a new society (the Soviet people),[115] the Muslims are perceived as a seriously disturbing factor for achieving all three goals.

Lenin himself attributed considerable significance to Pan-Islam;[116] Stalin condemned it, early in 1921, in a speech to Muslim Communists,[117] along with Pan-Turkism (Soviet

[110] Reported in the *Revue du Monde Musulman*, 14/4 (Apr. 1911), 137.

[111] 'K' voprosu o Panislamizm'', esp. 12. For Russian apprehensions in the same year, see also Halil Halid, 'Panislamische Gefahr', esp. 308.

[112] 'Panislamizm' i Pantyurkizm'', 556–7.

[113] This continued even as late as 1923: 'The Bolsheviks have definitely identified themselves in Turkestan with a Pan-Islamic policy'. See FO 371/10404, file 814/421, British Consul-General C.P. Skrine's memorandum no. 107 on events in Eastern Bukhara and Ferghana, dated in Kashgar, 23 Nov. 1923, 10.

[114] T.C. Young, 200. Bennigsen and Lemercier-Quelquejay, *Les Musulmans oubliés*, 279–81. Nissman, 52.

[115] Larson, 196–7.

[116] Quoted by L.I. Klimovich, *Islam v Tsarskoy Rossii*, 218–19. Cf. R. Pipes, 'Muslims in the Soviet Union', 13.

[117] Stalin, *Sochinyeniya* (Moscow, 1946–51), v. 1–4, referred to by

sources often refer to them together or interchangeably). The Tenth Congress of the Communist Party of the Soviet Union, in March 1921, adopted a resolution defining both Pan-Islam and Pan-Turkism as sources of deviation from Communism towards bourgeois democratic nationalism[118] and their accusation was oft repeated subsequently. The charges of deviation were meant to justify the subsequent liquidation of Pan-Islamists.[119] This condemnation was frequently juxtaposed with accusations that Pan-Islamists were supporting the revolt of the Basmachis,[120] Sultan-Galiyev, and later the Nazis. Denunciations in a similar vein were forthcoming at numerous professional conventions, during and after the end of World War II.[121] In the years since then, the Soviet authorities have suspected (particularly from the beginning of the war in Afghanistan) that Islam was becoming increasingly political and that 'their' Muslims were feeling solidarity with those in Afghanistan.[122]

Much of the official approach has been reflected, not surprisingly, in the press and other publications, as a few examples will demonstrate. Many criticized Islam, too,[123] but we shall keep solely to Soviet writings on Pan-Islam. In 1925, a pamphlet, innocuously entitled *The National Demarcation of Central Asia*,[124] focused on the evils of both Pan-Islam and Pan-Turkism, accusing them of a separatism initiated by feudal reactionaries committed to regional nationalisms and supported by the worst elements in society. In subsequent years, these accusations were tied up with similar ones against Islam, which was characterized as being 'the antithesis of science and progress'[125] or 'preserving the remnants of the past in Central

S. Blank, 'Soviet Politics and the Iranian Revolution of 1919–1921', 190, 194, n. 127.

118 Hostler, 118.

119 Bennigsen and Wimbush, *Muslims of the Soviet Empire*, 33.

120 H.C. d'Encausse, 'Civil War and New Governments', 234–5.

121 W.S. Vucinich, ed., *Russia and Asia*, 70, 112. Hostler, 191–2.

122 Cf. Broxup, 'Islam in Central Asia since Gorbachev', esp. 285, 288 ff.

123. For which see Baymirza Hayit, *Sovyetler Birliğinde'ki Türklüğün ve İslamın bazı meseleleri*, 226–40.

124 *Natsional'noye razmyedzhyevaniye Sryednyey Azii*. Cf. Landau, *Pan-Turkism in Turkey*, 18, 26, n. 78.

125 e.g. L.I. Klimovich, *Islam, yego proiskhodzhdyeniye i sotsial'naya sushchnost'* (Moscow, 1956), esp. 21 ff.

Asia'.[126] These attacks and others were obviously intended to break up the unity of the Muslim community (*umma*) and its attachment to the supra-national Muslim nation (*millet*).[127]

Of course, some Soviet works, scholarly and otherwise, were specifically concerned with Pan-Islam, generally attacking it, as the following examples indicate. A book of *Essays on Pan-Islam and Pan-Turkism in Russia*, published in 1931,[128] attacked the Pan-Islamists in the Tsarist Empire and their international, unpatriotic connections. A 408-page volume on *Islam in Tsarist Russia*,[129] published in 1936 by L. Klimovich, for many years the foremost official commentator on Islamic affairs,[130] included a long chapter on 'The *Provokator* in the Role of the Ideologue of Pan-Islam'[131] and another on 'The Religious Organization of Islam during World War I'.[132] An article by N.A. Smirnov, published in 1950, in a collection of essays issued by the Institute of History in the Soviet Academy of Sciences, and entitled 'Turkish Secret Service under the Banner of Islam',[133] was an attempt, based on Russian archival sources and other materials, to prove deeply committed Ottoman involvement in the Northern Caucasus, as early as the eighteenth century, under the guise of Islam and Pan-Islam. More recently, in 1961, Yu. V. Marunov, writing on 'The Pan-Turkism and Pan-Islam of the Young Turks', in a periodical published by the Institute for the Peoples of Asia in the Soviet Academy of Sciences,[134] passed severe judgement on Pan-Islam in those years and on its promoters in the Ottoman Empire.

Two examples from the 1980s complete the picture. A hand-

[126] e.g. A. Altmishbayev, *Nekotoriye pyeryedzhitki proshlogo v soznanii lyudyey v Sryednyey Azii i rol' sotsialistichyeskoy kulturi v borb'ye s nimi* (Frunze, 1958).

[127] Cf. Bennigsen and Lemercier-Quelquejay, *Les Musulmans oubliés*, esp. 279 ff.

[128] Arsharuni and Gabidullin.

[129] Klimovich, *Islam v Tsarskoy Rossii*.

[130] For a Muslim Arab opinion of Klimovich, see ʿĀshūr, ch. 7.

[131] Klimovich, *Islam v Tsarskoy Rossii*, 216–67.

[132] Ibid. 268 ff. Both chapters make a frontal attack on Pan-Islam.

[133] N.A. Smirnov, 'Turyetskaya agyentura pod flagom Islama', 11–63.

[134] Marunov, 38–56. For a different view, see however, J.M. Landau, 'Ideologies in the Late Ottoman Empire', *Middle Eastern Studies*, 25/3 (July 1989), 387–8.

book on Islam, published in 1983, contained a brief, but violent article on Pan-Islam,[135] condemning it of duplicity in having combined its anti-colonialist struggle with support for the local rulers, the landowners, and the mullahs; it then accused it of becoming a champion of reactionary trends and an ally of the anti-Communist bloc of states. The charges against Pan-Islam as both anti-social (as well as anti-Socialist) and anti-national (and harmful to Soviet interests) was repeated in a Russian work on Islam in the contemporary politics of Eastern lands published in 1986.[136] This work, which focused on the 1970s and early 1980s, considered early Pan-Islam as both spiritual and political, but even more so a bourgeois ideology, which has now become an ally of imperialist governments and a means of helping bourgeois rulers of their states to stem the advance of scientific Socialism.

It is no accident that nearly all the above publications, published in the Soviet era, discussed Pan-Islam almost exclusively in the Tsarist period. While an occasional invective against Pan-Islam was printed in the Soviet Union, this seems to have been limited to official pronouncements, in the main. Otherwise, writing about it was generally considered as almost taboo, no doubt out of a desire not to reawaken Pan-Islam by renewing public debate on it. To know more about attitudes towards it, one has to find out how Soviet encyclopaedic works defined and described it, but before doing so the definition offered by Marunov merits mention. It is as follows:

> Pan-Islam—a reactionary religious–political current, which came into being in the second half of the nineteenth century, Pan-Islam preached the union of all peoples believing in Islam, into one state. It was used widely, in the early twentieth century, by reactionary circles . . . to achieve the union of the Muslims of Afghanistan, Iran, Egypt, North Africa, Tsarist Russia, and other lands, under the dominion of the Ottoman Empire.[137]

This definition was, understandably, not too different from that of the official sources. Special interest attaches to the articles on Pan-Islam in the *Great Soviet Encyclopaedia.* The

135 *Islam kratkiy spravochnik*, 95–6.
136 *Islam v sovryemyennoy politikye Vostoka*, esp. 44–9.
137 Ibid. 39 n. 10. English transl. mine.

articles, entitled 'Panislamizm' in all three editions,[138] published, respectively, in 1939, 1955, and 1975, reflected then current views on the matter. The first article defined Pan-Islam in true Marxist terms, as 'a religious–political doctrine, advocating the inevitable governmental union of all Muslims under the supreme authority of the Muslim Caliph'. It further asserted that Pan-Islam was born at the end of the eighteenth century as an expression of the struggle led by the most reactionary Muslim feudal and clerical circles against the penetration of European capital into the lands of the Muslim East. Pan-Islam reflected the feudal Muslim teachings of *Jihad* and Caliphate and its slogans were employed in various anti-feudal movements of liberation during the nineteenth century. al-Afghānī was connected with Sultan Abdülhamid and, through him, with German imperialism. In 1911–12, a Pan-Islamic centre was established in Germany, distributing extensive propaganda in Turkey, Iran, and the Muslim-inhabited regions of Russia. A part of Pan-Islam's programme was taken up in Russia by such organizations as *İttifak-ul-Muslimin* and Jedidism in Bukhara, both exploited widely by imperialist groups during World War I, when clerical personalities and religious associations distributed Pan-Islamic leaflets, prepared in Turkey and Germany, through North Africa and India. After the abolition of the Caliphate in Turkey, in 1924, Pan-Islam was mainly active in Muslim lands which were still colonized, in the Khilafat movement in India, and in various international conventions. After the October Revolution, Pan-Islam was used against the Soviet Union by emigrants from it and by foreign Intelligence services.

Barring a few inaccuracies (al-Afghānī had no connection with the so-called German imperialism, for Abdülhamid's close relations with the Germans developed after al-Afghānī's death), the article is remarkable for weaving a good number of correct facts into a web of conspiratorial tales—characteristic of Stalin's days—intended to stigmatize Pan-Islam and its collaborators, past and present. The same spirit pervades the shorter article in the second edition. This started the story in

[138] *Bol'shaya Sovyetskaya Entsiklopyediya*, 1st edn., xliv, cols. 62–3; 2nd edn., xxxii. 3–4; 3rd edn., xix. 146 (cols. 424–5).

the second half of the nineteenth century and, in a studied attempt to place it in a social context, attributed the origins of Pan-Islam to the landed, bourgeois, and clerical groups in Turkey, from where it was said to have spread to the propertied classes in other Muslim countries. Its aims were said to be the same as in the first edition, but—in the spirit of the times—it was added that Pan-Islam was being exploited by the ruling classes in Muslim countries, in order to stimulate nationalist and religious differences, and so strengthen their own position and suppress the revolutionary workers' movement of the peoples of the East. The Ottoman involvement in Pan-Islam was given somewhat wider consideration, Germany and Great Britain being allotted a part of the blame for having used it for their own imperialist ends. In the Soviet Union, it was alleged further, Pan-Islam was used at the time of the Revolution and immediately afterwards by the hostile Muslim bourgeoisie in Central Asia. On the eve of World War II, the Nazis—preparing for their onslaught on the Soviet Union, set up and financed Pan-Islamic associations in the Near and Middle East; subsequently, the American and British imperialists strived to use Pan-Islam in order to fight the national liberation movement in the countries of the East.

Essentially this is the same article, in many details, but there is an obvious change of emphasis in the second edition, where blame was put less on the Ottomans and more on the Muslim bourgeoisie of Central Asia, that is, in the Soviet Union itself, branding them as criminally anti-revolutionary elements. In addition, Pan-Islam's supra-national character was emphasized and its alleged connections with Nazi Germany, the United States, and Great Britain presented as incontrovertible facts. The third edition's article reverted to some of these charges and updated them. While the earlier ones were anonymous, this one was signed by L.R. Gordon-Polonskaya, a historian who has written on Islam in India and Pakistan. A short bibliography, also, was appended. Gordon-Polonskaya repeated the outdated concept of Islamic union under a Caliph, but correctly pointed to the ever more significant political role of Pan-Islam, presented in the first two editions as a monolithic ideology. As she perceived it, al-Afghānī's conception of Pan-Islam had been anti-colonialist, but reactionary bourgeois circles got the upper hand as soon as Abdülhamid and the Young Turks succeeded in diverting Pan-Islam to their own aggressive objectives. No wonder that, after the October Revolution, Pan-Islam became one of the major slogans of the anti-revolutionary and nationalist forces in the Caucasus and Central Asia. It assisted imperialist forces in Muslim lands before and during World War II as well; afterwards, Pan-Islam joined the struggle against the national liberation movement in Asia and Africa and attempted to block in Muslim lands the development of popular thinking (by which, in general, Communism is implied).

The article in the third edition, while repeating some of the earlier accusations against Pan-Islam, condemned it roundly as the foe of progress, not in Muslim countries alone but everywhere in Asia and Africa, and as the determined ally of imperialism, nationalism, and anti-revolutionism (all pejorative terms in Communist phraseology). However, what is missing here is no less suggestive: Gordon-Polonskaya's article did not attack the Muslims in the Soviet Union—indecd refrained from mentioning them—and left alone their religious establishment too. This seems to toe the party line of striving to shape Soviet Muslims into the body politic and Soviet society. However, the implication that any supporter of Pan-Islam is in league with the enemies of the Soviet Union runs through the whole of Gordon-Polonskaya's article, as in those of previous editions of the *Great Soviet Encyclopaedia*. This is obvious in other official pronouncements on Pan-Islam, such as the *Soviet Encyclopaedic Dictionary*[139] in 1954 and the *Historical Encyclopaedia*[140] in 1967 (in the latter, the article was also written by Gordon-Polonskaya). These accusations were even more explicit in a *Little Scientific-Atheist Dictionary*, published in Russian in 1964, where Ye. A. Byelyayev, a prolific writer on Islamic and Arabic matters, accused the Pan-Islamists of 'actively carrying on a pro-imperialist policy ... following the success of the anti-imperialist struggle of the peoples of the East'.[141]

139 'Panislamizm', *Entsiklopyedichyeskiy slovar'* (Moscow, 1954), ii. 598.

140 'Panislamizm', *Istorichyeskaya Entsiklopyediya* (Moscow, 1967), x. 787–8.

141 Ye. A. Byelyayev, 'Panislamizm', 419–20.

It is not difficult to understand the disappearance of practically all Pan-Islamic activities in the Soviet Union (except, perhaps, in the Muslim fraternities), considering this strongly worded propaganda of 'guilt-by-association' against them, combined as it was with disparagement of traditional Islam and unrelenting supervision by the anti-subversion organs of the State. What is less understandable is the continuing passionate condemnation of Pan-Islam, after it had ceased to have any importance (if it ever had) in the Soviet Union. Indeed, whatever variations in the official Soviet image of Islam and changes of policy towards Soviet Muslims have occurred,[142] there has been little, if any, modification in the negation of Pan-Islam. Probably the explanation lies in the tendency to suspect all ideologies competing with Communism, and most particularly international ones, whose potential supporters are very numerous both within the Soviet Union and just across its own borders. These suspicions may well have received a boost, during the 1980s, by broadcasts from the Islamic Republic of Iran and from Pakistan (after the Soviet Union's invasion of Afghanistan), beamed at Soviet Muslims and appealing for Islamic solidarity.[143] Some other Communist governments, too, express their unease occasionally at what they consider the identification of their Muslim communities with Pan-Islam rather than with Socialism; for example, Yugoslav officials condemned Pan-Islam in 1979.[144] Indeed, Communist misgivings about Pan-Islam proved not entirely unfounded in recent years, which have witnessed a wave of reforms. On 10 May 1987, S. Syeutov (who called himself 'the Official Religious Representative of the Muslim Believers of the Crimean Tatar Nation') published a letter which he had sent to the President of the Council of Ministers of the Soviet Union, demanding that the Tatars be allowed to return to the Crimea. The letter ended with the following lines, 'If the Soviet government cannot or is unwilling to solve our national and religious problems without delay satisfactorily by internal

[142] B.R. Bociurkiw, 'The Changing Soviet Image of Islam', 59–80.

[143] See e.g. Bill Keller, 'Allah and Gorbachev Mixing in Central Asia', *The New York Times*, 12 Feb. 1988, A1 and A14.

[144] Cf. Zdenko Antic, 'Pan-Islamic Nationalism Condemned by Yugoslav Official', 1–4. This report is based on the Yugoslav press.

measures and possibilities, then our million strong Muslim Crimean nation will consider itself entitled to ask for the help of the 800 million strong Muslim World abroad.'[145]

[145] English transl. in *The Central Asian Newsletter* (London), 7/5–6 (Dec. 1988–Jan. 1989), 4.

IV. *Turkey Opts Out, while India's Muslims Get Involved*

1. *Abolition of the Caliphate: Catalyst for India's Muslims*

The Muslims in India, even more numerous than those in Russia, had expressed Islamic solidarity long before World War I. However, it was the abolition of the Caliphate soon after the end of that war, which served as a catalyst for their feelings and activities in the cause of Pan-Islam.[1] It ended the political legitimacy of Pan-Islam and compelled Islamists to search for alternative loyalties.

It has already been mentioned above[2] that, in Abdülhamid's reign, the Sultanate and Caliphate came under scrutiny, chiefly in Europe, where scholars, journalists, and statesmen (irked at Abdülhamid's despotism and support of Pan-Islam)[3] wondered about the applicability and suitability of those institutions in modern times. The arguments continued even after he had been deposed in 1909, with some expressing doubt about the legitimacy of a Caliph who was neither a descendant of the Prophet Muḥammad, nor even of his tribe, the Quraysh.[4] These views were rarely shared by Muslims, who usually maintained in no uncertain terms that the Caliphate was their own affair; a few even wrote to this effect in European languages for a European audience.[5] The arguments continued and

[1] The most recent work to consider this is Mohammad Sadiq's *The Turkish Revolution and the Indian Freedom Movement*.

[2] See ch. 1.

[3] e.g. 'Panislamism and the Caliphate' (1883).

[4] See, among others, Stanley Lane-Poole, 'The Caliphate', 162–77. For a contradictory opinion, Halil Halid, *The Diary of a Turk*, 207–9.

[5] e.g. Halil Halid, *Diary*, 207–10. Ameer Ali, 'The Caliphate', esp. 694; Ameer Ali was a distinguished Shiite Indian.

assumed an increasingly political character, chiefly after the defeat of the Ottoman Empire in 1918. An Orientalist like C.A. Nallino, Professor of Islamic Studies at the University of Rome, in a pamphlet written in Italian in 1917 and translated into English two years later,[6] maintained that Islam *was* conceivable without a Caliph[7] and that, anyway, the one in Istanbul was merely a pretender.[8] Arguing against the commonly accepted view of the Caliph's spiritual power over the Muslims, Nallino concluded that 'the Caliphate is nothing else than the universal monarchy of Islam, nothing else than political Pan-Islamism'.[9] An apparently anonymous[10] tract, published in 1919, bore the title *Aperçu sur l'illégimité du Sultan Turc en tant que Khalife.*[11] That this tract was distributed in Paris, probably at the Peace Conference, by the Greek delegation, gave it a special flavour. In true Pan-Hellenic spirit, it combined a condemnation of Turkish designs on Constantinople with an impassioned plea against the Sultan having any right to the Caliphate.[12]

It was only after the Caliphate had been abolished in Turkey that excitement cooled off and more balanced evaluations of the Caliphate were published, in either book form (such as T.W. Arnold's deservedly renowned study)[13] or articles.[14] However, the international stage had already been set for doing away with the Caliphate, even though the final act, played out in Turkey, was obviously dictated by the ideological approach

[6] C.A. Nallino, *Notes on the 'Caliphate' in general and on the Alleged 'Ottoman Caliphate'.*

[7] Ibid. 13.

[8] Ibid. 14 ff.

[9] Ibid. 32.

[10] Its introduction is signed C.G. Bello, Advocate in Constantinople, but there is no hard proof that he authored this tract.

[11] *Aperçu sur l'illegitimité du Sultan Turc en tant que Khalife: Quelques remarques relatives aux prétendues visées turques sur Constantinople.*

[12] Sample phrase, ibid. 14; 'Il n'y a pas de place en Europe pour l'usurpateur du Khalifat'.

[13] T.W. Arnold's *The Caliphate* appeared in print right after the abolition of this office, and mentioned it on p. 180.

[14] Such as D.S. Margoliouth, 'The Caliphate Yesterday, To-day, and Tomorrow', publ. in 1925. For a later work, cf. A.H. Lybyer, 'Caliphate'. For a recent study by an Egyptian Muslim, see Muḥammad ʿAmāra, *Naẓariyyāt al-Khilāfa al-Islāmiyya*, publ. in 1980.

and the practical needs of the new state, as its leaders perceived them.

Mustafa Kemal (1881–1938), later called Atatürk by a grateful people, planned and carried out a series of revolutionary reforms, whose main objective was the shaping of a new nation in a modernized state. This he achieved thanks to his charisma, talents, and the prestige acquired as the leader of the War of Independence, which brought the Turks, defeated in World War I, to military victory over the Greeks and renewed self-esteem. One of the first aims of Mustafa Kemal, as president of the popularly elected parliament, the Grand National Assembly, was to ensure that this Assembly alone would make the final decisions in matters of public interest; hence the steps he would persuade it to take against the office of the Sultan–Caliph.[15] However, he had to act prudently, since he did not wish to antagonize the Islamists within the Assembly and the Turkish population, nor to alienate the potential support of Muslims abroad (chiefly in India), offered to Turkey during its War of Independence and still very much needed during the protracted negotiations for international recognition of Turkey in a new peace settlement with the powers.

Hence, Mustafa Kemal temporized, showing respect for Islam. Consequently, public opinion in Europe in those years erroneously considered him a Pan-Islamist. In this, they did not differentiate between him and the leaders of the Committee of Union and Progress[16] who had fled to Germany, Italy, and Switzerland and were attempting to reorganize and recruit Muslim students for a programme committed to Pan-Islam.[17] British Intelligence, in particular, was convinced in the early 1920s that Mustafa Kemal was striving to promote the cause of an Islamic union by reviving Pan-Islam, basing it in Ankara.[18]

[15] As recently demonstrated by Halil İnalcık, 'The Caliphate and Atatürk's inkılâp'.

[16] FO 371/6345, file 13559, an inter-departmental letter, written on behalf of Montagu, Secretary of State for India, to Earl Curzon, dated in London, 9 Dec. 1921: 'There would appear to be little real difference between the Pan-Islamic policy of Mustafa Kemal and that of Enver Pasha'.

[17] FO 371/9002, file 618/618, being a report of the Inter-Departmental Committee on Eastern Unrest, 'Most Secret', dated Jan. 1923.

[18] e.g. FO 371/6497, file 52, Summary of Intelligence reports for the week ending 9 Dec. 1920, para. (d). FO 371/9129, file 1577/199, 'Secret', dated 3 Feb.

The misconception, at least within the British Intelligence community, was due in no small part to the Inter-Departmental Committee on Eastern Unrest, which had been set up at the Foreign Office. Headed by the Secretary of State for India—that is, the official concerned with the heavy involvement of Indians in Turkey's affairs in the early 1920s and their support of Mustafa Kemal—the committee could not but assume the worst about Mustafa Kemal's Pan-Islam. It tended to give credence even to the most fantastic rumours, for example, that in Iran, late in 1921, a group of Kemalists—described as the Pan-Islamic Kemalist Party—had connections in Turkey;[19] or to accept, a while later, that an article by Eşref Edib, editor of the Islamist *Sebilür-Reşad*[20] (The Right Way), advocating a Pan-Islamic convention, expressed Mustafa Kemal's own opinion.[21] A report of the committee characteristically sums up its conclusions as follows:[22]

> The information on the subject of the Pan-Islamic movement of the Angora Government ... demonstrates *a*. That the Pan-Islamic campaign was taken up with enthusiasm by the Angora Government, sanctioned by the Great National Assembly, and forms an integral part of Turkish Nationalist policy. *b*. That serious efforts are being made to establish and maintain connection with all other Moslem communities, whether independent or otherwise. *c*. That the Pan-Islamic policy is directed to encouraging all other Moslem communities to obtain their complete independence, and become self-contained units in a Federation or Union of Moslem communities under the religious and political leadership of Turkey. *d*. In order to advance their Pan-Islamic scheme, representatives and delegates have been called together in Anatolia, in order to produce concerted action, and harmonize the activities of the other communities with those of the Turkish nationalists.

This way of seeing things both reflected and, to a degree,

1923, and file 1912/199, telegram of 12 Feb. 1912—both on plans to convoke a Pan-Islamic convention in Ankara.

[19] FO 371/7803, file 2517/6, English transl. of an intercepted letter, dated 13 Dec. 1921.

[20] *Sebilür-Reşad*, 13 Apr. 1922.

[21] FO 371/7883, file 5949/27, 'Eastern summary', 'Secret', dated 8 June 1922, based on Intelligence reports.

[22] FO 371/7790, file 5336/402, 'Most Secret', Interim report of the Inter-Departmental Committee on Eastern Unrest, dated 24 May 1922, 5–6.

influenced public opinion in Western Europe, which was largely anti-Turkish. Mustafa Kemal may well have taken this into consideration when deciding to reply to accusations of Pan-Islamism. Although his victory over Christian armies, in the War of Independence, had given new hope to Pan-Islamists,[23] he was no Pan-Islamist himself. On 30 November 1921, speaking in the Grand National Assembly, Mustafa Kemal unequivocally disclaimed any intention of Pan-Islam, emphasizing that while the Turks wished all Muslims well, they were convinced that it was fantastic to desire the material union of all Muslim governments in one empire, since this was bound to arouse the hostility of the world; he concluded by declaring that the Turks were just a nation wishing to live independently.[24] Mustafa Kemal was to reassert this position subsequently.[25] Meanwhile, the course of events confirmed his statements.

Soon after his 1921 speech, Mustafa Kemal persuaded the Grand National Assembly to dismiss the Sultan, who was competing with Kemal's revolutionary government for the right to negotiate with the Powers on Turkey's behalf; and to abolish the Sultanate altogether, on 11 November 1922. Another member from the House of Osman was appointed by the Assembly as Caliph, on the understanding that his authority would be religious only and essentially resemble that of the Pope. In the Turkish context, this implied being a figurehead. On 3 March 1924, the Grand National Assembly abolished the Caliphate too,[26] thereby stirring a controversy in various parts of the Muslim world, where people had wanted to believe that the last Caliph's powers were quite considerable.[27]

[23] W.H. King, 'The American Treaty with Turkey at Lausanne and the Kemalist Pan-Islamic Adventure', 7. This article, by a senator from Utah, was originally written in 1926.

[24] Reported in the press. Cf. Bayur, *Türk inkilabı tarihi*, 331–72. FO 371/6537, file 13622, enclosures, for a summarized English transl. See also A.L. Bianchini, 'I movimenti nazionalisti nei paesi maometani', 510–11.

[25] In his 1927 *Nutuk* (1959 edn. publ. by the Türk Devrim Enstitüsü), iii (*vesikalar*), 1189–90, document 120.

[26] For the various stages, see İnalcık, 25 ff. S.G. Haim, 'The Abolition of the Caliphate and its Aftermath'. A.G. Chejne, 'Pan-Islamism and the Caliphal Controversy', esp. 685–8. Hamid Enayat, *Modern Islamic Political Thought*, 55. Necdet Öklem, *Hilâfetin sonu.*

[27] For the reactions of Muslim public opinion to the abolition of the Cali-

For pragmatic purposes, obviously, the support of Muslims outside Turkey had become less important, once the Peace Treaty of Lausanne was signed with the Powers in 1923. However, just as the abolition of the Sultanate had been intended to vest all state authority in the Grand National Assembly, that of the Caliphate was meant to conform with the new political ideology with which Mustafa Kemal wished to endow the young Republic of Turkey. Of the old ideologies, Ottomanism had died a natural death with the defeat and disintegration of the Ottoman Empire; Pan-Islam and Pan-Turkism were regarded as risky for the peace with its neighbours that Turkey so badly needed. So the Kemalists 'abandoned the concept of a universal Muslim community'.[28] There remained secular modernism, which Mustafa Kemal adopted determinedly, putting it in a context of nationalism focused on the Turks and Turkey within the boundaries of the new republic.[29]

For a generation or more, Pan-Islam lay dormant in Turkey. Some Turkish writers, such as M. Halil Halid, still praised Pan-Islam in 1925,[30] but subsequently few, if any, did. Only with the revival of Islam in Turkey during the 1950s, and its re-entry into general politics in the 1960s and 1970s, were there signs of renewed interest in Pan-Islam there. An example is a book by Hasan Tahsin Başak, published in 1957; the translation of the title of the Turkish version was *The Theory of Agreement and Unity of the Islamic Union*,[31] and of the Arabic version, *Muslims Have to Agree and Unite*.[32] While most of the author's arguments were on the religious level, he reminded his readers that 'united, Muslims ruled the world', for 'unity is the ultimate power'. A union of Muslims would be beneficial for them all, in both military and commercial matters; the alternative would be overall decline.[33] A generation later, in 1986, a 112-page book by Mustafa Talip

phate, see Madīḥa Darwīsh, 'The Caliphate and its Revival in the Twentieth Century'.

[28] D.H. Khalid, 'The Kemalist Attitude towards Muslim Unity', 31, 38.

[29] Landau, *Pan-Turkism in Turkey*, 72–3.

[30] In his book *Turk hâkịmiyeti ve İngiliz cihangirliği*.

[31] Hasan Tahsin Başak, *Islâm birliği ittifak ve ittihat nazariyesi*.

[32] *Yalzam ʿala al-Muslimīn al-ittifāq wa-'l-ittiḥād*.

[33] Ibid. 8–10.

Güngörge, *Islam Wishes Unity: National Unity and Islamic Brotherhood against Separatism*,[34] rated divisiveness as a grave danger for both the national existence of the Turks and the Islamic faith.[35] The panacea suggested was national and Islamic unity.[36]

The official attitude of the republic's leaders seems to have been summed up aptly in 1926 by Ahmed Cevdet, editor of the Istanbul *Iqdam* and a well-known intellectual. Cevdet was commenting on an article by Basil Matthews in the *Journal de Genève* which, shortly before the meeting of a Pan-Islamic convention in Cairo, had argued that a Muslim League of Nations was not only feasible, but also probable and a danger to be reckoned with by Western civilization.[37] In response, Cevdet maintained that the entire conception of Pan-Islam had existed merely in the minds of a handful of writers, confined to the offices of one Istanbul newspaper, and was, anyway, totally impracticable.[38]

Contrary to the case of the Republic of Turkey, however, Islam remained a topical issue, in the years immediately following the end of World War I, among India's Muslims, particularly on the issue of the Caliphate.

2. *From Solidarity to Active Involvement*

India's Muslims in the nineteenth century[39] were in the unenviable situation of being a very large minority group under foreign, non-Muslim domination, while memories of the reign of the Moguls (interrupted by the deposition of the last Mogul emperor, Bahadur II, in 1857) lingered on. For pious Sunnites, it was natural to turn to other Muslim centres for moral support,

[34] Mustafa Talip Güngörge, *Islâm birlik ister.*

[35] Ibid. 8 ff.

[36] Ibid. 111–12.

[37] *Journal de Genève*, 3 May 1926; English transl. in NA, RG 59, 867.404/165, US Consul-in-Charge C.E. Allen's no. 6246, to the Secretary of State, dated in Istanbul, 18 May 1926.

[38] English transl. in NA, RG 59, 867.406/166, Allen's no. 6258, dated in Istanbul, 5 June 1926.

[39] There is, of course, a vast literature on India's Muslims at that time. For an excellent essay on materials about Pan-Islam, see M.N. Qureshi, 'Bibliographic Soundings in Nineteenth Century Pan-Islam in South Asia'.

chiefly to those not ruled by Unbelievers.[40] While Pan-Islamic sentiment, including appeals to Muslims all over the world to work in unison for the glory of Islam, had been expressed in the writings of Shah Waliullah (1703–62)[41] and others in the eighteenth century,[42] these were, first and foremost, an expression of religious Pan-Islam, which is not in the province of our study.[43] Cultural and political shocks were needed for India's Muslims to respond with an organized, politically minded Pan-Islam.

Regarding the cultural shock, in the second half of the nineteenth century an increasing number of Muslims in India acquired reading fluency in foreign languages, mainly in English, and became acquainted with a steadily growing European Orientalism which, in the area of Islamic research, appeared irreverent and offensive to them, for example, the treatment of the Prophet Muḥammad, the oral tradition, or Islam in general. They responded vigorously,[44] basing themselves on Muslim sources and studies published in India and elsewhere, and this increased not only their contacts with co-religionists in India, but interest in Muslims abroad. A few of India's Muslims travelled to Muslim countries and many more, of course, undertook the Pilgrimage, visiting on their way other Muslim places. Such contacts worked both ways, evidently, and one of the best-known Muslim visitors to India was Jamāl al-Dīn al-Afghānī, who went there four times, of which the longest and most important visit was in 1880–2, after his expulsion from Egypt.[45] His discussions, speeches, and newspaper articles, at the time, however, were mainly focused on rethinking the entire system of Islam, without

[40] F.W. Buckler, 'The Historical Antecedents of the *Khilafat* Movement', 604.

[41] About whom see Aziz Ahmad, *Studies in Islamic Culture in the Indian Environment*, 201–17.

[42] Naimur Rahman Farooqi, 'Pan-Islamism in the Nineteenth Century', 285.

[43] See above, the introduction to our study.

[44] Examples in P. Hardy, *The Muslims of British India*, 175–6.

[45] For which see Aziz Ahmad's well-documented 'Afghani's Indian Contacts'. Cf. Syed Murtaza Ali, 'Saiyed Jamal al-Din Afgani'. Anwar Moazzam, 'Jamāl al-Dīn al-Afghānī in India'. Syed Moizuddin Ahmad, 'The Political Ideas of Jamal-ud-Din al-Afghani', esp. 54–5.

breaking with the past,[46] highlighting the contradictions between Islam and the West and, in this context, bitterly attacking British rule in India and rousing the local Muslims against it.[47] He deeply impressed some of India's Muslims, in Hyderabad, Calcutta, and other towns,[48] but this was before al-Afghānī became immersed in his Pan-Islamic projects and activities (although he seems to have nurtured hopes, then and subsequently, of Indian financial support for these activities).[49] Indeed, al-Afghānī had been impressed by the potential significance of India's Muslims for Pan-Islamic activity. His friend and admirer, ʿAbd al-Qādir al-Maghribī, who spent a year with him in Istanbul, when the latter lived there in gilded captivity, reported[50] that al-Afghānī had told him that the Muslims in India ought to be the focus of activity for the union of Islam.

Political Pan-Islam in India seems to have been a local product, however. Before 1857, there were few signs of it (such as marked sympathy for the Ottoman Empire during the Crimean War).[51] However, after the so-called Mutiny of 1857 (in which Muslims had had a notable share in certain parts of India)[52] and the deposition of the last Mogul emperor, the latter's office, even if devoid of real power, remained vacant. A symbol for Muslim solidarity and the legitimacy of the *sharīʿa* was needed; the Caliph in Istanbul, as the one remaining independent Sunnite leader, seemed the best possible head for an entire symbolic structure of authority.[53] His name was proclaimed in the Friday sermons and a feeling of loyalty to him gradually evolved—not without some help from Ottoman emissaries.[54]

[46] Munawwar Ahmad Anees, 'End of Empire', 47.

[47] Aziz Ahmad, 'Sayyid Aḥmad Khān, Jamāl al-Dīn al-Afghānī and Muslim India', esp. 63 ff.

[48] Acc. to the evidence of W.S. Blunt, who visited there in 1883. See Blunt's 'Note', in E.G. Browne's *The Persian Revolution of 1905–1909*, 402. For a list of al-Afghānī's friends and followers, cf. Moazzam, 87–8.

[49] M. Stepaniants, 'Development of the Concept of Nationalism', 29. Sharif al-Mujahid, 'Pan-Islamism', 105, maintains that a number of Indians financed the publication of *al-ʿUrwa al-wuthqā* in Paris.

[50] ʿAbd al-Qādir al-Maghribī, *Jamāl al-Dīn*, 72.

[51] M.N. Qureshi, 'Mohamed Ali's Delegation to Europe', 79, based on the correspondence of Dalhousie, then Governor-General of India.

[52] Asghar Ali Engineer, *Indian Muslims*, ch. 1.

[53] G. Minault, *The Khilafat Movement*, 5.

[54] Hardy, 177. Aziz Ahmad, *Islamic Modernism in India and Pakistan 1857–1964*, 124. There is no definitive evidence as to when these activities

Indian Ulema implicitly acknowledged the Ottoman Sultan's claim to a universal Caliphate.[55] This gradual trend towards substituting a foreign loyalty for commitment to India was not left unchallenged. The All-Indian Muslim League,[56] first headed by Sir Sayyid Ahmad Khan (1817–1898),[57] a scholar and a prominent representative of Islamic liberalism in India, then led by Aga Khan,[58] adopted a supportive attitude to British rule and a reserved one towards Pan-Islam (and was upbraided for this by al-Afghānī in Indian newspapers).[59] Its main concern was the interest of India's Muslims.

In the meantime, however, there had been signs of a trend amongst India's Muslims from the 1850s (approximately since the Crimean War), towards sympathy with the Caliph in Istanbul and Muslims everywhere.[60] Those who revered the Caliph and liked the Turks did not know—or did not care—about the despotic regime in the Ottoman Empire.[61] Respect and affection might have been a side-effect of the British having persuaded the Ottoman Sultan to appeal to India's Muslims in their favour during the Crimean War and the Indian Mutiny.[62] However, it seems that the Turco-

started; the claim that they began as early as 1877 is refuted by Badger, 'The Precedents and Usages Regulating the Muslim Khalîfate', 281.

[55] Aziz Ahmad, *Islamic Modernism*. Cf. Chirol, 'Pan-Islamism', 15–17, 20 ff.

[56] On the Muslim League see Jamil-ud-Din Ahmad, 'Foundation of the All-India Muslim League'. Yu. A. Ponomariyev, *Istoriya Musul'manskoy ligi Pakistana*, esp. part 1, chs. 1–2. W.C. Smith, *Modern Islam in India*, 246–92. Ram Gopal, *Indian Muslims*, 97–120, 129 ff.

[57] W.C. Smith, 15–28. L. Hubert, *Avec ou contre l'Islam*, 21. A.C. Niemeijer, *The Khilafat Movement in India 1919–1924*, 29 ff. Prem Narain, 'Political Views of Sayyid Ahmad Khan', 114 and *passim*. Hafeez Malik, *Sir Sayyid Ahmad Khan and Muslim Modernization in India and Pakistan*, 218 ff. and *passim*. M.Y. Abbasi, 'Sir Syed Ahmad Khan and the Reawakening of the Muslims' in: Ahmad Hasan Dani, ed., *Founding Fathers of Pakistan*, 1–44. Aḥmad Amīn, *Zuʿamāʾ al-iṣlāḥ fī al-ʿaṣr al-ḥadīth*, 129–48.

[58] On whom see S. Razi Wasti, 'The Role of the Aga Khan in the Muslim Freedom Struggle', in: Dani, ed., 101–8. For Aga Khan's own views on Pan-Islam, cf. Aga Khan and Zaki Ali, *L'Europe et l'Islam*, 13–21 (the chapter by Aga Khan, entitled 'Le Panislamisme').

[59] For this controversy, see Aziz Ahmad, *Studies*, 55–62. ʿAbd al-Munʿim Nimr, *Kifāḥ al-Muslimīn fī taḥrīr al-Hind*, 48–50.

[60] D.E. Lee, 'The Origins of Pan-Islamism', 283.

[61] Mohammad Habib, 'Recent Political Trends in the Middle East', 134–5.

[62] A. Vambéry, 'Pan-Islamism and the Sultan of Turkey', 9. Farooqi, 'Pan-Islamism', 286–8. Anees, 47. Aziz Ahmad, *Islamic Modernism*, 124.

Russian War of 1877–8 was a watershed in this respect. Shortly before, the Ottoman war in Serbia had provoked large meetings of sympathy with the Ottomans among Muslims in Bombay and Calcutta.[63] But the Russian victories in 1877–8 and their advance on Istanbul aroused a very real excitement and feeling of solidarity among those of India's Muslims who worried about the continuous reduction in the lands of Islam.[64] These started pressuring the British to assist the Ottoman Empire, agitating in the Urdu press, and collecting and sending to Turkey substantial sums.[65] Sayyid Ameer Ali (1849–1928), a distinguished Indian Muslim and author of studies on Islam,[66] described the Islamic impact of the war in his country, as follows:[67]

> Few observers can have forgotten the extraordinary outburst of sympathy among the Mussulmans of India with the wrongs of Turkey and the afflictions to which their coreligionists were subjected in consequence of that war ... I am in a position to speak of the enthusiasm that prevailed among all classes to help the Ottoman nation and to relieve the universal suffering and distress among the stricken people of Turkey. Even women in the humbler walks of life sent their earrings, bracelets and anklets to be sold and the proceeds remitted to the Turkish Compassionate Fund; and one Province alone was able to forward three lakhs of rupees, or nearly £20,000, whilst many Mussulman soldiers offered their services[68] to the Ottoman Government.[69] The feeling aroused in India in 1877 was a revelation to most observers whose prejudices and antipathies did not blind them to the realities of national life, and from that day forth few competent Europeans have felt inclined to underrate the solidarity that exists among the Moslem races of the world.

[63] As noted and reported by the Viceroy of India, quoted by Farooqi, 286. See also Mohammad Sadiq, *Turkish Revolution*, 18.

[64] Details in Feroz Ahmad, 'The Kemalist Movement and India', 113–14.

[65] See Çetinsaya, 'II. Abdülhamid döneminin ilk yıllarında "islâm birliği" hareketi', 34–46, 96–101, 133 ff. Cf. Aziz Ahmad, *Islamic Modernism*, 124.

[66] On whom see W.C. Smith, 49–55. M.Y. Abbasi, 'Syed Ameer Ali', in: Dani, ed., 45–82. Aḥmad Amīn, 149–57.

[67] Ameer Ali, 'Moslem Feeling', 101–2.

[68] It seems that volunteers from India for the Ottoman Red Crescent did arrive in Istanbul. Cf. Larcher, *La Guerre turque dans la guerre mondiale*, 483.

[69] One may wonder, indeed, if this spontaneous impressive show of Muslim solidarity did not suggest to Abdülhamid or his advisers the Pan-Islamic campaign which they undertook.

This informs us of the strengthening of sentiment with other Muslims, and of a growing feeling of solidarity, but not in the framework of any organized Pan-Islamic movement in the 1870s. The situation changed somewhat in the following years and the British were partly responsible. The British annexation of Cyprus in 1878 and Occupation of Egypt in 1882, along with the verbal attacks on the Sultan–Caliph by leading British statesmen, such as Gladstone, challenged the loyalty of many Muslims in India who had found it convenient to express an allegiance to both the British and the Ottomans while these were allied. According to W.S. Blunt, the British friend of Arabs and Muslims, who visited India, local Muslims regretted the end of this alliance.[70] They expressed their attachment to the Sultan–Caliph in various ways, including the publishing in London of a new Muslim semi-monthly in Persian, entitled *Ghayret* (Zeal). Thanks to an enterprising Turkish journalist, named Salih, who summarized the first issue, we know its contents.[71] Its contributors deplored in no uncertain terms what they considered as undeserved attacks on Islam and its Caliph, and appealed for inter-Muslim amical relations along with full obedience to the Caliph. They highly praised Abdülhamid and his efforts to defend Muslim interests—enjoining all Muslims, as an article of faith, to love and respect him. The union of Muslims was presented as the main factor of their past grandeur and the *sine qua non* of their future progress. Russia was described as coveting Islamic lands, while Great Britain had ceased to be the ally of the Ottoman Empire since its takeover of Cyprus.

Others among India's Muslims even suspected the British of attempting to drive a wedge between them and the Muslims in the Ottoman Empire.[72] The controversy among these Muslims was fed by continuing debates between the local Ulema as to whether India was to be considered as *Dār al-Islām*, implying acceptance of British rule, or *Dār al-ḥarb*, enjoining all Muslims

[70] W.S. Blunt, *India under Ripon*, 293–4. In 1908, Mushir Hosain Kidwai, a dedicated Pan-Islamist, even wrote in support of an Ottoman–British alliance; see his *Pan-Islamism*, 31.

[71] Salih, 'İttihad-ı İslam' (in French: 'L'Union Islamique'). *Ghayret* must have been published in late 1880 or early 1881.

[72] XX, 'La Solidarité islamique et l'Angleterre', part 2, 526–7.

to drive out the Unbelievers.[73] The arguments of a Civil Servant of the Nizam of Hyderabad, who published a tract maintaining that India was in neither of the two Abodes,[74] seems to have made no impression on the Ulema.

Here is a brief selection of instances of increasing activity for Pan-Islamic solidarity among India's Muslims in the late nineteenth and early twentieth centuries. Rafiüddin Ahmad, writing in 1895, reported that Muslims in India had been horrified and indignant at rumours that the Aya Sophia Mosque in Istanbul was to be changed into a church.[75] A year later, others tried to defend the Sultan against accusations in Europe and warned that any measures taken by Great Britain against the Ottoman Sultan would deeply offend the 50 million Muslims in India.[76] *The Moslem Chronicle* of Calcutta was, at the time, one of the more articulate newspapers in India in shielding the Sultan.[77] In 1897, Anthony MacDonnell, Lieutenant-Governor of the United Provinces, reported great sympathy with the Ottoman Empire, the publication of a book preaching *Jihad* and a leaflet describing the Ottoman Sultan as Commander of the Faithful,[78] and the holding of public speeches congratulating him.[79] Although MacDonnell did not realize this, the excitement may well have been due to the brief Ottoman campaign and victory in Crete in that year. In 1901, hundreds of notables in Bengal petitioned Abdülhamid to appoint an Ottoman Consul in Calcutta (in addition to the ones in Bombay, Madras, and Karachi).[80] More significant, organizationally no less than ideologically, was the gradual veering of the All-India Muslim League, previously pro-British (or at least neutral), towards a Pan-Islamic attitude, by increasingly ex-

[73] Vambéry, *Western Culture*, 230–1. Further details in Niemeijer, 32–4.

[74] Moulaví Cherágh Ali, *The Proposed Political, Legal, and Social Reforms in the Ottoman Empire and Other Muhammadan States*, 27.

[75] Rafiüddin Ahmad, 'A Muslim's View of Abdul Hamid and the Powers', 157, and cf. p. 162, for the author's perception of Abdülhamid's immense moral influence and popularity.

[76] H.A. Salmoné, 'Is the Sultan of Turkey the True Caliph of Islam?', 173–4.

[77] Reported by M. MacColl, 'The Musulmans of India and the Sultan', 280–1.

[78] *Amīr al-Mu'minīn* and *Pādshāh-i Mosalmānān*.

[79] Hardy, 177–8. Cf. Feroz Ahmad, 'The Kemalist Movement', 114–15.

[80] *Kölnische Zeitung* (Köln), 25 Mar. 1901.

hibiting its solidarity with Muslims abroad.[81] Further, India's Muslims donated generously to Abdülhamid's project of the Hijaz Railway, which they correctly interpreted as having Pan-Islamic significance. At the very start, *Anjumān-i Islām* (Association of Islam) in Bombay collected £2,000,[82] a large sum, in 1900–1. Other substantial sums were collected and dispatched to Istanbul, for the same purpose, in the following years.[83]

Mushir Husain Kidwai, who has already been mentioned as the secretary of 'Pan-Islam' in London, was actively writing for the cause. In a 1906 article, he argued that Pan-Islam was neither a fanatical nor a secret movement, but one dedicated to defending Islam from Christian calumnies.[84] Two years later, he published an interesting tract, before leaving Great Britain for India. Entitled *Pan-Islamism*, it was intended for Indian and other Muslims no less than for the British readers. In it he complained that 'the Muslim Powers never made a common cause, and unfortunately even to-day the weakened and ill-used Muslim kingdoms do not combine together to present a strong united front to the merciless blows of united Christendom'.[85] In his opinion, 'the brotherhood of Musalmans and the Islamic spirit in them is still a living force and it requires only to be organized'.[86] He had several practical suggestions, too, for this purpose:

For the success of Pan-Islamism a perfect organization is required and branch societies in every Muslim country needed to make the people as well as their Governments realize their condition and their

[81] Hubert, 22–3.

[82] H. Norman, *All the Russias*, 290.

[83] XX, 'La Solidarité islamique et l'Angleterre', 537. Rouire, 'La Jeune-Turquie et l'avenir du Panislamisme', 262. Larcher, 483. The French Consul-General in Calcutta, Coutouly, was aware of these facts and of the pro-Ottoman sentiments among India's Muslims, but expressed doubts whether these could be construed as an advance for Pan-Islam. See AE, NS, Indes, vol. 10 (1906–1917), fos. 6–11, Coutouly's report to S. Pichon, French Minister for Foreign Affairs, dated in Darjeeling, 27 Dec. 1906. For the feeling of loyalty to the Sultan–Caliph in India, see also PA, Orientalia Generalia 9, vol. 2, A. Quadt's no. 680, to von Bülow, dated in Simla, 20 July 1906.

[84] Mushir Hosain Kidwai, 'Pan-Islamism', *The Morning Post*, 20 Aug. 1906 (the letter had been written on 15 Aug.).

[85] Id., *Pan-Islamism*, 11–12.

[86] Ibid. 39.

backwardness as compared with advanced nations. The central Pan-Islamic Society should be placed on a firmer footing. London is the place best suited for the headquarters of the Society.[87]

Then Kidwai expressed warm support for the Ottoman Sultan as the leader of the Islamic movement (he had himself been to Istanbul and had reached this conclusion there).[88] Later, in India, Kidwai joined a Committee of Subscription for the Ottoman navy and went to Istanbul as the delegate of this committee in 1910. In the Istanbul periodical *Sıratul Müstakim* (The Straight Way), Kidwai published an open letter to India's Muslims, praising the Committee of Union and Progress and its alleged intention to preserve the power of the Caliphate. Upon his return to India, he became a correspondent for the Committee of Union and Progress, writing under the pen-name of Mullah Aftab Hussain Saheb.[89] Kidwai continued writing for the cause of Pan-Islam during and after World War I.

A much more powerful voice in support of a political conception of Pan-Islam, in the years immediately preceding World War I, was that of a highly esteemed activist in the ranks of India's Muslims, Abul Kalam Azad (1888–1958), a noteworthy scholar of Islamic theology who, as a young man, had ardently supported Pan-Islam in his important Urdu newspapers *al-Hilāl* (The Crescent) and *al-Balāgh* (The News), both published in Calcutta from 1912.[90] The influence of al-Afghānī on Azad's writing is noticeable.[91] Like al-Afghānī, Azad looked beyond the nation state and maintained that Islam would open the Muslims' way to politics, since for them the words 'Allah' and 'Islam' caused all their hearts to beat in unison.[92]

In his Pan-Islamic stance, which developed over the years,[93] Azad was joined by one of the most energetic political leaders

[87] Kidwai, *Pan-Islamism*, 72.

[88] Ibid. 21, 36–7, 46–7.

[89] Acc. to British reports from India, summarized in Shukla, 'The Pan-Islamic Policy of the Young Turks and India', 305–7.

[90] For Abul Kalam Azad, his life and writings, see L.S. May, *The Evolution of Indo-Muslim Thought after 1857*, 185–95. Mushir U. Haq, *Muslim Politics in Modern India 1857–1947*, chs. 4–5 and *passim*.

[91] Anwar Moazzam, 'Jamāl al-Dīn al-Afghānī in India', 95.

[92] Hardy, 179–80. Aziz Ahmad, *Islamic Modernism*, 129. Anees, 48.

[93] Shaukat Ali, *Pan-movements in the Third World*, 215–8. Rajat Ray, 'Revolutionaries, Pan-Islamists and Bolsheviks: Maulana Abul Kalam Azad and the Political Underworld of Calcutta', in: Mushirul Hasan, ed., *Communal and Panislamic Trends in Colonial India*, 85–108.

of India's Muslims, Muhammad Ali (1878–1931),[94] who, in his Delhi newspaper *Comrade,*[95] adopted the policies of Pan-Islam which he would subsequently champion in the political arena. Although he conceded, in an article on 'The Future of Islam', in February 1912, that Pan-Islam was essentially a defensive instrument,[96] he later became more extreme in his views. His article supported Muslims everywhere[97] and, more particularly, the Ottoman Caliph and the Turks.[98] Muhammad Ali included in the *Comrade* other forceful articles on Pan-Islam, for example, one by Zafar Ali Khan, the editor of the Lahore newspaper *Zamīndār* (Landholder),[99] in an English translation.[100] Entitled 'Indian Muslims and Pan-Islamism', it maintained that:

> To the man in the street Pan-Islamism was synonymous with a gigantic union of the Moslems of the world, having for its cherished object the extermination of Christianity as a living political force ... The bombardment of Meshed by the Russians,[101] the descent of Italy on Tripoli, the onslaught of the Balkan Allies on Turkey, with all their attendant horrors, have made the Moslems of India a changed people. They are not what they were two years ago ... The brotherhood of Islam, or Pan-Islamism if you will, transcends all considerations of race and colour and is of an extra-territorial type in which all the Muslim populations of the world merge their geographical identity and become one nation.[102]

[94] For whose career see May, 195–201. S. Moinul Haq, 'Maulana Mohamed Ali'. M.N. Qureshi, 'The Ali Brothers' (discusses both Muhammad Ali and his brother Shaukat), in: Dani, ed., 109–36.

[95] Aparna Basu, 'Mohamed Ali in Delhi', in: Mushirul Hasan, ed., 109–25.

[96] *Comrade,* Feb. 1912, repr. in Afzal Iqbal, ed., *Select Writings and Speeches of Maulana Mohamed Ali,* 45–62.

[97] *Comrade,* 8 June 1912, subsequently repr. in Rais Ahmad Jafri Nadvi, ed., *Selections from Mohammad Ali's Comrade,* 467, 484.

[98] e.g. *Comrade,* 27 Apr. 1912 and later, cf. Nadvi, ed., 489–524. For Muhammad Ali's Turcophile sentiments, see Kramer, *Political Islam,* 81.

[99] This article has also appeared, in a Turkish transl., in Celal Nuri [İleri]'s above-mentioned *Ittihad-ı Islam,* 396–401.

[100] *Comrade,* 14 June 1913, repr. in Nadvi, ed., 297–9.

[101] The Russians were suspected of invading Iran with the intention of conquering Tehran. A large meeting in the Jumʿa Mosque in Bombay, on 8 Dec. 1911, expressed the sympathies of India's Muslims for their co-religionists in Iran. See, AE, NS, vol. 10 (1906–1917), fos. 226–8, report no. 81 by the French Acting Consul-General, to de Selves in the French Ministry for Foreign Affairs, dated in Bombay, 14 Dec. 1911.

[102] These quotes cast some doubt on Mohammad Sadiq's interpretation of Zafar Ali Khan's views on Pan-Islam, in his article; cf. Sadiq's 'The Ideological Legacy of the Young Turks', 206 n. 54.

During May 1913, *Comrade* issued a supplement entitled 'Macedonian Atrocities', in three parts, each of which was four pages long.[103] The title served as a delayed reply to Gladstone's condemnation of the 'Bulgarian atrocities' attributed to the Ottomans. Obviously inspired by an Istanbul-based committee for the publication of documents relating to the atrocities of the Balkan allies, this was a glowing account of the behaviour of the Ottoman troops and a horror story of atrocities perpetrated by the Balkan allies and the local Christians. These referred chiefly to the persecution of the Muslims by the Greeks and Bulgarians in Salonica, Üsküb (Skopje), Dedeağaç, Kavalla, Serres, and many other localities in Macedonia—complete with data, photographs, and quotations from documents. The presentation expressed identification with the suffering of Muslims in the recently lost territories.

The relatively aggressive tone of such articles was not fortuitous. They tended to reflect, more than earlier ones, the frustration felt by a growing number of India's Muslims over foreign encroachment on and invasion into Islamic territory—and their increasing involvement in Pan-Islam. The general view was that the Ottoman Empire was the only state capable of stemming European aggression and, therefore, ought to be helped.[104] During the Turco-Italian War in Libya[105] and the Balkan War (when the entire Muslim press in India supported the Ottoman Empire),[106] events were highlighted in an increasingly outspoken press. Its most notable organs were: (i) *al-Hilāl*, in Urdu, published in Calcutta from 1912 and reaching a circulation of 11,000 within six months and 25,000 during World War I; edited by Azad, in radical style, both religious and political, it was very influential. One of its recurring themes was that there was no profit in local and national movements so long as the entire world of Islam was not united.[107] (ii)

[103] 'Macedonian Atrocities', supplement to *Comrade* (Delhi), 13, 24, and 31 May 1913. A copy can be consulted in IO (EPP 2/15–17).

[104] Examples in Nallino, *Notes*, 31–2.

[105] For the reaction of India's Muslims to this war, cf. Sadiq, *Turkish Revolution*, 26–9.

[106] Ibid. 30–1. For a Muslim demonstration of support in Calcutta, see PA, Orientalia Generalia 9, vol. 4, German Consul-General Heinrich Reuss's no. 377, to Bethmann Hollweg, dated in Simla, 4 Oct. 1911.

[107] Shaukat Ali, *Pan-movements*, 217.

Zamīndār, in Urdu, published in Lahore as a weekly since 1903 and as a daily since 1910; it had been edited since 1909 by Zafar Ali Khan (the son of *Zamīndār*'s founder) in a violently anti-British and Pan-Islamic tone, which served to increase its circulation to about 20,000 copies. (iii) *Comrade*, in English, published first in Calcutta in 1911, then in Delhi from 1912. (iv) *Hamdard* (Comrade) in Urdu, in Delhi from 1913.[108] Both *Comrade* and *Hamdard* were edited by Muhammad Ali, with increasing emphasis on Muslim protest everywhere.[109] All four had marked Pan-Islamic tendencies; this, along with criticism of British rule in India, brought about their suppression.[110] Others, however, carried on the struggle, notably *The Moslem Chronicle*, in English (which translated articles from the *İkdam* of Istanbul) and *Ḥabl-ol-Matīn*, in Persian; both were published in Calcutta.[111]

In the pre-war years, many new organizations were set up to collect funds and send medical relief to the Ottomans and to defend the Holy Places.[112] The Ottoman–Italian War in Libya caused considerable excitement amongst India's Muslims, many of whom expressed themselves in Pan-Islamic terms (though some dissented).[113] In 1911, Red Crescent groups were dispatched;[114] in the following year, an all-India medical mission reached Istanbul.[115] According to a report, prepared by P.C. Bamford, Deputy Director of the Intelligence Bureau in the British Government of India and published in 1925,[116] this mission was in close contact in Istanbul with public supporters of Pan-Islam, such as Enver Pasha and ʿAbd al-ʿAzīz Shāwīsh. Zafar Ali Khan also visited Istanbul in 1913. Upon his return to India in mid-1913, Dr Muhktar Ahmad Ansari (1880–1936),

108 On *Comrade*, see Aparno Basu, 109–25.

109 Shukla, 306. W.C. Smith, 196–7. May, 196.

110 Sharif al-Mujahid, 'Pan-Islamism', 110.

111 Ibid. 111, also mentions that an Indian Muslim published, in Tokyo, a short-lived anti-British, Pan-Islamic newspaper, entitled *The Islamic Fraternity*. For the newspaper and its editor see also below, nn. 135–9.

112 Minault, *The Khilafat Movement*, 10–11.

113 Details in Koloğlu, *Islamic Public Opinion*, ch. 2, B, section 5, and ch. 5, C, section 9.

114 Larcher, 483.

115 Aziz Ahmad, *Islamic Modernism*, 131–2.

116 P.C. Bamford, *Histories of the Non-co-operation and Khilafat Movements*, 113.

the head of the medical mission, lectured the students at Aligarh College about the importance of a union between Turkey and India.[117] Pertinently, the Pan-Islamic *Hamdard* devoted a complete issue[118] to the mission's visit to Turkey. However, in addition to these Pan-Islamic gestures and similar ones (such as Azad's appeal in *al-Hilāl*, early in 1913, for a boycott of European goods and a *fatwā*, in the *Aligarh Institute Gazette* of March that year, cursing all oppressors of Muslims),[119] there was very strong criticism, among India's Muslims, of Great Britain's reluctance to assist the Ottoman Empire in its wars.[120] The All-India Muslim League, once pro-British, passed resolutions to this effect, too.[121]

One may hypothesize that it was these and other expressions of deep sympathy by many of India's Muslims with the Ottoman Empire and its Muslims,[122] as well as a certain admiration for the Germans,[123] that led the Turks and Germans to hope for widespread revolt among India's Muslims, upon the declaration of war and proclamation of *Jihad*. The British in India, however, still had considerable support. At the end of 1914, Aga Khan (1877–1957), leader of the Ismaili community and head of the All-India Muslim League, announced his support of the British Empire, avowing that neither Islam nor the Ottoman Empire was in danger—and he was a religious Pan-Islamist.[124] Then, the Council of the All-Muslim League reversed its position and announced its support of the British Government.[125] However, the latter was still concerned about the potentially disruptive power of Pan-Islam, especially in war-time.[126] Abul Kalam Azad was interned during World War I, as were Muhammad Ali and his brother Shaukat

[117] Bamford, *Histories*, 113.

[118] *Hamdard*, 1/35 (12 July 1913).

[119] Bamford, 112, 114.

[120] The British were even suspected of helping the Ottoman Empire's enemies. See ibid. 112. Ameer Ali, 'Moslem Feeling', 103–8 (with examples of meetings and demonstrations). Cash, *The Moslem World in Revolution*, 71 ff. Shaukat Ali, *Pan-movements*, 223–4.

[121] Hubert, 23–5. Aziz Ahmad, *Islamic Modernism*, 131–2.

[122] See Bamford, 118–19.

[123] Cf. Afzal Iqbal, *The Life and Times of Mohamed Ali*, 162–3.

[124] Muḥammad ʿAmāra, *al-Jāmiʿa al-Islāmiyya* . . ., 49–50.

[125] AE, Guerre 1914–1918, vol. 1651, fo. 64.

[126] Sareen, 176.

Ali (1873–1938),[127] as well as several other prominent anti-British, pro-Ottoman Pan-Islamists. Few publications with Pan-Islamic content were available at that time in India, and most of these seem to have been smuggled in, possibly from the Ottoman Empire. Examples are Urdu pamphlets bearing such titles as *The Message*[128] (two essays, of which the second discussed the Islamic position concerning the war and the Caliphate and called on all Muslims to unite in the cause of Islam); *The First Year of the War*;[129] *Friendly Advice to Indian Brothers*;[130] and *Jihad for Right*.[131] Among the few Pan-Islamic works published in India itself was the 168–page collection of Urdu poems on the Balkan Wars by Vajahat Husayn, a former assistant editor of *Zamīndār*, entitled *Naẓam-i-Vajahat* (The Poetry of Vajahat) and published in Lahore in 1914.[132]

Several Muslim intellectuals and students left India before or during the war, some to Afghanistan (hoping to join the *Jihad*),[133] others to continue their Pan-Islamic, nationalist, or revolutionary activity. One of the most colourful of these was Muhammad Barakatullah (1859–1927), of Bhopal, who had taken part in Pan-Islamic activities as early as 1892. As chairman of a Committee of Indian Mohamedans in London, he wrote (together with three of his colleagues in the committee) a letter to *The Standard*, protesting at calumnies of the Sultan–Caliph and reaffirming the loyalty of all India's Muslims to Abdülhamid II: 'From every Mohamedan Indian who has visited Constantinople we hear that His Majesty the Sultan is devoting all his time to the prosperity of his Empire and the happiness of his subjects, which had made him gain their love and their respect'.[134]

[127] Hardy, 185. Aziz Ahmad, *Islamic Modernism*, 132ff. Anees, 48. For the two brothers, to whose public Pan-Islamic role we shall revert presently, see also Shan Muhammad, ed., *Unpublished Letters of the Ali Brothers*.

[128] Sulaymān Ashraf, *al-Balāgh* (1914?). A copy can be consulted in IO.

[129] *Dunyā kī jang kā pahlā sāl* (1915?). A copy is available in IO.

[130] Muḥammad Jamāl, *Hind* (1917). A copy is available in IO.

[131] Id., *al-Jihād fī sabīl al-ḥaqq* (1917). A copy can be consulted in IO.

[132] N.G. Barrier, *Banned*, 194.

[133] Sareen, 176–7.

[134] *The Standard* (London daily), 9 Sept. 1892. The letter had been written a day earlier.

In 1910, Barakatullah went to Japan, with the intention of opening an anti-British and Pan-Islamic office.[135] He became a teacher of Hindustani languages at Tokyo University. Soon after his arrival, he took over the publication of an English monthly, *The Islamic Fraternity*, whose object had been to inform non-Muslims of the character of Islam. Barakatullah turned it into a Pan-Islamic, anti-British organ (which he smuggled into India)[136] with a determinate stand on Islamic union. For example, one of its issues[137] included an unsigned lengthy article on 'Turkey to be Ousted from Mecca', which not only blamed Great Britain for fostering Arab nationalism, particularly in the Arabian Peninsula, but warned Muslims everywhere that their only remaining hope was in uniting and rallying to Istanbul's support: 'The present delicate condition of the Muslim community in the world makes it imperative for all Muslims to rally round the throne of Muhammad V—the Commander of the Faithful—who personifies the symbol of the unity of Islam today ... The Khalif ... protects rights, property and honor of the Believers'. After the Balkan Wars, the publication advocated a Pan-Islamic alliance led by Afghanistan, which Barakatullah considered 'the future Japan of Central Asia'.[138] Barakatullah also wrote several pamphlets expressing the same views,[139] and later a book on the Caliphate.[140]

In his evaluation of Afghanistan, Barkatullah was not

[135] For his career there and subsequently, see Sareen, 146–7.

[136] See, for 1912, PA, Orientalia Generalia 9, vol. 5, German Consul Saunier's no. 83, to Bethmann Hollweg, dated in Bombay, 16 Sept. 1912.

[137] *The Islamic Fraternity* (Tokyo), 1/5–6 (15 Sept. 1910). This issue, as well as 2/2 (15 May 1911), can be consulted in IO (under EPP 2/19–20).

[138] Sareen, 146. It appears that this is the monthly referred to by E. Ronssin, French Consul-General in India, who reported on its distribution in India and elsewhere in 1912. See AE, NS, vol. 10 (1906–17), fos. 283–4, Ronssin's no. 136, to Raymond Poincarré, French Prime Minister and Minister for Foreign Affairs, dated in Calcutta, 13 Aug. 1912.

[139] After Tokyo University had not renewed his contract (possibly due to British protests), Barakatullah left Japan to pursue a revolutionary course, during which he co-operated with Lenin, too. During World War I, he published the Pan-Islamic magazine *Ghadr* in San Francisco. See Feroz Ahmad, '1914–1915 yıllarında İstanbul'da Hint milliyetçi devrimcileri', 8. Brown, *Har Dayal*, ch. 4, esp. 181.

[140] This work is known to me only from PA, Politik, Abt. III, Türkei, Politik 16, vol. 1, Barakatullah's letter to the German Ministry for Foreign Affairs, dated in Zurich, 26 Sept. 1924.

entirely mistaken—at least in so far as Pan-Islam was concerned. Since the late nineteenth century, the Emirs who ruled Afghanistan had been partial to Pan-Islam, which they considered as a potentially supportive force against the pressures of Russia and Great Britain.[141] Moreover, they encouraged a newspaper, *Sirāj ol-akhbār* (Candle of the News), first published 10 October 1911, with clearcut Pan-Islamic tendencies.[142] Written in Persian, it was regularly sent to India (until the British seized it) and Iran.

3. *The Organizing of Pan-Islam: Khuddām-i Kaʿba*

As had been pointed out by Kidwai and others, in the Indian context Pan-Islam was a potentially significant political force if those Muslims supporting it could be organized. The terrain for such organization, in India no less than elsewhere, was obviously the numerous Islamist circles which not only took Islam seriously, but considered it as pivotal to their private and public life. One example of these, *Anjumañ-i Islām*, which collected donations for the Hijaz Railway, has already been mentioned. The trouble with them, from a Pan-Islamist's point of view, was that they were local groups with overriding concern for India's Muslims. A more comprehensive association, such as the All-India Muslim League, overturned its own support for Pan-Islam, as has been stated. The enormous size of the Indian subcontinent and its huge masses of Muslims raised, indeed, both a challenge to Pan-Islamists and a set of grave problems in any attempt to provide organization.

An association committed to Pan-Islam existed in Calcutta at the end of the nineteenth century. Its promoter and moving spirit was Haji Nur Muhammad Zakariya, a wealthy Muslim trader,[143] whose prestige among Calcutta's Muslims increased further after he built a mosque and became involved, on behalf of poor Muslims, in a communal riot in 1897. Zakariya was one of the earliest and most dedicated Pan-Islamists in Calcutta from the 1870s. He held frequent meetings at his

141 L.W. Adamec, *Afghanistan, 1900–1923*, 24–6, 81–2.
142 Ibid. 101–3.
143 For Zakariya, see Dipesh Chakrabarty, 'Communal Riots and Labour', 154 ff.

home or in the mosque, to present lectures (even Ameer Ali gave one there), pray for the welfare of the Ottoman Caliph, and raise money for the Turkish Relief funds; in 1878, between 40,000 and 60,000 rupees were collected at each of these meetings.[144] In 1897, following the Ottoman victory over the Greeks, Zakariya and other Pan-Islamists in Calcutta founded a 'party'. It consisted chiefly of traders and professionals and it was 'active in holding up the Sultan as the head of Islam, in representing him as being unjustly harassed by Great Britain and the European Powers, and in magnifying his might as manifested by victories over the Greeks'.[145] It was from this association and its likes that Pan-Islamic ideas filtered into the poorer population groups of Calcutta's Muslims, for example, the view that the Ottoman Sultan was an all-powerful Muslim ruler, or that the British regulations against the plague were actually intended to damage that pillar of Islam and Pan-Islam, the Pilgrimage.[146]

Zakariya's 'party' and similar associations were, however, only local groupings, foreshadowing the organization of Pan-Islam on a wider and more politicized scale. This occurred only in 1913, following the shock waves in India's Muslim public opinion after the Ottoman defeats in Libya and in the Balkan Wars. Its most forceful expression was in the *Anjumān-i Khuddām-i Kaʿba* (Association of Servants of the Kaʿba).

According to scattered information in the press and in works on India's Muslims,[147] and more detailed and reliable reports in the records of the British Foreign Office and India Office,[148]

[144] Ibid. 161, based on the Indian National Archives, Delhi.

[145] From a British report, now in the Indian National Archives, quoted by Chakrabarty, 162.

[146] Ibid. 162–5.

[147] One finds only a few lines, at most, in Bamford, 113; Shaukat Ali, *Pan-movements*, 219; Kohn, 47; Afzal Iqbal, *Life and Times*, 168; M. Mujeeb, *The Indian Muslims*, 400; Shafique Ali Khan, 'The Khilafat Movement', 36; S. Razi Wasti, 'The Khilafat Movement in the Indo-Pakistan Subcontinent', 18.

[148] IO, L/P&S/20/H137, Criminal Intelligence Office's report, entitled 'Anjuman-i-Khuddam-i-Kaaba 1913–1914', 'Secret', dated 1914. FO 882/15, Vivian's memorandum, dated 30 July 1917 (in the files of the Arab Bureau, Egypt), 164. FO 371/4204, file 9152, no. 64090, confidential report, dated in 1919, enc. in a letter, no. 86, from J.H. DuBoulay, Secretary to the Government of India, to Sir T.W. Holderness, Under-Secretary of State for India, dated in Delhi, 22 Jan. 1919.

the association was decided upon in December 1912[149] and set up in May 1913. It had been initiated in Lucknow[150] by Mushir Husain Kidwai and Qayyamul Din Muhammad Abdul Bari[151] (1879–1926), an ʿālim from Lucknow, who was active in the promotion of both Muslim education and Pan-Islam.[152] Muhammad Ali and Shaukat Ali came together to confer with them in Lucknow and the four of them decided to set up the *Anjumān-i Khuddām-i Kaʿba* as a religious association whose object would be to protect Mecca and Medina from non-Muslim aggression. Its rules, subsequently published in the press, specified that its chief aim was to preserve the honour and safety of the Kaʿba and defend Islam's Holy Places from non-Muslim aggression, and as subsidiary objectives to spread Islam, provide for Muslim schools and orphanages, and assist the Pilgrimage. All members were to pay, equally, one rupee per year, of which a third would be sent to the independent Muslim power responsible for safeguarding the Kaʿba, another third donated to schools and orphanages, and yet another third reserved for commercial undertakings to benefit the Kaʿba or kept to assist its defence in case of need.

Members were to be either regular ones, or votaries (*shaidai*)[153] who would undertake special missions (of propaganda, perhaps of martyrdom as well). New members had to take an oath upon joining the association (later, the oath was replaced by a promise of intent). Not only would the Muslims in India be encouraged to enrol, but also Indians living in Mecca, Medina, and the Holy Places in Iraq. A pyramidal structure was envisaged, with its basis being the mosques (they considered drafting all imams and muezzins into the association) and its apex a Central Committee of six members: a president, modestly titled *Khādimul Khuddām* (servant of the servants), Maulvi Abdul Bari; two secretaries, Mushir Husain Kidwai and Shaukat Ali; and three additional members, Muhammad Ali

[149] See also Sadiq, *Turkish Revolution*, 69 n. 4.

[150] Acc. to S.M. Zwemer, *Across the World of Islam*, 330, Shiites had an important role in the early organization in Lucknow.

[151] Self-styled as a *Maulvi*, that is a *Mawlawī* (Turkish, *Mevlevi*) dervish. This interconnection, again, of the new association with a Muslim fraternity is suggestive.

[152] Cf. D. Page, 'Prelude to Partition', pp. xxx ff.

[153] Probably from the Arabic *shahīd*, 'martyr'.

and two Muslims from Lucknow (a notable and a barrister).[154] The rules were only slightly modified when issued in November 1913, and again, early in 1914—chiefly with a view to increasing the recruitment of members and enlarging the Central Committee to fourteen.

During 1913, members of the Central Committee toured India, striving to increase membership by holding public rallies. By October, nineteen branches of the association had been set up in India and overall membership had increased from 23 to 3,431; it rose to about 17,000 by 1915.[155] Kidwai and Zafar Ali Khan went to London and opened a branch there in the summer of 1914. Another was opened in Istanbul, apparently with the co-operation of Shāwīsh; others seem to have been inaugurated, shortly afterwards, in Cairo and Singapore.[156] Enquiries regarding the aims and the rules of *Khuddām-i Kaʿba* were received at its secretariat from Muslims in Turkey, Egypt, and elsewhere. Shaukat Ali consistently replied that solely Indian Muslims, at home or abroad, might join, but that others could of course set up their own associations. At the same time, he had an Arabic translation of the rules prepared and printed, and looked for an opportunity to issue an Arabic newspaper in Egypt (which never materialized).

All this tells us something about the association's general character and intentions, which were not entirely within the purview of religion. Collecting money for the Ottoman Sultan (with no way of ensuring that the funds would be earmarked for the Holy Places) and establishing channels of communication with Muslim groups abroad undoubtedly had a ring of political Pan-Islam. Turks who visited India just before World War I, such as representatives of the Red Crescent, met members of the association.[157] British Intelligence in India reported that the association's founders intended to build ships for pilgrims (which was true) and maintain an army and a navy for the protection of the Holy Places (possible, but not fully

[154] See below, app. L.
[155] Weekly Report of the Director of Criminal Intelligence, 1 June 1915, mentioned by David Page, p. xxxi.
[156] M.N. Qureshi, 'The Khilafat Movement In India, 1919–1924', Ph.D. thesis (Univ. of London, 1973), 26.
[157] Sadiq, *Turkish Revolution*, 41.

proven).[158] Its founders and leaders, in particular the Ali brothers, Kidwai, Abdul Bari, and others in central positions, such as Zafar Ali Khan, were known for their dedication to Pan-Islam and several of them (such as Kidwai) for their Pan-Islamic contacts in Istanbul. Ḥāfiẓ Wahba, an Egyptian who arrived in India in 1913 to take a job as editor of the association's projected Arabic journal,[159] was well known for his connections in Istanbul, particularly with ʿAbd al-ʿAzīz Shāwīsh, the fiery Pan-Islamist, to whose monthly, *al-Hidāya*,[160] he had contributed.[161]

Lest some of this be considered guilt-by-association, Muhammad Ali's presentation of *Khuddām-i Kaʿba*[162] and its objectives is quite instructive. He asserted that the changing of the map of the Muslim world by the European powers ought to matter to every Muslim—especially since worse was to come. He then emphasized that the association was religious, non-violent, concerned about the future of the Holy Places, and desirous to assist the Ottoman Empire in defending them 'and in maintaining an independent and effective Muslim sovereignty over those lands'. Muhammad Ali, however, also expressed the interest of the association in other Muslim lands and its will to co-ordinate action to stem European aggression. He actually appealed to Muslims to organize themselves so as to repulse with their weapons an attack on the Holy Places of Islam. Later, in September 1916, Abdul Bari, still the association's president, gave an interview in which he stated that, 'when the Association was started, one of its chief aims was to send educated Muslims of importance, politically, from all parts of the world to Arabia in order to bring about union between the Turks and Arabs and to teach the latter the expediency of having one powerful Muslim Power by securing unity among themselves'.[163]

The founders and leaders of the *Khuddām-i Kaʿba* were very

[158] A resolution passed at a meeting of the association's Central Committee, on 15 Feb. 1914, spoke of 'building a fleet, etc. in defence of the Holy Places'. See IO (n. 148, above), app. D.

[159] Letter in *al-Shaʿb* (Cairo), 22 Jan. 1914, perhaps written by Ḥāfiẓ Wahba himself.

[160] For which see above, ch. II, sect. I of our study.

[161] e.g. *al-Hidāya* (Istanbul), 4/3 (Rabīʿ al-awwal 1331 H), 233–9.

[162] *Comrade*, 31 May and 7 June 1918.

[163] FO 37/4204 (as in n. 148, above).

conscious of the importance of organization and mobilization. In the New Delhi branch alone, membership increased from an initial 900 to 2,000 by June 1914.[164] In many other Muslim-populated areas growth was equally impressive.[165] Even in 1918, after the lean years of the World War, Shaukat Ali estimated overall membership at no less than 20,000.[166]

The association worried, on the eve of World War I, about the conflict of loyalty for India's Muslims, should the Ottoman Empire join Germany; its president, Abdul Bari, called on the Sultan, on 31 August 1914, to remain neutral.[167] *Khuddām-i Kaʿba* slowed down its activities during the war, particularly following the internment by the British authorities of the Ali brothers and several other Pan-Islamists. This resulted, also, in the closing of the association's monthly, *Khuddāmul Kaʿba*, which had started publication in June 1914, with Shaukat Ali as its editor. A minimal activity was kept up, however, during the war, in various parts of India, and a branch appears to have been established then, in Kashgar, for Pan-Islamic propaganda.[168] Significantly, however, a number of Ulema joined in these Pan-Islamic activities[169]—signalling their growing political awareness, as well as their future involvement in India's Islamic and Pan-Islamic politics at the end of World War I. For indeed, as soon as the war ended, the *Khuddām-i Kaʿba* resumed their Pan-Islamic activity (and the Ali brothers had signed many of their letters, in internment, as 'Servants of the Kaʿba).[170] After the war's end, the association opposed the dismemberment of the Ottoman Empire and the endangering

[164] Indian National Archives, reported by Basu, 118.
[165] Cf. M. Kar, 'Khilafat and Non-cooperation Movements in Assam', in: Mushirul Hasan, ed., 126–40.
[166] FO 371/4204 (as in n. 148, above). However, this figure applied more strictly to the All-India Muslim League, see Larcher, 483, based on the *Revue du Monde Musulman*.
[167] Text of this cable in Sadiq. *Turkish Revolution*, 41–2.
[168] FO 371/3057, file 103481, being a memorandum, dated in Jedda, 25 Mar. 1917, by Captain N.N.E. Bray, on Pan-Islam. Based on Intelligence sources, research, and numerous conversations, it is entitled 'A Note on the Mohammedan Question'. Its main conclusion is that Pan-Islamic agitation was continuing throughout India and fast becoming a political movement there.
[169] Mujeeb, 400–1. Aziz Ahmad, *Islamic Modernism*, 133 ff. Mushirul Hasan, 'Religion and Politics in India', esp. 10–11.
[170] Examples in Shan Muhammad, ed., 171–2.

of the Caliph's office.[171] In this it was joined by other groups in the more popularly supported Khilafat movement, whose most prominent leaders included the leaders of *Khuddām-i Kaʿba*.

4. *Popular Involvement: The Khilafat Movement*

The popular character assumed by the Pan-Islamic movement in India during the years immediately following the end of World War I may be explained by several factors. Firstly, on the eve of this war, the new middle classes of India's Muslims, better off and more educated, outgrew their dependence on British imperial rule and increasingly voiced their involvement in politics.[172] Secondly, the Ulema, as has been mentioned, came out of their political isolation and assumed a more participatory role in Islamic and Pan-Islamic politics;[173] indeed, every important move of the Khilafat movement was preceded by a *fatwā*[174] and the Ulema, *Piri*s (heads of Muslim fraternities), and other religious dignitaries became political agents for recruiting support. Thirdly, it seems that a growing sentiment pervaded politically aware Muslims in India that ensuring Muslim power and sovereignty abroad was a guarantee for their own religious and national survival as a minority group; in other words, Pan-Islam assumed for them a nationalist significance.[175] One may surmise that, but for the outbreak of war in 1914 and the resulting measures instituted by the British (including the internment of political leaders), the above factors would have combined even then in a popular Islamic and Pan-Islamic movement, probably led by *Khuddām-i Kaʿba*. This was perforce postponed to the end of the war, when more suitable conditions developed again.

In late 1918, upon the defeat of the Ottoman Empire, Abdul Bari, president of *Khuddām-i Kaʿba*, sent a strongly worded cable to the Viceroy of India, expressing his concern over the

[171] Afzal Iqbal, *Life and Times*, 168 ff. Hardy, 185 ff. Shafique Ali Khan, 36–7.

[172] Smith, 195. N.R. Keddie, *Sayyid Jamāl ad-Dīn 'al-Afghānī'*, 26 n. 30.

[173] M.N. Qureshi, 'The Indian Khilafat Movement', 155–7.

[174] Mushir-ul Haq, 'The Authority of Religion in Indian Muslim Politics', in: Mushirul Hasan, ed., esp. 361 ff.

[175] Qureshi, 'The Indian Khilafat Movement', 155, 161.

situation of the Holy Places in Arabia. The Council of the All-India Muslim League at Lucknow followed suit. Subsequently, at the annual session of the same league, in Delhi, on 30–1 December 1918, in the presence of several eminent Ulema, serious worry about the Caliphate was expressed. In September 1919, M.A. Jinnah and several other leaders of this league forwarded to Lloyd George a memorandum, asserting that the Ottoman Sultan–Caliph alone 'is the sovereign fit to be the defender of the faith and the custodian of the holy places'.[176] Further moves were then undertaken to secure the Caliph's position in Istanbul and contacts established with Muslims abroad, for instance in Kabul.[177] However, popular involvement was yet to come.[178]

Angry at what seemed to them a wish to reverse the results of the Crusades, traditional[179] and secular Muslims, young and old, organized in November 1919 a Khilafat Conference (i.e. a Conference for the Caliphate),[180] where an All-India Central Khilafat Committee was set up.[181] Subsequently, regional and local committees were established.[182] This was to become the main framework for many—although by no means all[183]—of

[176] Text of the memorandum in *The Daily Telegraph* (London), 17 Sept.1919.

[177] Afzal Iqbal, *Life and Times*, 168–72.

[178] Private involvement of Muslims from India in Pan-Islamic propaganda continued, e.g. in North Africa, reportedly in 1920. See FO 371/5467, file 217, British Vice-Consul Basil S. Cave's no. 112, to Earl Curzon, dated in Algier, 20 Nov. 1920.

[179] Numerous Indian Ulema had meanwhile organized themselves in local and all-India associations.

[180] For this convention and the early stage of the movement, see Gopal Krishna, 'The Khilafat Movement in India', 37ff. For the the movement itself, see Ram Gopal, 137–51. Sadiq, *Turkish Revolution*, chs. 3 and 5. Engineer, ch. 2. S. Moinul Haq, 'The Khilafat Movement'. S. Razi Wasti, 'The Khilafat Movement'.

[181] For many of the documents concerning this movement, see K.K. Aziz, compiler, *The Indian Khilafat Movement 1915–1933*.

[182] Shafique Ali Khan, 43. For the main course of events, cf. NA, RG 59. 845.00/236 to 845.00/403, including reports and newspaper clippings for 1920–3.

[183] There were some, like an active Indian Pan-Islamist, Abdur Rab, who went to the Soviet Union, but never joined the Communist Party there. For a while, he was Enver Pasha's secretary, then, in 1921, became 'India's representative' in the June 1921 convention of Muslim delegates who met in Moscow and formed a sort of Pan-Islamic committee to co-operate with the Third International in propaganda in Muslim lands. See Amba Prasad, 'Pan-Turanian and Pan-Islamic Movements and India 1908–1922', esp. 254–9.

India's Pan-Islamists. The movement was joined not only by Sunnite political leaders (with increased prestige after their internment) such as the Ali brothers, Abdul Bari, Azad, Ansari, and many others, but also by Shiites and Ahmadis. A leader of the latter, Mirza Mahmud Ahmad, from the Punjab, acknowledged that, while he could not recognize the Sultan in Istanbul as his Caliph,

> The conference should take its stand upon the position, that the complete extinction or curtailment of the sovereign power of a Moslem state, the head of which is considered by a large section of the Moslems of the world to be their Khalifa, will be an act which cannot but be disliked by all sections of Moslems, to whom even the contemplation of such an eventuality caused the deepest pain. In such a case the motion could be heartily and universally supported by all sections of Mussalmans.[184]

The Khilafat was being rapidly transformed from an agitational alliance[185] into a religio-political mass movement[186] with a perceptibly romantic component.[187] Poems on Pan-Islam and Turkey, in Urdu, were published then.[188]

The adherence of Mahatma Gandhi himself[189] to the Khilafat movement was a significant event. In a series of articles in *Young India*, collected in book form[190] in 1921, he maintained that he had joined the movement out of a sense of moral responsibility and deep feeling for the just cause of India's and other Muslims.[191] His participation created the comprehensive alliance which the British had long been hoping to prevent. Divergent views among India's Muslims[192] and between them and the Hindus notwithstanding, some of their most influential leaders had joined forces in a common organization. The Muslims were more willing than previously

[184] Mirza Mahmud Ahmad, 'The Future of Turkey', 276.

[185] D. Page, pp. xxvii ff.

[186] To borrow an expression from G. Minault, 'Islam and Mass Politics', 176.

[187] Engineer, 76–7.

[188] Barrier, 194, for examples.

[189] For Gandhi's share in the Khilafat movement, see Muhammad Munawwar, 'Khilafat Movement'.

[190] Mahatma Gandhi, *Freedom's Battle*.

[191] Ibid, 3 ff. and *passim*.

[192] Examples in Hardy, 190–7 and *passim*.

to support the Hindus' policies of non-cooperation with the British, while the Hindus were ready to support the Muslims' campaign for preventing a total dismemberment of Turkey and injury to the prestige of the Sultan–Caliph in Istanbul. Of course, Turkey and the Caliphate were matters of pivotal significance around which the entire Khilafat campaign turned, for political opinion among the Muslim élites (secular no less than religious) was increasingly devoted to the Caliphate as a free institution and the Caliph as a free agent (without real attachment to the person of one caliph or another). Azad, the ideologue of the Khilafat movement, maintained that all Muslims owed *political* loyalty to the Caliph.[193] While not all India's Muslims, naturally, were dedicated to this premise, the Khilafat Conference and its Central Khilafat Committee soon drove more moderate Muslim organizations, such as the All-India Muslim League, into the periphery of the political arena in India. It seems that a sizeable majority of India's politically aware Muslims supported the following paragraph in the Central Khilafat Committee's manifesto of January 1920: 'Islam has ever associated temporal power with the Khilafat. We, therefore, consider that to make the Sultan a mere puppet would add insult to injury and would only be understood by Indian Muslims as an affront given them by a combination of Christian Powers'.[194]

The main activities of the Central Khilafat Committee can be summed up under two headings, internal and external. Under the former, general conventions, starting in 1919, and continuous propaganda in well-attended mass meetings[195] drummed up funds.[196] The response of India's Muslims to these Pan-Islamic appeals was very generous: in 1921–2, a total of 36.5 lakhs of rupees (about US $1,180,000 at that time) were collected (of which only 19 lakhs, however, were trans-

[193] Aziz Ahmad, *Islamic Modernism*, 134–7. Voll, 225.

[194] Quoted by May, 204.

[195] e.g. a large meeting of Muslims and Hindus in Bombay, on 5 Mar. 1920, report in the local press enclosed in FO 371/5142, file 5632, North West Frontier Intelligence Bureau Diary no. 14, for the week ending 1 Apr. 1920.

[196] Some funds were used for expenses, other were sent to the Red Crescent in Turkey, yet others seem to have reached Pan-Islamists in Europe. See FO 371/9002, file 618/618, report of the Inter-Departmental Committee on Eastern Unrest, 'Most Secret', dated Jan. 1923.

ferred to Turkey).[197] An important role was fulfilled by the publishing of two periodicals by the Committee, in English and Urdu, entitled respectively, *Khilafat Bulletin* and *Khilāfat-i 'Uthmāniyya.*[198] Regarding external activities, delegations were sent abroad and offices opened there to promote its demands. The external activities were, on the whole, less successful than the internal. A first delegation of India's Muslims, headed by Muhammad Ali (whose brother, Shaukat Ali, was now the most powerful politician in the Central Khilafat Committee) went to Europe in February 1920.[199] Mushir Hosain Kidwai had preceded them to London, to inaugurate an Islamic centre there,[200] the most important component of which was an Islamic Information Bureau. The latter started, on 23 October 1919, publishing in English a weekly, *Muslim Outlook,*[201] later continued by the *Islamic News,*[202] and subsequently by *The Muslim Standard.*[203] Staffed by Muslims from India, all three journals were committed to Muslim unity, the Khilafat movement, the position of the Sultan–Caliph, and the independence of Turkey. Kidwai was very active in them, maintaining[204] that, as a *rapprochement* between Germany and Russia was likely, Great Britain's best bulwark against this peril would be 'to secure the goodwill of Islam and to strengthen the position of Islamic countries and empires'.

London remained the main centre of activity for the delegation to Europe, which strived to sway the British Government in favour of Turkey, despite a strong wave of anti-Turkish public opinion. In addition to press propaganda, leaflets stating

[197] NA, RG 59, 845.00/392, US Consul A.M. Warren's no. 800, to Dept. of State, 'Strictly confidential', dated in Karachi, 19 July 1923.

[198] Chejne, 685.

[199] On this delegation's activities in Europe, see M.N. Qureshi, 'Mohammed Ali's Delegation to Europe, 1920', based on his unpubl. Ph.D. dissertation.

[200] Kidwai had already published, in 1919, two booklets, in preparation for the delegation's trip abroad, entitled, respectively, *The Future of the Muslim Empire Turkey* and *The Sword against Islam or a Defence of Islam's Standard Bearers.* Both advocated maintaining the Ottoman Empire as a Muslim unit.

[201] 54 issues, 23 Oct. 1919–28 Oct. 1920.

[202] 38 issues, 4 Nov. 1920–21 July 1921.

[203] 33 issues, 18 Aug. 1921–25 Jan. 1923. A set of all three magazines can be consulted at the New York Public Library.

[204] *Muslim Outlook,* 2 (30 Oct. 1919), 5.

its case, and letters to the editor,[205] the delegation had several interviews, including one with Prime Minister Lloyd George. The main points expressed by Muhammad Ali on behalf of the delegation, in this interview, were that, since the Caliphate combined spiritual with temporal authority, it was vital for all Muslims to preserve the institution in its irreducible minimum; also, that it was essential to maintain the Caliph's control over the Holy Places in Arabia. This stand, which combined religious and political Pan-Islam, was very moderate —perhaps too moderate to have any real impact. Lloyd George and other statesmen rejected the delegation's demands *in toto*; after all, the British Government not only had to consider imperial interests (as it understood them) and its own public opinion, but was naturally reluctant to acknowledge the right of Muslim Indians to speak on behalf of the Ottoman Empire, which would be a recognition of Pan-Islam as such. Two interviews with Edwin Montagu (1879–1924), Secretary of State for India (1917–1922) were more pleasant, for he was a strong supporter of the Turkish case[206]—but equally unproductive. Nor did interviews with Labour leaders, nor public meetings addressed by members of the delegation and their friends, produce concrete results.

Considerable efforts went into organizing the public meetings, in which Muhammad Ali was usually the main speaker. For example, in a speech in London, on 22 April 1920, he asserted that the victorious Allies ought to respect the sanctity of Muslim lands, for the Muslims felt united with one another and it was inconceivable that Islam should be deprived of its Caliph and that Arabia should pass under foreign control.[207] At another meeting, also in London, on 2 July 1920, he denounced the peace treaty imposed on Turkey, maintaining that, should Muslim majorities be submitted to the rule of Christian minorities, Greek or Armenian, India's Muslims might refuse to fight in the next war.[208]

In desperation at their total failure, the delegation repeatedly

[205] Examples in Shan Muhammad, ed., 186 ff.

[206] For Montagu's views, cf. H. Armstrong, *Turkey in Travail*, 136–7. Armstrong had been, in 1919, Acting Military Attaché to the (British) High Commissioner in Istanbul and on the Headquarters' Staff of the Allied Army of Occupation there.

[207] See the 1920 pamphlet *Justice to Islam and Turkey*, esp. 6–15.

[208] Afzal Iqbal, *Select Writings and Speeches*, 195–204.

visited France, Italy, and Switzerland, attempting to influence political leaders and public opinion there.[209] Even though the mood in Italy was more favourably disposed, the delegation failed to make a political impact in all these countries. France, however, was apparently considered sufficiently important for the delegation to establish contact with a magazine, which it probably sponsored, too, at least in part. Entitled *Echos de l'Islam*, it was published irregularly in Paris between 20 February 1920 and early 1924.[210] Muhammad Ali, Kidwai, and other Khilafat activists wrote in it. *Echos de l'Islam* promoted the cause of Turkey and its struggle for independence, identifying it with the interests of Muslims everywhere —with a particularly high profile for Islamic agitation in India and special emphasis on the Khilafat movement's activities. An obvious aim was the mobilization of French public support for the Turkish and Pan-Islamic causes.

One of the delegation's boldest acts was its cabling the Sultan to tear up the peace treaty imposed on the Ottoman Empire; it explained its views further to the Sultan in a lengthy letter from Paris, on 28 May 1920.[211] In a way, the delegation's emphasis throughout on Muslim solidarity was double-edged. It was its entire *raison d'être* and justified its action; on the other hand, it raised again the spectre of Pan-Islam. The delegation's members, however, were so committed to religious and political Pan-Islam, that a change in approach, on this count, was never considered. While there is no factual basis for British suspicions that Muhammad Ali intended to bring about Muslim India's independence, in order to unite it with a revived Islam,[212] the delegation established numerous and close contacts with other Muslims then in Europe; it tried (unsuccessfully) to mediate between Turks and Arabs, in order to promote some sort of Muslim union;[213] and, most particularly, kept continuously in touch with the Turks—first with

[209] For the delegation's speeches in Paris, see e.g. Comité 'La France et l'Islam', *Recueil de discours en faveur de l'Islam et de la Turquie*, 5–9.

[210] A set may be consulted at the Bibliothèque de Documentation Internationale Contemporaine, Paris.

[211] The letter, signed by Muhammad Ali and three other members of the delegation, was reprinted in *Comrade*, 14 Nov. 1924, then repr. by Bamford, 244–50, and Shan Muhammad, ed., 191–201.

[212] Qureshi, 'Mohamed Ali's Delegation', part 2, 172.

[213] Ibid. 172–3.

the leaders of the Committee of Union and Progress in Europe and with Mustafa Kemal in Ankara, then solely with the latter.[214] As a sign of solidarity of India's Muslims with Turkey's, Muhammad Ali promised Mustafa Kemal to strive to stop the sending of Muslim troops outside India, most especially if these were intended to fight Turkey.[215]

Upon the delegation's return to India, the Khilafat movement expressed disappointment at its failure through anti-British tactics, chiefly by joining with the (Hindu) National Congress in a countrywide movement of non-cooperation with the British. Fiery speeches landed the Ali brothers and other leaders in gaol, again, and it was the Khilafat movement's new president, Seth Chotani, who led another delegation to Great Britain in 1921, with equal lack of success. By this time, the British Government had less of a say in Turkey's affairs, since the peace treaty had been signed and Mustafa Kemal was leading a successful campaign in Turkey. Ultimately, it was neither the Khilafat movement, nor any other external factor, that determined the course of Turkey's history, but rather the Turks, led by Mustafa Kemal. The last chapters of the Khilafat movement's attempts to influence developments in Turkey in a Pan-Islamic direction were ultimately no more successful than its earlier efforts.

Meanwhile, the Khilafat movement had been weakened by the arrest of its prominent leaders and then torn apart by internal ideological strife and tensions between Muslims and Hindus.[216] Aghast at the abolition of the Sultanate in 1922 and the appointment of a powerless Caliph by the Grand National Assembly in Ankara, two noted personalities, the Aga Khan and Ameer Ali, tried to intervene in favour of the Caliph with İsmet İnönü, Turkey's Prime Minister.[217] This angered Mustafa

[214] Details in M.N. Qureshi, 'The Rise of Atatürk and its Impact on Contemporary Muslim India, the Early Phase'.

[215] Cf. M. Ali Asgar Khan, 'The Turkish Nationalists and the Indian Khilafatists', esp. 42. The letter was dispatched via the Italian diplomatic pouch, but was later intercepted by the British.

[216] See Shaukat Ali's letter in Shan Muhammad, ed., 263–4. The Ulema seemed to be chafing at Gandhi's prominence in 'their' movement, once the Muslims leaders had been arrested, cf. Mushirul Hasan, 'Religion and Politics in India', 11–20.

[217] Their letter was subsequently published in Istanbul in late November 1923 and in *The Times* (London), 14 Dec. 1923.

Kemal to such a degree that, although he replied courteously to the Indians, explaining his own point of view,[218] it seems he felt that his hand was being forced by this intervention; this may indeed have hastened the abolition of the Caliphate in 1924.[219] The move shocked many of India's Muslims who, when they finally believed it, considered it a political blunder.[220] The Khilafat movement was left in confusion, refusing however to renounce or modify its Pan-Islamic conception. This found an expression in the sending of emissaries to Turkey[221] and in moves to secure Muslim control of the Holy Places in Arabia, in 1923, culminating in the dispatch of a delegation, in late 1924,[222] which, however, proceeded only to Jedda, due to the hostilities which had started in Hijaz. Consequently, it confined itself to delivering to the Government of Hijaz a statement on the 'Aims of the Indian Khilafat Committee', of which the two first paragraphs were a suitable conclusion to the Committee's Pan-Islamic involvement:[223]

a. To set up a lawful[224] republican government in the Hejaz which shall be independent internally and whose foreign policy shall be such as to satisfy the Muslim world and meet its views in regard to the complete and absolute independence of the country—an independence free from foreign influence, whether open or concealed.

b. To call a Muslim Conference for the formation of this republic, in which there shall participate delegates from admittedly independent-minded Muslim societies in Muslim lands which are under domination, and representatives of the independent Muslim Governments, and delegates of the Hejaz.

Following Ibn Saʿūd's conquest of Hijaz, the Committee's concern over the Holy Places was superfluous. Yet, even though it announced in 1925 that it was turning its attention

[218] Shafique Ali Khan, 59–60.

[219] Enayat, 54. V. Chirol, 'The Downfall of the Ottoman Khilafat', 235. For a different view, see N.A. Ayyubi, *Kemal Ataturk*, 24.

[220] See e.g. *The Times* 5 Mar. 1924, based on information from New Delhi. Cf. NA, RG 59, 867.404/90, US Consul-General A.W. Weddel's no. 920, to Secretary of State, dated in Calcutta, 27 Mar. 1924. See also Öklem, 29–31.

[221] NA, RG 59, 845.00/403, US High Commissioner Rear Admiral M.L. Bristol's no. 945, to Secretary of State, dated in Istanbul, 8 Oct. 1923.

[222] See the letters of Shaukat Ali, then President of the Central Khilafat Committee, during 1924, in Shan Muhammad, ed., 237–61.

[223] Bamford, 209.

[224] i.e. in accordance with the *sharīʿa*.

to the communal welfare of India's Muslims,[225] that is, shifting from Pan-Islam to nationalism,[226] it interested itself in the all-Muslim congresses in the 1920s[227]—prompting Mustafa Kemal to state, in his lengthy 1927 speech, that those who continued to occupy themselves with the chimera of the Caliphate were misleading the Muslim world.[228] The Khilafat movement petered out in the late 1920s and the early 1930s, despite Muhammad Ali's efforts to revive it in 1925–6.[229] It existed, without much further impact on Indian public opinion, until December 1933, when it held its last annual conference, in Lucknow. Its resolutions,[230] in a true Pan-Islamic spirit, ended as follows, 'To get in touch with Musalman rulers and Muslims of the world so that with unanimity they can carry on propaganda'.

In its heyday, the Khilafat movement had immense prestige among circles inclined towards Pan-Islam not solely in India, but also abroad, chiefly in Turkey and Egypt. In July 1923, ʿAbd al-ʿAzīz Shāwīsh sent letters to it, urging the continuation of its struggle for Pan-Islam and Caliphate.[231] A newspaper in Afghanistan, *al-Mujāhid,* later in the same year, claimed that this movement was promoting the unity of Islam and deserved the support of all Indians in the North West Frontier Province in its effort to unite them with Afghanistan.[232] Still later, in August 1925, when a joint congress of two Islamic organizations in the Dutch East Indies, *al-Islām* and *Sarikat al-Islām,* met at Djokja, it decided to send a delegate to British India, to get in touch with the Central Khilafat Committee, in order to join forces in convening a Caliphate Conference.[233]

[225] May, 206.

[226] Anees, 48.

[227] About which see in the next chapter. Cf. also G. Young, 544, concerning the interest of India's Muslims in these congresses.

[228] Mustafa Kemal's *Nutuk* (1959 edn.), iii (*vesikalar*), 1189–90, doc. 120.

[229] M.S. Jain, 'Mohamed Ali and the Khilafat Committee, 1925–1926', in: Mushirul Hasan, ed., 164–9.

[230] K.K. Aziz, compiler, 336.

[231] FO 371/10110, file 3657/2029, App. to Inter-Departmental Committee on Eastern Unrest, 'Pan-Islamism and the Caliphate, July–March 1924', 3.

[232] FO 371/10396, file 897/21, Intelligence Bureau North West Frontier Province Diary no. 46, 'Secret', for the week ending 20 Dec. 1923, para. 657.

[233] FO 371/11924, file 252/252, British Consul-General J. Crosby's no. 151, 'Confidential', entitled 'Notes on the Native Movement and on the Political

The failure of the Khilafat movement in its Pan-Islamic moves came as a surprise to many. It lacked real political power to impose its will on Great Britain and its allies,[234] but this was only one of the more obvious causes, along with the clash of views among the leaders of what was far from being a monolithic movement. A more inherent weakness of the movement was the widespread ignorance of the precise relations between Mustafa Kemal and the Caliph; those Khilafatists who knew or suspected the truth were reluctant to analyse those relations rationally.[235] At the very time when the Khilafatists were striving to preserve the Caliphate, the Turkish leaders were taking steps to finish it off. This is why its abolition proved to be such a shock and dealt a death-blow to the movement. Even after the event, some of India's Muslims hoped that the Caliphate could be re-established in Istanbul.[236] When it was discovered that the Caliph, exiled in Europe, had but little money at his disposal, it was the Nizam of Hyderabad, although critical of the Khilafat movement (he had banned it in his dominions, on 22 May 1920),[237] who provided him with an annual life pension of £3,600.[238]

The Khilafat movement was never merely a spiritual expression of Pan-Islam, interested in preserving the Caliphate on religious grounds. When examined closely for objectives and organization, it bears definite marks of political Pan-Islam. Its interests, like those of *Khuddām-i Kaʿba,* covered a wide range of Islamic affairs. The Khilafatists were concerned not solely with Ottoman Muslims, but also with those in Iran (India numbered quite a few Shiites), Iraq, Libya, Morocco, and elsewhere.[239] As the single largest Muslim concentration in the world, India's Muslims—and, chiefly, the Pan-Islamists—

Situation in the Netherlands East Indies generally', dated in Batavia, 30 Nov. 1925.

[234] Thus it was not quite 'a Muslim replica of the Crusades', as Buckler, 610, defined it.

[235] Acc. to Moin Shakir, *Khilafat to Partition,* 74, even Muhammad Ali was unaware of the secularist trends in Turkey.

[236] e.g. Mohammad (Maulavie) Barakatullah, *The Khilafet,* publ. in 1924.

[237] For his *firman* in the matter, cf. K.K. Aziz, compiler, 132.

[238] Cash, 72–3.

[239] Prabha Dixit, 'Political Objectives of the Khilafat Movement in India', in: Mushirul Hasan, ed., 45, 50.

constantly felt that they ought to help Muslim groups everywhere.[240] But there was more than this in their political approach. They needed an alternative supportive centre to the British rulers of India—whose former special relationship with the Muslims there had been diminishing—as well as allies in their rivalry with the huge Hindu majority. Such a centre could be provided only by the Sultanate–Caliphate in Istanbul; hence the insistence of India's Pan-Islamists (as expressed by Azad[241] and others) on preserving the Caliph's spiritual and temporal powers.[242]

Upon the abolition of the Caliphate, a significant part of the ideological message of India's Pan-Islamists was flawed, as their link with Turkey's and other Muslims received a rude blow.[243] Organized activities seemed to dwindle or, at best, became infrequent and sporadic. As a movement, Pan-Islam could be considered dormant in India from the late 1920s.[244] Ideologically, it was still topical, at least in prompting plans for a great league of Muslim nations.[245] The best-known exponent of this view was Muhammad Iqbal (1873–1938), the intellectual leader of India's Muslims since the 1920s.[246] One of his poems reads as follows: 'From the banks of the Nile | To the soil of Kashgar | The Muslims should be united | For the protection of their sanctuary'.[247] However, Pan-Islam does not seem to have been central to Iqbal's thinking, for he could not but perceive the growing importance of state nationalism (he is sometimes credited with being the spiritual father of Pakistan).[248] To repeat Gail Minault's thesis, Pan-Islamic sym-

[240] Prabha Dixit, 'Political Objectives ...', in Mushirul Hasan, ed., 53–4.

[241] e.g. ibid. 55–6.

[242] Cf. ibid. 54 ff.

[243] Sadiq, *Turkish Revolution*, 15–16.

[244] M.J. Steiner, *Inside Pan-Arabia*, 49. Sirdar Ikbal Ali Shah, 'Ferments in the World of Islam', 137, referring to 1927.

[245] Sirdar Ikbal Ali Shah, 'Ferments', 133.

[246] On Muhammad Iqbal's thoughts on Pan-Islam and a league of Muslim nations, see Aziz Ahmad, *Studies*, 68–70, 268 ff. J.L. Esposito, 'Muhammad Iqbal and the Islamic State', 188. Stepaniants, 31–2. Shaukat Ali, *Pan-movements*, 219–21.

[247] Transl. by Sharif al-Mujahid, 'Muslim Nationalism', 33 n. 15.

[248] Probably the most steadfast of India's Muslims in his Pan-Islamic commitment was Mushir Hosain Kidwai who, in 1937, published his last book on the subject, *Pan-Islam and Bolshevism*. This is a large, 500-page work, in

bols were increasingly employed to forge a Pan-Indian Muslim constituency.[249]

In retrospect, as the first popular, modernly organized agitation for Pan-Islam among India's Muslims, with notable success in mobilizing them, the Khilafat movement attempted to combine a universalist cause with a nationalist one.[250] Its failure to weld meaningfully these two elements ultimately led to the breakdown of the movement. This development unintentionally served as a script for subsequent rivalries between Pan-Islam and nationalism in India and elsewhere.

which he says, *inter alia*, that the need for an Islamic power persists (p. 17) and that a coalition and federation of Muslim states is imperative (p. 19). However, Pan-Islam meant to him, as it did for Iqbal, a world federation, based on Muslim brotherhood (p. 211).

[249] Minault, *The Khilafat Movement*, 2 ff. and *passim*.

[250] Kushwant Singh, 39.

V. *Between Two World Wars: The Convention Age*

1. *The Post-World War I Period*

The era between the two World Wars or, more precisely, the twenty years between the breakdown of the Khilafat movement in India and the end of World War II, was a low point for Pan-Islam. It had been a 'losing' ideology during World War I, apparently confirming pre-war evaluations of it as a mere utopia;[1] moreover, its failure to recruit Muslim support during this war, the absence of any co-ordinated protest at the harsh treatment of Muslim peoples under the Versailles Peace Treaties, and the abolition of the Caliphate in 1924—all these seemed to signal the death of Pan-Islam,[2] or, at least, its being put out of action.[3] The few who thought that Pan-Islam was still a potent force were, in general, foreign officials and military officers whose duty was to forestall a Pan-Islamic threat,[4] or politicians who considered it from a narrow perspective. For instance, at the end of World War I, a French *député*, de Menzie (1876–1947), argued that a Pan-Islamic union was not noxious *per se*, provided that it was not directed against France; hence France should take the initiative in organizing it.[5] In subsequent years, some officials believed that the Muslims were still awaiting the reinstatement of a Caliphate to reunite them,

[1] See e.g. V.V. Bartol'd, 'Panislamizm', 400–2.

[2] Cf. R.G. Torre, 'Panislamismo, Panarabismo y acción ibérica', 257–64. F. Bellotti, *Arabi contro Ebrei in Terrasanta*, 19–20.

[3] 'Panislamismo' (in *Enciclopedia Universal Illustrada Europeo-Americana*), xli. 795. R. Hartmann, *Die Krisis des Islam*, 33 ff. Elie Salem, 'Nationalism and Islam', 277: 'The fate of Pan-Islam was sealed with the fall of the Ottomans'.

[4] See e.g. Général Bajolle, 'Le Panislamisme et la paix mondiale', 28–36. Cf. above, ch. III, and below, section 3 of the current chapter.

[5] De Menzie, *Rome sans Canossa*, 238–50.

and it was even suggested that the Sultan of French-ruled Morocco would be appropriate for the role.[6]

Everywhere, Pan-Islam seemed at its nadir, with no prominent political leadership to develop it ideologically or organize its followers on a long-term basis. Some Pan-Islamic propaganda went on among pilgrims at Mecca and Medina,[7] but this was apparently little planned and irregular.[8] Moreover, there was no official support for Pan-Islam immediately available. The Ottoman Empire had been partitioned, the leaders of the Committee of Union and Progress were self-exiled and scattered, and the Republic of Turkey, one of the few independent states with a preponderantly Muslim population, had renounced Pan-Islam publicly. Other Muslim independent states, such as Saudi Arabia, Iran, and Afghanistan, were little interested in it at that time. All other Muslim populations were under foreign domination—mostly by the British, French, Dutch, or Italians—and their own politically aware élites were struggling in other directions, most notably for the ending of foreign rule and to follow independence with the establishing of nation states (as demonstrated by the examples of Iraq and Egypt).

State nationalism has obviously competed with other pan-ideologies and pan-movements (such as Pan-Slavism and Pan-Africanism) and has hampered their course.[9] In the case of Pan-Islam, however, such concepts as nationalism, parliamentarianism, and the nation state—all imported from the West—seemed to oppose its basic aims and to postulate that modern states are not built on religion.[10] Worse still, from Pan-Islam's point of view, local grievances and objectives seemed more concrete than a universal ideology,[11] while another

[6] e.g. Robert Bos (a municipal councillor in Paris), in *Le Matin* (Paris), 24 Aug. 1933, repr. in E. Jung, *L'Islam se défend*, 21–4.

[7] For a report on Algerian pilgrims returning from Mecca and Medina with Pan-Islamic notions and spreading them, cf. FO 371/17294, file 5391/84, Consul-General G.P. Churchill's no. 22, to the British Ambassador in Paris, Lord Tyrrell of Avon, dated in Algier, 9 May 1933, enclosing a French circular, dated 16 Feb. 1933.

[8] Although, to use Feduchy's words in *Panislamismo*, 61, meetings there were a sort of 'permanent Panislamic congress'.

[9] Cf. Shaukat Ali, *Pan-movements in the Third World*, 209 ff.

[10] Nallino, 'Panislamismo', 196.

[11] H.A.R. Gibb, *Modern Trends in Islam*, 119–20.

ideology, Pan-Arabism, was starting to compete with it from within.

However, forecasts about Pan-Islam's disappearance proved premature. The outcome of World War I and the contents of the peace treaties served as a rude shock to many Muslims,[12] increasing their concern about the troubles of their co-religionists elsewhere.[13] A spirit of Muslim solidarity, possibly based on enthusiastic romanticism,[14] gradually evolved, instead of the former idea of a united Muslim state (difficult enough to conceive without a caliph at its head),[15] and was to reach its practical follow-up only after World War II. In the inter-war period, Pan-Islam had to come to terms with the existing situation by modifying both its ideology[16] and its activities—sometimes under the impact of leftist politics, more often of right-wing ones.[17] Clearly, a rethinking was necessary, about both objectives and means. This found expression in ideological debate, sporadic activities, and—more strikingly—in the bolstering of Muslim solidarity via comprehensive international gatherings.

2. *The Ideological Debate*

Before World War I and Turkey's abolition of the Caliphate, devotees of political Pan-Islam took it for granted that a worldwide union of Muslims was both necessary and feasible; their writing and preaching concentrated on persuading others of the need for such a union and agreeing among themselves as to the ways of achieving it. In the following generation, disheartened Pan-Islamists who still believed in a union of Muslims[18] sought new answers to the old questions. Considering the failure of the proponents of political Pan-Islam, emphasis was laid, in the inter-war period, on the religious and moral bonds

[12] A. Ghirelli, *El renacimiento musulmán*, 55–6.

[13] Un Africain, *Manuel de politique musulmane*, 32 ff. This book was publ. in 1925.

[14] To use an expression of Torre, 259.

[15] A. Gouilly, *L'Islam devant le monde moderne*, 67.

[16] For the dilemma of choosing between a single Muslim state and a universal, Pan-Islamic one, see Tunaya, *İslâmcılık cereyanı*, 91–3.

[17] Some examples in B. Lewis, 'The Return of Islam', 22.

[18] Cf. T. Reichardt, *Der Islam vor der Toren*, 326–7.

essential for Muslim unity, with less attention being given to political union. Since Islam is both religion and state, *dīn wa-dawla*, and the *umma* is both a religious and a political entity, the idea of political union was never renounced officially, but received less consideration. This was a natural response to the secular nationalism which was becoming increasingly fashionable in many Muslim lands.[19]

None the less, pronouncements in favour of political Pan-Islam recurred from time to time, such as the call for a 'Muslim League of Nations' with a joint army for the defence of the Holy Places, in *al-Mujāhid*, a Muslim magazine distributed in India's North West Frontier Province in 1923,[20] and variously repeated since. Moreover, Islamists, while stressing religion and morals, had serious reservations about the concept of the nation state[21] and, in some of their attacks on it, denounced its exponents for damaging Muslim unity. Some of the latter, indeed, while refraining from attacking Islam directly, were very critical of Pan-Islam which, they argued, was much too broad to be practically feasible. Saṭiʿ al-Ḥuṣrī (1882–1968), a Syrian nationalist, writer, and educator, was one of those who accepted Pan-Islam as a concept of brotherly affection, but rejected its political content.[22] The emergence of Muslim activism in the late 1920s[23] and its gathering momentum as a political force could not but find affirmation (albeit not a central one) in the politics of Pan-Islam. We shall attempt to set out several representative statements of this.

Muḥammad Rashīd Riḍā has already been mentioned as a distinguished disciple of Muḥammad ʿAbduh, who presented a propaganda plan to the Committee of Union and Progress in Istanbul.[24] Editor of the influential Cairene monthly *al-Manār*, Riḍā concerned himself largely with Muslim theology and

[19] See B. Lewis, *The Middle East and the West*, 108.

[20] *al-Mujāhid*, 10 Nov. 1923, reported in FO 371/10396, 897/21, Intelligence Bureau North West Frontier Province Diary, no. 46, for week ending 20 Dec. 1923, 'Secret'.

[21] Cf. J.P. Piscatori, *Islam in a World of Nation-states*, 42.

[22] See Satiʿ al-Husri, 'Muslim Unity and Arab Unity', transl. into English by S.G. Haim, 145–53. For al-Ḥuṣrī's other views on Pan-Islam, see B. Tibi, *Nationalismus in der Dritten Welt am arabischen Beispiel*, 149–66.

[23] W.L. Cleveland, 'The Role of Islam as Political Ideology in the First World War', 85.

[24] See above, ch. II, section 7.

jurisprudence. In so far as he was interested in politics, he (a Syrian by birth and education) favoured Syrian nationalism.[25] However, after the 1922 abolition of the Sultan's powers in Turkey and the limitations imposed on the Caliph's authority, Riḍā wrote in the following year an important Arabic book on the matter. Entitled *The Caliphate or the Greatest Imamate*,[26] it criticized the Turkish decision to abolish the Sultanate as well as the claims of various pretenders to the Caliphate.[27] His main argument, however, was that the Caliphate had been—and ought to be—a combination of spiritual and temporal authority.[28] While the reassertion of this familiar view was not a sign of support for Pan-Islam *per se*, it did start a spate of publications on the issue of the Caliphate, which did touch on Pan-Islam. Certainly, Riḍā's contention[29] that 'the main reason for Great Britain's being the greatest and strongest of all those militating against the Islamic union is the fear of a Muslim renewal and the fruition of Pan-Islam', as well as his avowed preference for a 'Muslim Government, free from Western traditions and laws',[30] are indications of his views. Also, it is probably no mere coincidence that Riḍā chose, five years later, to publish another book, in which he reprinted earlier articles of his own from *al-Manār*, under the title *Islamic Unity and Religious Brotherhood*.[31] In his later years, indeed, Riḍā tended to resist Pan-Arab ideology less and to consider it as a step in the achievement of, for him, the immeasurably more important Islamic unity.[32]

[25] Haim, 'The Abolition of the Caliphate and its Aftermath', 229 ff., for an analysis of Riḍā's work.

[26] Muḥammad Rashīd Riḍā, *al-Khilāfa aw al-imāma al-ʿuẓmā*. This 144-page book was repr. from *al-Manār*, 42–3. An annotated French transl. was publ. by Henri Laoust, in 1938, entitled *Le Califat dans la doctrine de Rašīd Riḍā*.

[27] *al-Khilāfa aw al-imāma al-ʿuẓmā*, 52–7.

[28] See analysis of this central theme by M.H. Kerr, *Islamic Reform*, 153–85. Cf. M. Colombe, 'Islam et nationalisme arabe à la veille de la première guerre mondiale', 92. W. Caskel, 'Western Impact and Islamic Civilization', 339.

[29] *al-Khilāfa aw al-imāma al-ʿuẓmā*, 115–16.

[30] Ibid. 118–19.

[31] Muḥammad Rashīd Riḍā, *Kitāb al-waḥda al-Islāmiyya wa-'l-ukhuwwa al-dīniyya*. This 168-page book is repr. from *al-Manār*, 3, 4 and 5.

[32] Cf. I. Gershoni, *Mitsrayim beyn yiḥūd lě-aḥdūt*, who emphasizes Riḍā's continued opposition to the particularist nationalism of individual states.

Not all Muslims who wrote on the Caliphate in the 1920s advocated the reinstatement and continuation of the office. In 1925, ʿAlī ʿAbd al-Rāziq (1888–1966), a professor at the ancient al-Azhar Academy, a very respected Islamic centre of learning, published a book entitled *Islam and the Foundations of Government*.[33] In it he insisted on the need to distinguish between spiritual and temporal power in Islam and expressed doubts about the worthiness of the Caliphate. He was dismissed from his position and his book was repeatedly attacked in publications by other Ulema.[34]

Perhaps the most original book on the Caliphate published in the 1920s, connecting it to Pan-Islam, was A. Sanhoury's *Le Califat: Son évolution vers une société des nations orientales*, which was printed in Paris in 1926, based on this Egyptian's dissertation at the University of Paris. A voluminous work,[35] the author discussed, in the main, the legal and historical role of the Caliphate. However, since the book was published soon after the abolition of the Caliphate, Sanhoury could hardly not address himself to the issue of Islamic unity too. From the fact that the Ottoman Empire had not comprised all the Muslims in the world but, on the contrary, had also included non-Muslim areas, he deduced that the union of the Muslim world could not be achieved within the parameters of a centralized empire.[36] In his discussion of Pan-Islam,[37] Sanhoury considered its doctrines to have both religious and political bases and that, due to the latter, Pan-Islam was a significant modern movement, whose proponents wished to bring out of Islamic solidarity a political co-operation between

[33] ʿAlī ʿAbd al-Rāziq, *al-Islām wa-uṣūl al-ḥukm*. On its impact, see Enayat, 62–3. Haim, 'Abolition', 235–6. M. Fakhry, 'The Theocratic Idea of the Islamic State in Recent Controversies', 456–9. On the writer, cf. Fatḥī Raḍwān, *Mashhūrūn mansiyyūn*, 97–113.

[34] Haim, 'Abolition', 236–7. Hartmann, 30–1. B. Lewis, *Middle East*, 109. Chejne, 'Pan-Islamism and the Caliphal Controversy', 694–5. A recent criticism of of ʿAbd al-Rāziq's work is Muḥammad Ḍiyā' al-Dīn al-Rīs, *al-Islām wa-'l-Khilāfa fī al-ʿaṣr al-ḥadīth*.

[35] It comprises xvi + 627 pages.

[36] A. Sanhoury, *Le Califat: Son évolution vers une société des nations orientales*, 326. A copy of this book is available in the library of the Middle East Centre, St Antony's College, Oxford.

[37] Ibid. 504–9.

Muslim peoples, to enable them to strive together for their own liberation and eventual union.[38]

In the concluding chapters of his book, Sanhoury argued for an Islamic union, based on an organization headed by a caliph to be elected by a General Assembly of the Caliphate. This was to be made up of delegates from all Muslim communities, as well as minorities in non-Muslim states, sent, proportionately to population figures, to Mecca at Pilgrimage time. A smaller Supreme Council would comprise one or more delegates from each Muslim state, meet several times a year, and be divided into specialized committees. The Caliph would preside over both bodies and thus play an important role, particularly if these institutions were recognized internationally.[39] This would be an intermediate step to instituting an association of all independent Oriental nations, within whose framework Muslim society was to play not a religious, but a political, role. Affiliated to the League of Nations, this association would be headed by the Caliph, again combining spiritual and temporal powers.[40]

While not entirely new (ʿAbd al-Raḥmān al-Kawākibī and Mehmed Murad had already made some of these suggestions),[41] their timing was suggestive, after the foundation of the League of Nations and the abolition of the Sultanate and the Caliphate. Some of the ideas were to be considered by Muslim conventions, to meet soon afterwards. Such views found an echo readily enough. An example is offered by an interview granted, on 15 May 1926, by a Muslim, Maḥmūd Sālim al-ʿArafātī, formerly a judge in Egypt and subsequently the publisher of a journal, *ʿArafāt*.[42] He expressed himself incisively, asserting that the only way for Muslim countries to solve their problems was for them to unite under the leadership of a caliph freely elected by the various Muslim governments. In these United States of Islam, each unit would preserve its

[38] A. Sanhoury, *Le Califat*, 509–13.

[39] Ibid. 574–7.

[40] Ibid. 584–606.

[41] See above, ch. 1. On al-Kawākibī, see also Aḥmad Amīn, *Zuʿamāʾ al-iṣlaḥ*, 267–301.

[42] The interview was recorded by E. Jung, *L'Islam sous le joug (la nouvelle croisade)*, 75–95.

current administration, while the central government would direct defence and foreign policy. The Caliph would supervise the application of the Koranic laws and the Pilgrimage. al-ʿArafātī insisted that the re-establishment of the Caliphate was essential, despite the opposition of the British and the reluctance of the nationalists, Christians, and Jews in Muslim countries[43] (the classification of the nationalists with the infidels was probably not fortuitous).

Perhaps the most sophisticated approach to the ideology of Pan-Islam, in the new situation of nationalism-bound Egypt of the 1920s and the 1930s, was developed by Ḥasan al-Bannā (1906–49). The founder, in 1928, and General Guide of the Muslim Brethren (*Jamʿiyyat al-Ikhwān al-Muslimīn*),[44] al-Bannā was chiefly responsible for formulating the politics of this association. Although these were largely concerned with the domestic affairs of Egypt, and subsequently of the other countries in which branches were established, the idea of a universalist Muslim state was constantly implied, due to the association's focus on Islam. While strongly disapproving of local brands of nationalism, particularly if they were Western-inspired and secularly minded, al-Bannā developed his own version of Pan-Islamic nationalism,[45] insisting that Islam and nationalism were complementary, especially when the latter operated within the parameters of the Islamic faith—since, for the Muslim Brethren, Islam was, of course, both religion and state. His approach aimed at neutralizing local nationalism by considering all areas inhabited by Muslims to be one Islamic fatherland (*waṭan*),[46] if not in one Islamic state, then in an association of Muslim nations (*Hay'at al-umam al-Islāmiyya*). This attitude was paralleled by al-Bannā's striving to play down the significance of differences among Islamic groups and

[43] Ibid. 81–8.

[44] Numerous works about the Muslim Brethren have been published in recent years. Among those in English, see R.P. Mitchell, *The Society of Muslim Brothers*. Abd al-Monein Said Aly and M.W. Wenner, 'Modern Islamic Reform Movements'. I. Gershoni, 'The Emergence of Pan-Nationalism in Egypt: Pan-Islamism and Pan-Arabism in the 1930s', 71 ff. Caskel, 344 ff. E.I.J. Rosenthal, *Islam in the Modern National State*, 116–22.

[45] To use a term of Gershoni, 'Emergence', 70.

[46] Rosenthal, 116–17. Heyworth-Dunne, *Religious and Political Trends in Modern Egypt*, 64.

schools.[47] He even devised a prayer, for the use of his followers, combining the sentiments of Egyptian nationalism and Islamic solidarity.[48] While the idea of political Pan-Islam was less central to his thinking than that of religious Pan-Islam, al-Bannā did recommend the union of Muslim nations around the precepts of the Koran,[49] and he held in high esteem political organization, propaganda, and active involvement.[50]

In the fifth general meeting of the association, held in Cairo on 11 October 1938, al-Bannā addressed himself directly to these issues, by lecturing on 'The Stand of the Muslim Brethren towards Union[51]—Nationalist, Arab, and Islamic':[52]

Islam does not recognize geographical boundaries, nor does it acknowledge racial and blood differences, considering all Muslims one *umma*. It considers the Islamic fatherland as one, despite distances and frontiers. The Muslim Brethren consider this unity as holy and believe in this union, striving for the joint action of all Muslims and the strengthening of the Brotherhood of Islam, declaring that every inch of land inhabited by Muslims is their fatherland . . . The Muslim Brethren do not oppose everyone's working for one's own fatherland. They support Pan-Arabism, which they consider a second step in their advance. They strive for Islamic unity,[53] which they perceive as the complete structure for the general Muslim fatherland . . . they believe that the Caliphate is a symbol of Islamic union and an indication of the bonds between the nations of Islam . . . the Muslim Brethren see the Caliphate and its re-establishment as a top priority, although they believe that preparatory work for this is absolutely needed: cultural, social, and economic co-operation between all the Muslim peoples is necessary, to be followed by alliances, treaties, and the convocation of meetings among these lands . . . Subsequently, an association of Muslim peoples should be set up, which would elect the Imam.

[47] Mitchell, 217.

[48] Transl. into English by Elie Kedourie, in his 'Pan-Arabism and British Policy', 106.

[49] Aly and Wenner, 340.

[50] Gabriel Baer, 'Islam and Politics in Modern Middle Eastern History', 20–1. For the early political views of the Muslim Brethren, see also Zakariyya Sulaymān Bayyūmī, *al-Ikhwān al-Muslimūn wa-'l-jamā'āt al-Islāmiyya fī al-ḥayāt al-siyāsiyya al-Miṣriyya 1928–1948*.

[51] Or 'unity' (Arabic, *waḥda*).

[52] 'Mawqif al-Ikhwān al-Muslimīn min al-waḥda al-qawmiyya wa-'l-'Arabiyya wa-'l-Islāmiyya', repr. in 'Amāra, *al-Islām wa-'l-'Urūba wa-'l-'almāniyya*, 171–4. English transl. of extracts mine.

[53] Or 'union'.

The Muslim Brethren were the best organized and, politically, the most important Muslim organization outside the official Islamic establishment in Egypt. Other Muslim groups, however, were similarly active, several of them connected to the Muslim Brethren by personal ties. Among these were the Young Men's Muslim Association (*Jamʿiyyat al-shubbān al-Muslimīn*), the Islamic Guidance Association (*Jamʿiyyat al-Hidāya al-Islāmiyya*), and the Society of Islamic Brotherhood (*Jamāʿat al-Ukhuwwa al-Islāmiyya*).[54] Established in the late 1920s and the 1930s, all these groups attempted to define their respective positions to Islam, Pan-Islam, and nationalism. All reacted in characteristic Pan-Islamic fashion to crises. For instance, the Young Men's Muslim Association published a ringing declaration, when clashes occurred in East Jerusalem in 1929, that 'Palestine is the Islamic world and the Islamic world is Palestine'![55] It is interesting to note that the author of this declaration, Muḥibb al-Dīn al-Khaṭīb, was simultaneously the secretary of this association and the editor of the Muslim Brethren's newspaper.[56]

A year later, it was apparently the same association[57] which protested energetically about a 1930 French decree reputedly preventing the Berbers of Morocco from following the precepts of Islam.[58] It is noteworthy that the protest was addressed to Muslim sovereigns and peoples, the Ulema of the Holy Places, al-Azhar, and the mosques in Fez, Najaf, and India, as well as to Muslim organizations in Bombay, Delhi, Sumatra, Java, Jogjakarta, Jerusalem, Beirut, and China.[59] This is evidently an appeal to Muslim solidarity with the Berbers. Later, the association established branches in Palestine, which were active in Pan-Islamic agitation during the 1930s.[60]

[54] Gershoni, 'Emergence', 70–1.

[55] Quoted by Cleveland, 'The Role of Islam', 98.

[56] Ibid.

[57] Its protest was signed La Jeunesse Musulmane au Caire.

[58] The text, in French, was repr. by E. Jung, *Les Arabes et l'Islam en face des nouvelles croisades*, 56–62. It is signed by Abdel-Hamid Said, Président Général de l'Association de la Jeunesse Musulmane au Caire.

[59] Ibid. 56–7.

[60] Cf. FO 371/16009, file 753/87, undated memorandum—prob. Dec. 1931—enc. in Sir Arthur Wauchope's, British High Commissioner in Palestine, dispatch dated in Jerusalem, 24 Dec. 1931.

In 1938 the Society of Islamic Brotherhood was set up in Cairo. Its Pan-Islamic character was evident in its objectives, organization, and leadership. It aimed at overcoming doctrinal differences and increasing co-operation between Muslims throughout the world, at all levels, by strengthening the bonds among them. In addition to its Egyptian headquarters and branches, it soon claimed to have opened forty-six branches abroad. A strong emphasis was laid on Islamic and Pan-Islamic propaganda. The directors were Muslims from different Muslim countries, as were the various committees. The president was a well-known Egyptian Pan-Islamist, ʿAbd al-Wahhāb ʿAzzām.[61] Although the society's activities were soon curtailed by World War II, it made its mark, at least in Egypt.

It is not surprising that Egypt was the centre of Pan-Islamic debate in the inter-war years; it fitted well into the country's developed cultural and religious life—without, however, leaving a profound impression. A similar interest in Pan-Islam was evinced, in the period under discussion, among certain groups of India's Muslims, even after the failure of the Khilafat movement. Two of their most notable thinkers took a definite stand in the favour of Pan-Islam, Muhammad Iqbal and Abu Aʿla al-Maududi.

Muhammad Iqbal (1875/6–1938), already mentioned as one of the most significant of India's Muslim thinkers,[62] was convinced by his stay in Europe in 1905–8 that solidarity among Muslim peoples provided the only chance for the survival of Islam.[63] His works, published after his return to India, recommended such solidarity.[64] Even his *Stray Reflections* referred to this, as follows: 'Our solidarity rests on our hold on the religious principle. The moment this hold is loosened we are nowhere. Probably the fate of the Jews will befall us'.[65] Iqbal's travels among Muslim communities outside India, in the 1920s, convinced him that there existed support for Pan-Islam even among the Shiites in Iran, most particularly for

[61] Heyworth-Dunne, 106–7.

[62] A. Schimmel, *Gabriel's Wing*. Schimmel (p. 35), maintains that Iqbal was born in 1876 (other dates, too, have been suggested).

[63] Javid Iqbal, ed., *Stray Reflections*, introduction, p. xviii.

[64] Ibid., p. xxiv.

[65] Ibid. 21.

the idea of a confederation of Islamic states.[66] By 1929, he was calling for its establishment.[67] It was only later that Iqbal moved towards the support of nationalist Muslim entities, in India and elsewhere,[68] arguing then that Muslims ought to concentrate on the struggle for independence in their own respective countries.

Abu A'la al-Maududi (1903–79) was a prolific journalist and editor of several Muslim newspapers in Lahore. An indefatigable writer and speaker, in India and later in Pakistan, for the cause of Islam and Pan-Islam, al-Maududi had many followers.[69] His characteristic approach was expressed in a booklet on *The Patriotic Appeal and the Islamic Union*, first published in Urdu in 1932.[70] In it he attacked patriotism, based on race, territory, language, and culture, recommending in its stead Islamic nationalism, which for him meant the bond forged by the first Muslims. This bond, leading to an Islamic union, ought to supersede all other ties,[71] since these, along with the imitating of the West, were the present danger for Islam. Unlike Iqbal, al-Maududi continued to support political Pan-Islam throughout his life, even advocating a bloc of Muslim countries and a Muslim International Court of Justice to resolve differences between them.[72]

al-Maududi joined Iqbal (in his earlier period) and other Muslim thinkers in India and elsewhere who argued that Pan-Islam, on both the religious and political levels, was essential

[66] Sirdar Ikbal Ali Shah, 'Les États-Unis d'Islam', 146–7. He was a distinguished Indian Muslim who warmly supported this idea, considering Pan-Islam 'the grand factor of union in all Muslim countries'.

[67] Sir Mohammad Iqbal, *The Reconstruction of Religious Thought in Islam*, 159. See also Piscatori, 174. Cf. Hamid Enayat, *Modern Islamic Political Thought*, 59–60, for Iqbal's response to the abolition of the Caliphate.

[68] Javid Iqbal, ed., p. xxi.

[69] On his life and thought, see Muḥammad 'Amāra, *Abū al-A'lā al-Mawdūdī wa-'l-ṣaḥwa al-Islāmiyya*, and Aḥmad Idrīs, *Abū al-A'lā al-Mawdūdī ṣafaḥāt min ḥayātih wa-jihādih*. See also Aziz Ahmad, *Islamic Modernism*, 208–23. F. Abbott, *Islam and Pakistan*, 172–82. For an obituary, see *al-Ahibba* (Lahore), 9/3–4 (July–Dec. 1979), 1.

[70] This work is known to me only in its Arabic transl., entitled *Bayn al-da'wa al-qawmiyya wa-'l-rābiṭa al-Islāmiyya*, a copy of which can be found in the Bobst Library at New York University.

[71] Ibid. 61–2.

[72] S. Abu A'la Maududi, 'The Task before the Muslim Summit', in Maududi's *Unity of the Muslim World*, also mentioned by Piscatori, 174.

for maintaining Islam in its position and making it more powerful and more effective.[73] Support for this idea was offered by some poets, too, such as Sayyid Ali Husayn, an Indian Muslim who published in 1931 a book of poems in Persian, *The War Cry*, supporting Pan-Islam.

3. *Sporadic Pan-Islamic Activity*

Except for international conventions, to be discussed below,[74] Pan-Islamic activity in the inter-war era appears incidental and disorganized. Much of what one knows about it is derived from British and French confidential reports. The following is a representative sample of this activity which, although of limited scope and effect, continued sporadically for several years.

There seems to have been very little Pan-Islamic activity in the Ottoman Empire between its defeat in 1918 and the establishment of the Republic of Turkey in 1923. What there was, existed mostly in Islamist circles and was largely a part of their propaganda for Islam. An example is provided by an association entitled *Teâli-yi İslâm cemiyeti*, or the Society for the Elevation of Islam. Founded in Istanbul on 19 February 1919, it was headed by a number of teachers, most of them in Islamic or Arabic studies.[75] The society opened several branches in Anatolia and set about promoting Islamic education and publications, offering counselling and organizing public sermons. These activities continued in 1920, but petered out afterwards. What distinguished this society from other Islamist groups of that time, however, was its overt commitment to Islamic unity and support for the Caliph's leadership, in a Pan-Islamic tenor, as was evident from its programme[76] and its manifestos.[77]

There were more Pan-Islamic moves, however, outside the vanquished Ottoman Empire, in which military units of the

[73] Maududi's Pan-Islamic teachings will also be discussed below, in ch. VI.

[74] See the next section in our study.

[75] T.Z. Tunaya, *Türkiye'de siyasal partiler*², ii. 382 ff., based on the Ottoman press.

[76] Published in *İkdam* (Istanbul daily) and repr. by Tunaya, ibid. 386–7.

[77] Repr. ibid. 387–97.

victorious Entente Powers were watching developments closely. Abroad, the erstwhile leaders of the Committee of Union and Progress and some of their steadfast emissaries and propagandists had found refuge, at the end of World War I, in Germany (usually in Berlin), Switzerland, and, occasionally, Italy.[78] Hunted by the Entente Powers and shadowed by their secret agents, these refugees could not return immediately to Turkey, first occupied by the Entente armies, then liberated by Mustafa Kemal (no friend of the *ancien régime*). Some support could conceivably be found in the Soviet Union—and a few tried for it—but the new regime there was too busy with consolidating its own power. Hence, Pan-Islamic involvement remained the only practical way for some of these refugees to continue their political activity. Some of them, indeed, mainly the leadership of the Committee of Union and Progress, had prepared themselves for Islamic and Pan-Islamic moves even before the empire's defeat.[79] It is not quite clear what they expected to achieve in real terms; but even nuisance value has its uses.

Although the suspicions of the British and French secret services of a widespread international plot by Pan-Islamists, masterminded in Berlin since the end of the war, seem far-fetched,[80] certain moves were probably undertaken. Involved were Talat, Cemal, and Enver,[81] the defunct Ottoman Empire's top rulers, as well as such veteran Pan-Islamists as ʿAbd al-ʿAzīz Shāwīsh and Ṣāliḥ al-Sharīf. If an intercepted letter from

[78] See the detailed 'Note sur le Panislamisme', prepared by the press service of the French High Commissioner's office in Syria and Lebanon, dated 18 July 1921, in AE, Archives Diplomatiques, Levant 1918–1940, Turquie, Série E, vol. 561, fos. 195–202. For Italy, see FO 371/8967, file 917/85, British Ambassador Sir R. Graham's report to Curzon, 'Confidential', dated in Rome, 17 Jan. 1923: Rome 'has become a centre of Pan-Islamic agitation'.

[79] Details in A.A. Cruickshank, 'The Young Turk Challenge in Postwar Turkey', esp. 16–19.

[80] Particularly, as a not disinterested party, the Greek Minister for Foreign Affairs, Politis, supplied some of the details. See the memorandum, dated Feb. 1920, in AE (as in n.78), fos. 151–2, in which R. de Billy, French Minister to Greece, forwarded it to Millerand, French Minister for Foreign Affairs, with his report no. 37, dated in Athens, Mar. 1920. See also FO 371/5220, file 2233, for this memorandum.

[81] Cf. AE, ibid., esp. fo. 154, a report from the French Chargé d'Affaires in Berlin to the French Minister for Foreign Affairs, dated in Berlin, 21 Apr. 1920—where the Bolsheviks, too, are implicated.

Aleppo to Izmit merely referred to the need of Arabs and Turks to join forces in a Pan-Islamic union and rise together,[82] secret organizations in Istanbul for political Pan-Islamic action appeared more threatening. After all, Istanbul, although under the occupation of the Entente, was still the seat of the Caliphate. An association aptly named Union of Islam, or *İttihad-ı İslam*,[83] was busy sending propaganda to several other Muslim countries, both to buttress their support of the Caliphate and the Ottoman throne and to synchronize their steps for achieving and maintaining their independence.[84] This was paralleled—and possibly complemented—by Pan-Islamic organizations in Europe (centred in Berlin), striving to unite Muslim groups and individuals for joint action;[85] and another in the Soviet Union (located in Moscow), supporting Islam in various countries outside its borders, with the objective of urging them to rise, singly or jointly, against West European domination.[86]

Western suspicions notwithstanding, all these organizations produced only limited results in the years immediately following the end of World War I. Some agents carried Pan-Islamic propaganda into French-ruled North Africa,[87] Italian-ruled Tripolitania,[88] some areas of Asia Minor,[89] and parts of India[90]

[82] French transl. of this letter ibid., fos. 55–6.

[83] Along with a subsidiary, *Dār al-ḥukūma*, or 'the seat of government'.

[84] AE (as in n. 78, above), a secret report from Lieut. Rollin, head of the French Navy's Intelligence Service in Turkey, no. 1490–E–2, 'Secret', dated in Istanbul, 27 Dec. 1919. FO 371/5166, file 2784/262, Weekly Summary of Intelligence Reports, Istanbul, for the week ending on 11 Mar. 1920. Cf. also ibid. for the Intelligence Reports in file 2813/262 and in FO 371/5171, 12473/262.

[85] AE, ibid., vol. 562, fos. 2 ff., note of the French Secret Service, dated 1 Jan. 1922. NA, RG 59, 867.00/866, Samuel Edelman's Near Eastern Intelligence Report no. 624.T.285, dated in Geneva, 31 Mar. 1919.

[86] AE, ibid., vol. 561, fo. 154, dispatch from the French Chargé d'Affaires in Berlin to the French Minister for Foreign Affairs, dated 21 Apr. 1920. Ibid., vol. 562, fos. 65–6, report of Fernand Prevost, French Minister in Iran, no. 26, to Poincarré, dated in Tehran, 11 July 1922.

[87] Ibid., vol. 561, fo. 203, cyphered telegram no. 385, 'Urgent Secret', from Billy to the French Ministry for Foreign Affairs, dated in Athens, 22 Sept. 1921.

[88] Ibid., vol. 562, fos. 72–4, enclosing a copy of an Italian request in this matter, dated in Paris, 3 July 1922.

[89] FO 371/5065, file 13532, transl. of an undated memorandum presented to ʿAbd al-ʿAzīz ibn Saʿūd by a member of the Syrian deputation from Damascus which visited him in the spring of 1920.

[90] FO 371/5166, file 2813, Weekly Summary of Intelligence Reports for the week ending on 18 Mar. 1920, 9–10.

and Central Asia[91]—where it looked for a brief while as if the Emir of Bukhara[92] and the Emir of Afghanistan[93] were engaging in Pan-Islamic schemes of their own. However, with the deaths of Talat, Enver, and Cemal, the abolition of the Caliphate, and the decrease of both German and Soviet support for Pan-Islam, whatever organized activity had existed diminished in scope and tempo.

There was also, it seems, a change in its character. In the first years after the end of World War I, a modest amount of Pan-Islamic activity was carried out by a few groups with no central administration. Those which have already been mentioned as maintaining a semblance of organization had few, if any, relations with each other, except through occasional personal contacts. Most, possibly all, were active locally, rather than internationally. The only one which appears to have had an international scope was a hardly known association named the Pan-Islamic League which, in 1921, seems to have opened branches in Berlin, Geneva, London, and Rome,[94] as well as in Damascus.[95] However, its activity was scant and its life brief. Further information about the activities of other Pan-Islamic organizations, in subsequent years, is usually provided in bits and pieces—with each bit of evidence concerning a one-off move. For example, a 'Committee of United Muslims',[96] possibly inspired by Berlin, met in Scutari, Albania,

[91] FO 371/6630, file 226/226, British Consul-General in Kashgar, Lt. Col. P.T. Etherton's report no. 265/1920, to the Secretary to the Government of India in the Foreign and Political Dept., dated in Kashgar, 20 Oct. 1920.

[92] *The Times*, 12 Feb. 1919, 10, letter from its correspondent in Bombay.

[93] Ibid. Also 'Une ligue musulmane dans l'Asie Centrale', AAT, Archives de la Guerre, 7N2106, Ministère des Colonies, Revue de la presse et des questions musulmanes, Comte rendu analytique, 15 Mar. 1919. Cf. FO 371/8038, files 2073 and 2544, Etherton's report, dated in Kashgar, 3 Apr. 1922, esp. the last chapter.

[94] Much of the information about it is in AE, Archives Diplomatiques, Levant 1918–1940, Turquie, Série E, vol. 561, fo. 204, note by the French Secret Service, dated 10 Oct. 1921. Ibid., vol. 564, fos. 9ff., two Intelligence reports, dated in Berlin, 17 and 25 Oct. 1921. Ibid., vol. 564, fos. 47–8, cyphered telegram from Athens, to the French Minister for War, dated 18 Oct. 1921.

[95] FO 371/6463, file 14239, British Consul in Damascus C.E.S. Palmer's no. 253 P, dated in Damascus, 9 Dec. 1921, lists the members, mostly retired officers, in the Damascus branch.

[96] The FO documents (cited n. 95) call them 'Comité des Musulmans unifiés'.

and, calling itself 'The Voice of the Crescent', started devising plans of propaganda for political Pan-Islam.[97] Equally short-lived was a handbill, in Spanish, by the Asociación Pan-Islamismo in Buenos Aires, expressing its desire to assist Muslims everywhere.[98]

Subsequently, this ephemeral trend became the norm, highlighting the sporadic character of Pan-Islamic activity, which increasingly became a matter of personal initiative, lacking both co-ordination and any follow-up. This was even the case for Pan-Islamic conventions, generally convoked by private persons and without a permanent framework.

Not unexpectedly, our information about Pan-Islamic activities (the conventions apart) becomes more fragmentary, particularly as the interest of the secret services of the powers gradually diminished, even though these services did not desist from watching Pan-Islam.[99] On the other hand, the local Muslim press increasingly demonstrated its interest in Pan-Islam, sometimes being quite articulate in its support for it; an example is the Syrian press in late 1921.[100] This was true, also, of some newspapers in Egypt at the time and later, as well as in Afghanistan and elsewhere. For instance the Afghan *al-Mujāhid*[101] urged all Muslims to unite, emulating Christian devotion during the Crusades, in order to wrest Arabia from non-Muslim control. Similar matters were discussed, in a kindred spirit, by several Pan-Islamic European periodicals, such as the bi-monthly *Echos de l'Islam*, published in Paris in 1924 by the Bureau d'Information Islamique.[102] Some press

[97] FO 371/7558, file 3256/2390, secret report from Durazzo, supplied by the Italian Legation, dated in Scutari, 10 Feb. 1922.

[98] This is undated, but seems to be from 1923. See AE, Archives Diplomatiques, Levant 1918–1940, Turquie, Série E, vol. 563, fo. 12, report by the French Chargé d'Affaires in Buenos Aires no. 160, dated 15 Sept. 1923. The manifesto is enclosed in fos. 13–15.

[99] As late as mid-1938, the French Resident General in Morocco reported an intensification of Pan-Islamic activity there, which he attributed to the German authorities in Berlin. See a copy of his report in FO 371/22004, file 2269/2269, enclosed in Bentinck's dispatch.

[100] FO 371/6458, file 121, a dispatch from C.E.S. Palmer, dated in Damascus, 29 Oct. 1921, reporting on *Fatāt al-ʿArab* and other Arabic newspapers.

[101] Of 28 Sept. 1923, summarized in FO 371/9288, file 9685/153, in the North West Frontier Province Intelligence Bureau Diary, no. 42, 'Secret', for the week ending 15 Nov. 1923.

[102] Acc. to S.M. Zwemer, 'Present-day Journalism in the World of Islam', 145.

organs, however, closed down, while others shifted their interests to other matters.

The sporadic activity that continued was expressed in infrequent Pan-Islamic agitation in Sudan, reportedly by the Sanūsiyya, in 1921;[103] Iran in 1923;[104] Palestine in the same year;[105] and Morocco in 1925.[106] Speeches extolled it, even at official functions in Ankara in 1923 (at a dinner party offered by the Afghan Minister there, Sultan Ahmad Khan, in honour of Sayyid Shaykh Aḥmad Sanūsī).[107] Letters of all sorts were exchanged, some of which have survived, suggesting Pan-Islamic schemes. A typical one, sent from Berlin to Damascus in 1921, was authored by the Pan-Islamic League mentioned above, calling itself in Arabic *Jamʿiyya al-Islāmiyya.*[108] The letter[109] assumed that the Western powers were stepping up their campaign to enslave the entire Orient, including all Muslims. It appealed to the world's Muslims to unite for survival and organize in their own interests, letting each nation administer itself independently. Theirs should be a clandestine association and ought to set up its own army (for guerrilla warfare) and establish contacts with other revolutionary organizations.

Yet another opportunity for Pan-Islamic activity, chiefly propaganda, continued to be the Pilgrimage to Mecca and Medina, which served this and other objectives as well.[110] However, all these activities in political Pan-Islam dwindled

[103] FO 371/7746, file 3013/264, enclosing a copy of the Sudan Monthly Intelligence Report for Nov. 1921, dated in Khartoum, 11 Jan. 1922. See esp. 2–4.

[104] FO 371/9046, file 2944/2944, British Minister to Iran Percy Loraine's no. 54, 'Secret', dated in Tehran, 29 Jan. 1923.

[105] Two contradictory reports about Pan-Islamic agitation: FO 371/8999, file 10769/206, H. Young's no. 42908/1923, to Clayton, dated in London, 14 Sept. 1923; and FO 371/8999, file 10441/206, G.F. Clayton's letter, dated in Jerusalem, 5 Oct. 1923.

[106] *The Times*, 5 Jan, 1925, a dispatch from Harris, its correspondent in Morocco.

[107] FO 371/9141, file 1135/1135, Acting British High Commissioner in Istanbul Neville Henderson's no. 50, dated in Istanbul, 22 Jan. 1923.

[108] Or *Jamʿiyyat al-Islāmiyya*. One would have expected *al-Jamʿiyya al-Islāmiyya*.

[109] The original, intercepted, has been preserved in FO 371/6463, file 12485, enclosed in C.E.S. Palmer's no. 192 P, dated in Damascus, 21 Oct. 1921 (with an abbreviated English transl.).

[110] Cf. Nagazumi Akira, 'The Abortive Uprisings of the Indonesian Communist Party and its Influence on the Pilgrims to Mecca', 1–2.

when the conventions became a centre of public attention; from then on, practically all individual moves were connected with those conventions, in one way or another. Exceptions were generally on the lunatic fringe, such as the case of Sir Henri Wilhelm August Detering (1866–1939). This Dutch-born first managing director of the Royal Dutch/Shell consortium of oil companies was, in some respects, an eccentric. In August 1928, he attempted (unsuccessfully, as it turned out) to persuade the British Foreign Office of his political scheme for an anti-Bolshevik union of Muslim nations. With the oil funds at his disposal, he planned to persuade Turkey, the Soviet republics in the Caucasus, Iran, and Afghanistan to join in such a union, in both political (anti-Soviet) and economic (oil revenues) interests.[111] As far as one knows, Detering's project did not advance beyond the planning stage, but it remains, nevertheless, an interesting instance of a rare individual initiative by a non-Muslim to promote the cause of political Pan-Islam.

4. *Pan-Islamic Conventions*

The international convention was the single most striking phenomenon of Pan-Islamic activity in the inter-war era. Few of the associations and individuals attempting to promote Pan-Islam in those years refined their thinking into an ideology of political Pan-Islam or tried to carry it over into a serious consideration of organization, ways, and means. The Pan-Islamic conventions grappled with these lacunae, to a degree. While the conventions were less than a complete success due to the obstacles they faced,[112] they offered a substitute for joint action, in the absence of a Caliph.[113] They caused some excitement in government circles in Europe[114] and this, too, may have encouraged Pan-Islamists to reconvene.[115]

While the organizational model of these conventions was

[111] FO 371/13044, file 3939/3939, Sir R.C. Lindsay's memorandum, dated 3 Aug. 1928, and G.R. Clerk's private letter, to L. Oliphant at the Foreign Office, dated Istanbul, 21 Aug. 1928.

[112] For which see M. Guidi, 'Islām e Arabismo', 16–17.

[113] Aga Khan and Zaki Ali, *L'Europe et l'Islam*, 19–20.

[114] Cf. E. Jung, *Le Réveil de l'Islam et des Arabes*, 15–16.

[115] The most useful book on these conventions is Martin Kramer's *Islam Assembled*. My study will focus solely on their Pan-Islamic context.

European, their inspiration was largely local. Several proposals were advanced before World War I, of which Gasprinskiy's has been discussed above.[116] During the war, several Muslim conventions met, for example, in Stockholm in October 1917.[117] After the war, several of the leaders of the defunct Committee of Union and Progress toyed with the idea, as we have explained;[118] their activity might well have had a Soviet connection.[119] In 1919, the Egyptian daily *al-Ahrām*[120] advocated the convocation in Egypt of a congress of Muslims from Asia and Africa, to support the Caliph and Turkey.[121] These and other projects produced no immediate practical results. Even the tactical moves, sponsored in Turkey by Mustafa Kemal, to establish in 1919 a Pan-Islamic association,[122] were chiefly intended to attract local and foreign Muslim support for the Ottoman Caliphate—and even more so for Turkey's nationalist struggle. A congress in Sivas, in Eastern Anatolia, in February 1921, in which Arab delegates, too, were present,[123] had the same goals and remained without any follow-up, due to the Kemalist measures against the Sultan–Caliph, which soon followed, and the subsequently intensified secularization of the newly established Republic of Turkey. Nor did reports about a projected convention in Cairo in late 1922,[124] or a project by ʿAbd al-ʿAzīz Shāwīsh in 1923, for a 'grand conference of Muslims ... to co-ordinate political activities of Muslims',[125] fare any better, for much the same reasons.

[116] Ibid., ch. 4. Zwemer, *Across the World of Islam*, 306–8. See also above, ch. III, section 3.

[117] PA, Orientalia Generalia, 9, vol. 8, German Ambassador Johann-Heinrich Bernstorff's cable no. 1246, to the Ministry for Foreign Affairs, dated in Istanbul, 10 Oct. 1917; German Under Secretary for Foreign Affairs von dem Busβche-Haddenhausen's no. 697, to the German Ambassador in Stockholm, dated in Berlin, 30 Oct. 1917.

[118] See above, section 3.

[119] Kramer, *Islam Assembled*, 69–72.

[120] 20 Sept. 1919.

[121] See also AAT, Archives de la Guerre, 7N2106, Ministère des Colonies, Revue de la presse et des questions musulmanes, Comte rendu analytique, 31 Oct. 1911, 10–11.

[122] Kramer, *Islam Assembled*, 73 ff.

[123] B. Lewis, *Middle East*, 108.

[124] FO 371/8967, file 520/85, confidential Intelligence Report, dated 10 Jan. 1923.

[125] FO 371/10110, file 3657/2029, App. to the Inter-Departmental Committee on Eastern Unrest's 'Pan-Islamism and the Caliphate, July 1923–March 1924', 3.

Subsequent years, however, witnessed a number of conventions with Pan-Islamic contents, of which the most noteworthy were those in Mecca in 1924; in Cairo and in Mecca in 1926; in Jerusalem in 1931; and in Geneva in 1935.

(a) Mecca, 1924

Ḥusayn of Mecca (1853–1931), who claimed to be a *sharīf*, or descendant of the Prophet Muḥammad (or, at least, of his tribe),[126] had joined with his family in 1916 the British war effort against the Ottoman armies. As a reward, his sons were to rule Iraq and Transjordan, both under British Mandate. Ḥusayn himself was proclaimed Caliph in Shuna (Transjordan), on 5 March 1924, by his close followers, soon after the Ottoman Caliphate had been abolished by the Grand National Assembly in Ankara.[127] However, Ḥusayn needed more than the acclamation of his entourage, of course. Following certain preparations, a mini-convention inspired by him met in Mecca in July 1924, during the Pilgrimage season, to set about legitimizing his claims to the Caliphate.[128] The convention, at which Arabs predominated, agreed on a charter which specified that Muslim unity was its main goal, but emphasized the pre-eminence of Arab unity as the basis for a future union of all Islamic countries. No agreement was reached on investing Ḥusayn with the Caliphate, so that the main achievement of the convention was declaratory only, with Muslim unity as a secondary objective. Even these modest results were nullified when Ḥusayn's rival, ʿAbd al-ʿAzīz Ibn Saʿūd (*c*.1880–1953), conquered Mecca in October 1924 and thus laid another cornerstone to the Kingdom of Saudi Arabia. Ḥusayn himself abdicated the Caliphate[129] and fled to Cyprus, a non-Muslim country, thus practically forfeiting whatever claims he might still have had to the Caliphate.

[126] Haim, 'Abolition', 225.

[127] PA, Politik, Abt. III, Türkei, Politik 16, vol. 1, Ayétbey Dibokhova, of the Royal Hashimite Legation in Rome, to Constantin von Neurath, German Ambassador in Rome, dated in Rome, 15 Mar. 1924. For Ḥusayn's own proclamation, in Arabic, on assuming the Caliphate, see *al-Jazīra* (Jaffa), 16 Mar. 1924.

[128] Kramer, *Islam Assembled*, 84–5. Baer, 19. Chejne, 18.

[129] Haim, 'Abolition', 225.

(b) Cairo, 1926

After two years of preparation, a so-called 'Caliphate congress' was convoked in May 1926 by Egyptian Ulema, headed by the highest Muslim dignitaries in the country, including leading scholars from al-Azhar. Unofficially, Egypt's King Fu'ād (reigned 1923–36) was behind the convention, since he was widely thought to consider himself as a most suitable candidate for caliph and propaganda in Egypt served his ends.[130] The controversy which preceded the inauguration of the convention compelled the organizers to retreat from their original objective of electing a new caliph[131] instead of the one demoted and exiled from Turkey (the latter was one of those agitating against convoking the convention).[132] None of the 39 men attending was delegated by his own government; they were there on behalf of various institutions and organizations, or privately (nine of them had no mandates, but were just invitees). Most were from Egypt and Palestine, although a few came from Tripolitania (including an important figure, Idrīs al-Sanūsī), Tunisia, Morocco, Hijaz, Yemen, Iraq, Poland (where there was lively Islamic activity),[133] the Dutch Indies, the Sultanate of Johur, and South Africa.[134] Conspicuous by their absence were such important Islamic centres as India, Iran, Afghanistan, Turkey, and Algeria.[135] While most of the convention's time was taken up with procedural matters, a committee report on the Caliphate was debated and approved.[136]

[130] Elie Kedourie, 'Egypt and the Caliphate, 1915–52', 182–98. Haim, 'Abolition', 241 ff. Kemal A. Faruki, 'Approaches to Muslim Unity', 29. A.C. Eccel, *Egypt, Islam and Social Change*, 295 ff., 451.

[131] For these moves, see A. Sékaly, *Le Congrès du Khalifat (le Caire, 13–19 mai 1926) et le congrès du monde musulman (la Mekke, 7 juin–5 juillet 1926)*, 4 ff. Kramer, *Islam Assembled*, 86 ff.

[132] AE, Archives Diplomatiques, Levant 1918–1940, Turquie, Série E, vol. 563, fo. 131, Intelligence report, dated in Annemasse, 13 Mar. 1926.

[133] Ibid., vol. 570, fos. 129–31, French Ambassador's no. 9 to the French Minister for Foreign Affairs, dated in Warsaw, 8 Jan. 1926.

[134] Acc. to the minutes of the convention e.g. in Sékaly, 46–8.

[135] Cf. AE, Archives Diplomatiques, Levant 1918–1940, Turquie, Série E, vol. 571, fos. 95–115, French Minister to Egypt Henri Gaillard to the French Minister for Foreign Affairs, dated in Cairo, 25 May 1926 (including the proceedings of the convention).

[136] The minutes of the convention were published in the Arabic press and soon afterwards transl. into French by Sékaly, 29–122. On the convention itself, see Kramer, *Islam Assembled*, 100–5. L. Massignon, 'L'Entente

There was a widespread reluctance to make definitive commitments about the Caliphate, partly attributable to the fact that some of the delegates were soon to go to yet another 'Caliphate Congress' in Mecca.[137] After having discussed the definition of the Caliphate, the need for its continued existence, and the ways of nominating the Caliph,[138] it concluded that investing a new one was necessary but unfeasible in the then prevailing situation and that the convention's bodies, along with branches to be established abroad, ought to continue preparing for this eventuality.[139] This was an elegant way of conceding defeat on the political level, whilst paying lip-service to Muslim unity—it is briefly mentioned several times[140] but no specific method of achieving it is suggested. The absence of any sequel to the convention (despite the decision to reconvene) reinforced the absence of any substantial commitment to political Pan-Islam.

(c) Mecca, 1926

The convention in Mecca, in June–July 1926, was called by Ibn Saʿūd, ostensibly for consultations about the Pilgrimage and related matters, but really in order to obtain international Muslim legitimation of his conquest of the Holy Places and his rule over them. Controversy preceded the convention and continued during its meetings.[141] Nevertheless, an effort was made to give it a semblance of Muslim solidarity. The attendance was larger than the Cairo convention several weeks earlier, numbering about seventy delegates, some of whom arrived only near the convention's end. They were also much

islamique internationale et les deux congrès musulmans de 1926', 482–3. A.J. Toynbee, ed., *Survey of International Affairs*, i. 89–91. Hartmann, 35. Baer, 19–20. A.M. Goichon, 'Le Panislamisme d'hier et d'aujourd'hui', 29. Chejne, 692–4.

137 NA, RG 59, 883.404/6, J.M. Howell's no. 826, to Secretary of State, dated in Cairo, 21 May 1926.

138 Sékaly, 74–7.

139 Ibid. 103–9, for a French transl. of the report of the convention committee on the Caliphate; and Toynbee, ed., *Survey*, i. 578–81, for its English transl.

140 Sékaly, 39, 110, 114, 117, 121.

141 For the convention, see Sékaly, 11–25 (with minutes, pp. 123–219). Toynbee, ed., *Survey*, i. 308–19. FO 371/11433, file 4677/20, British Acting Consul in Jedda S.R. Jordan's no. 87, to Sir Austen Chamberlain, dated in Jedda, 15 July 1926, and enclosures. Kramer, *Islam Assembled*, 106 ff. Massignon, 'L'Entente', 483–5. Hartmann, 35–6. Baer, 20. Goichon, 29.

more varied in their make-up, including delegates from India, Turkey, Egypt, Afghanistan, and the Soviet Union (the most significant absentee was Iran;[142] nor were North Africa and China represented). Moreover, many of those present had been officially sent by their own governments. Among the most articulate delegates were Muḥammad Rashīd Riḍā, the Egyptian thinker and journalist;[143] Muhammad Ali and Shaukat Ali, the brothers who had distinguished themselves as Pan-Islamists in the Khilafat movement and other organizations in India;[144] and Amīn al-Ḥusaynī (1897–1974), Mufti of Jerusalem and President of the Supreme Muslim Council in Palestine.[145] With these and similarly minded participants, the convention—even if it had been meant to focus on religious affairs[146]—rapidly turned to the politics of Islam and Pan-Islam. Muhammad Ali wanted the convention to decide on Pan-Islamic guarantees for the independence of Arabia,[147] while Riḍā proposed an Islamic pact, by which Muslim governments would refer to the convention in Mecca for arbitration (thus leading to the establishment of a Pan-Islamic confederation).[148] Although both proposals were voted down, the very fact that they had been submitted and discussed was evidence of a trend towards political Pan-Islam. On the other hand, the convention's failure to take a joint stand on these and many other issues was indicative of the disunity which prevented Pan-Islam, at that time, from becoming a truly significant factor in Islamic politics. Neither of the conventions, in Mecca and Cairo, in 1926 had a sequel. King Ibn Saʿūd, disappointed at the failure of the Mecca convention to follow his guidelines, initiated no further conventions (certain rumours notwithstanding),[149] but adopted a more personal policy instead.

[142] Acc. to Iran's Prime Minister, what was needed was 'a general assembly of Muslims to regulate the holy shrines'. See Kramer, *Islam Assembled*, 11–12.

[143] See above, ch. v, section 2.

[144] See above, ch. iv.

[145] See below in the next section.

[146] See Ibn Saʿūd's speech in Sékaly, 128–31.

[147] Ibid. 163–5.

[148] Kramer, *Islam Assembled*, 115.

[149] AE, Archives Diplomatiques, Levant 1918–1940, Turquie, Série E, vol. 570, fos. 209–10, report of the Résidence Générale, dated in Rabat, Mar. 1928.

Regarding political Pan-Islam, he spoke repeatedly on the great need for Muslim unity, for instance in May 1930, and in the same year sought closer relations with the King of Afghanistan, the Shah of Iran, and leading Muslims in India (reportedly, to enlist their support in his claims to proclaim himself Caliph).[150]

(d) *Jerusalem, 1931*

Palestinian Muslims had been involved in the conventions of the 1920s, in an attempt to gain Muslim support in their struggle against the Jewish National Home in Palestine and its protector, the authorities of the British Mandate. In 1918, Amīn al-Ḥusaynī had convened in Jerusalem a 'general Muslim congress' which, however, was attended only by Palestinians and Arabs from neighbouring lands.[151] This served as a general rehearsal for the 1931 convention,[152] a much more comprehensive affair, in which its convener and president, Amīn al-Ḥusaynī, sought both to improve his own political standing and bargaining power[153] and to enlist external Muslim support.[154] The invitations specified three topics of debate: the conditions of Muslims, the holy Islamic shrines in Palestine, and other issues of interest to all Muslims.[155] Evidently, the last of these could refer to almost anything. About 133 delegates attended, more than at the previous conventions.[156] Except for Yemen's delegate, all were unofficial and several were exiles, coming from more than twenty countries, with

[150] FO 371/14481, file 3042/3042, Sir A. Ryan's dispatch no. 117, to the Secretary of State for Foreign Affairs, dated in Jedda, 18 May 1930. FO 371/14481, file 4318/3042, India Office's secret report no. P. 4991/30, dated 26 June 1930. The last suggestion is based on an interview with Maulvi Ismail Ghaznavi of Amritsar, allegedly an agent of Ibn Saʿūd.

[151] Kramer, *Islam Assembled*, 125.

[152] For a detailed survey of which see ʿĀdil Ḥasan Ghunaym, 'al-Mu'tamar al-Islāmī al-ʿāmm'. NA, RG 59, 867 n. 404/32, US Consul-General P. Knabenshue's no. 664, to Secretary of State, dated in Jerusalem, 17 Dec. 1931.

[153] IO, L/P&S/10/1314, file P.Z. 1449/1931: Panislamism, esp. 127–34, 'The Pan-Islamic Movement'.

[154] Voll, 270. Michael Assaf, 'Die muselmanische Konferenz in Jerusalem', 34–43.

[155] *al-Manār*, Feb. 1932, 117–18, also quoted by Thomas Mayer, 'Egypt and the General Islamic Conference of Jerusalem in 1931', 312.

[156] Acc. to the computation of U.M. Kupferschmidt, 'The General Muslim Congress of 1931 in Jerusalem', 141.

Palestine, India, and Yugoslavia especially well represented. Absent were Turkey, the Soviet Union, Afghanistan, and several Balkan states. More significantly, among the participants were some of those who had attended earlier conventions, including Muḥammad Rashīd Riḍā and Shaukat Ali (whose brother, Muhammad Ali, had died in the mean time). Among those who came for the first time were Muhammad Iqbal, from India, Ḍiyā' al-Dīn Ṭabāṭabā'ī, a former Prime Minister of Iran, and Shaykh Muḥammad al-Ḥusayn al-Kāshif al-Ghiṭā' (1877/8–1954), a well-known Shiite cleric from Iran, whose participation symbolized an all-Muslim effort to bridge over the Sunnite–Shiite controversy.[157] The convention itself was beset by disputes, regarding both its rules of procedure[158] and more substantive matters.[159] Shaukat Ali, who fought for the revival of Pan-Islam, wished to set up a consultative Caliphate body and establish a Muslim university in Jerusalem[160]—but the convention's majority (overriding his arguments) did not even debate the Caliphate issue[161] and insisted—ignoring Shaukat Ali's demands—that the language of instruction in a Muslim university ought to be Arabic only. The debates became more acrimonious and the convention more militant, inveighing against Zionist aspirations in Palestine, the British administration there, and colonialism in general. This occurred despite Amīn al-Ḥusaynī's earlier pledge to the British High Commissioner of Palestine that the convention would deal solely with religious matters. Many of the speeches and proposals were supportive of Muslim unity, referred to daily, either in general terms,[162] or in concrete

[157] On the significance of this *rapprochement*, cf. Enayat, 43.

[158] For which see FO 371/16009, file 87/87, dated 5 Jan. 1932, enc.

[159] For the convention and its debates see, in addition to FO files, A. Nielsen, 'The International Islamic Conference at Jerusalem', 340–54. Kupferschmidt, 142 ff. Kramer, *Islam Assembled*, 133 ff. Thomas Mayer, 318 ff. H.A.R. Gibb, 'The Islamic Congress at Jerusalem in 1931', 99 ff. Baer, 20. Goichon, 29–30.

[160] Shaukat Ali's interview with Ḥāfiẓ 'Afīfī, Egyptian Minister to London, reported in FO 371/15282, file 5495/1205, notes of Sir G.W. Rendel at the Foreign Office, dated 4 Nov. 1931. See also FO 371/16009, file 753/87, undated [prob. Dec. 1931] memorandum, entitled 'The Pan-Islamic Movement', enc. in Sir A. Wauchope's dispatch, dated in Jerusalem, 24 Dec. 1931.

[161] M. Ki., 228. Nielsen, 351. For the controversies, Nielsen, 346 ff.

[162] Examples in Kupferschmidt, 143, 148.

proposals (such as the drawing up of a statistical survey of the entire Muslim world),[163] or in the final resolutions (to establish a Muslim university in Jerusalem and an agricultural bank, both assisted financially by Muslims everywhere).[164] Moreover, the setting up of an organization, with a secretariat and committees, with the aim of convoking future conventions (these bodies continued to function for about five years, although the conventions failed to materialize, chiefly because of lack of funds)[165]—indicates that this was the convention most oriented to political Pan-Islam. The opening of branches in Palestine, Transjordan, and Syria,[166] and the personal relations established at the convention, also proved of importance to Pan-Islam, since some of the participants from Egypt, Syria, and Lebanon were to hold key political positions in their own countries in later years.[167] Pan-Islamic propaganda on the political level continued unabated in the 1930s;[168] and the Palestine issue was increasingly referred to in Pan-Islamic speeches and publications elsewhere, too.

(e) Geneva, 1935

Since the late nineteenth century, Western Europe had been the natural ground for the political activity of numerous Muslims. Driven away from their countries or self-exiled, these were busily involved in promoting the ideology of their choice. The somewhat earlier European moves of al-Afghānī, and of other Pan-Islamists soon after the end of World War I, have already been mentioned. These continued (albeit with a low profile) during the 1930s, in Switzerland and Germany, chiefly because the Muslims in Europe felt that their brethren in foreign-dominated lands lacked the freedom necessary for political activism. Even the Jerusalem convention of 1931,

[163] Examples in Kupferschmidt, 149.

[164] Ibid. 144–5. For the resolutions, cf. *Muqarrāt al-mu'tamar al-Islāmī al-ʿāmm fī dawratih al-ūlā*. For a French transl., see Jung, *Réveil*, 99–107.

[165] For an evaluation of the recommendations and the resolutions, see Gibb, 'Islamic Congress', 105 ff.

[166] Kramer, *Islam Assembled*, 139–40.

[167] The long-term impact was more significant than Z.R., 'Pan-Islam', 313, estimates.

[168] For examples, see FO 371/18957, file 1514/154, enc., Criminal Investigation dept.'s report no. 50/G/S, 'Secret', dated in Jerusalem, 13 Feb. 1935, entitled 'Pan-Islamic agitation'.

despite its decidedly political character, had no follow-up, while the second general congress of Oriental women—held in Tehran, in November–December 1932—shunned both politics and Pan-Islam.[169] Not surprisingly, various Pan-Islamic projects in Western Europe of the 1930s remained in the planning stage only. Such was the November 1931 announcement of ʿAbbās Ḥilmī, ex-Khedive of Egypt, who was living then in Geneva, of the establishment of an 'Alliance musulmane internationale', which would hold periodic all-Muslim conventions.[170] No practical results ensued. However, some progress was made due to the tireless efforts of an Egyptian journalist and lawyer, Maḥmūd Sālim al-ʿArafātī,[171] then a resident of Paris, who had participated in the Jerusalem convention of 1931. An enthusiast for political co-ordination,[172] his *idée fixe* was the convocation of a congress of Muslim intellectuals living in Europe.[173] Sālim had the good fortune to enlist the enthusiastic support of Shakīb Arslān,[174] who was then living in Geneva. A prolific historian[175] and a convinced Pan-Islamist, who had attended Kaiser Wilhelm II's speech in Damascus[176] and later opposed in the late Ottoman Empire a break between Turks and Arabs,[177] Arslān had already tried, in

[169] H. Massé, 'Le Deuxième Congrès musulman général des femmes d'Orient à Téhéran (novembre–décembre 1932)'. For other conventions at that time, see Goichon, 30.

[170] Details in Kramer, *Islam Assembled*, 138–9.

[171] The French Resident General in Tunisia called him by his full name, Maḥmūd Sālim al-ʿArafātī—see AE, Archives Diplomatiques, Levant 1918–1940, Turquie, Série E, vol. 690, fos. 229–30, French Resident General's no. 733, to the French Minister for Foreign Affairs, dated in Tunis, 1 June 1933.

[172] In 1916, signing Mahmoud Ben Salem El Arafati, he had written a 68-page book, entitled *La Coordination des forces alliées*, warmly advocating co-ordination of the Entente Powers' plans for the after-war years and especially their future policies towards Muslims.

[173] For his various moves, chiefly in 1933, cf. AE, Archives Diplomatiques, Levant 1918–1940, Turquie, Série E, vol. 690, fo. 215, of 9 May 1933, and fo. 218, of 11 May 1933; and vol. 692, fos. 23–5. See also Kramer, *Islam Assembled*, 143–4. I have been unable to consult Mahmoud Salem's (*sic*) work, *Le Congrès islamo-européen de Genève*, mentioned by Kramer.

[174] See also above, ch. II.

[175] See Sāmī al-Dahhān, *al-Amīr Shakīb Arslān ḥayātuh wa-āthāruh*. This work is more concerned with Arslān's literary than with his political activity. The same is true of Dahhān's *Muḥāḍarāt ʿan al-amīr Shakīb Arslān*.

[176] Cleveland, *Islam against the West*, 141.

[177] J. Bessis, 'Chekib Arslan et les mouvements nationalistes au Maghreb', 470.

1919, to organize Turks, Arabs, Kurds, and Muslims from the Caucasus to oppose the subjugation of Muslim lands by non-Muslims.[178] Moreover, he was a veteran of Arab and Islamic conventions[179] and a great asset in organizing and presiding over the one planned for Muslims in Europe. Just before the convention met in Geneva, in September 1935, Arslān's close associate, Ihsan el-Djabri, published in his own monthly, *La Nation Arabe*, a detailed article[180] about the convention's objectives. As he put it, these were 'to establish a social, economic and religious bond between the Muslims in the West and the Muslim world'.[181] However, he went on candidly:

> If the Muslims organize themselves, all their difficulties would disappear immediately. Those states having Muslim populations could not permit themselves to treat their Muslim subjects as they do nowadays. It is necessary that the initiatives of solidarity and of future convocations of the convention develop in view of resolving jointly the difficulties which the Muslims are facing in certain European countries.[182]

Due, however, to Arslān's pledge to the Swiss authorities, the invitations to the convention specified only the following aims: to develop ties between the Muslims of Europe and to foster among them a spirit of co-operation, Islamic virtue, and general culture (mention was made, also, of Muslim protection of the Holy Places in Palestine).[183] There were about 66 delegates,[184] from Yugoslavia, Poland, Hungary, Germany, Austria, the Netherlands, Switzerland, Romania, England, Italy, and France—with additional invitees from India, Afghanistan, Iraq, Iran, Syria, Lebanon, Egypt, Tunisia, and

[178] NA, RG 59, 867.00/866, Edelman's no. 624.T.285, dated in Geneva, 31 Mar. 1919.
[179] Arslān, *Sīra dhātiyya*, 108–12.
[180] Ihsan el-Djabri, 'Le Congrès islamique d'Europe', 369–74.
[181] Ibid. 369.
[182] Ibid. 373. Cf. ibid. 379–85, text of el-Djabri's own speech at the Geneva convention.
[183] AE, Archives Diplomatiques, Levant 1918–1940, Turquie, Série E, vol. 692, fos. 25–7, App. 4, report on 'Le Congrès musulman de Genève'; cf. fos. 28–38.
[184] Ibid. Acc. to el-Djabri, 376, seventy delegates. Acc. to. E. Lévi-Provençal, 'L'Emir Shakib Arslan (1869–1946)', 14, more than sixty.

Algeria.[185] As with previous Muslim congresses, moderate pleas for Islamic religious solidarity were soon superseded (Arslān's efforts notwithstanding) by impassioned political speeches on behalf of the Muslims in the Soviet Union, Palestine, and North Africa, as well as verbal attacks on colonialism (British, Italian, and French) and Zionism.[186] There was a strong current of Pan-Islamic pleading in the political speeches which, although they did not lead to further conventions in Geneva, impressed commentators that Pan-Islam was a political element to be considered.[187] They did not fail to notice, however, signs of dissension, for instance the non-attendance of the Muslims of Palestine and their criticism of the convention.[188]

5. *Towards World War II*

The Muslim conventions described above, and several gatherings of more limited scope,[189] served notice that Muslims were using European methods of organization to assemble, debate, and synchronize common policies in their own interest. Despite the failure to achieve tangible results in these conventions, the message was not lost on the Muslims themselves or on others. International politics gave increasing attention, in the late 1930s, to the potential force of joint Muslim action and to the element of political Pan-Islam involved. While in Great Britain and France certain officials were wondering

[185] AE, as in n.183. Also 'Der muslimische Kongreß von Europa Genf September 1935', 100; and 'Au congrès musulman d'Europe', 104, 111.

[186] For the debates, cf. el-Djabri, 379 ff. A Delegate, 'European Muslim Conference at Geneva', 396–7. FO 371/18925, file 5821/5696. Kramer, *Islam Assembled*, 149–52. 'Il congresso dei Musulmani d'Europa a Ginevra'.

[187] V.V. (= Virginia Vacca), 'Il congresso dei Musulmani d'Europa a Ginevra', 501–3. Id., 'Seguito dei lavori del congresso dei Musulmani d'Europa a Ginevra', 563–5.

[188] Id., 'Critiche palestinesi al congresso dei Musulmani d'Europa a Ginevra', 503–4.

[189] Such as a meeting of the Ulema of Palestine in 1935 and 1936 and a mini-convention at Mecca in 1937. See AE, Archives Diplomatiques, Levant 1918–1940, Turquie, Série E, vol. 693, fos. 253–6, de Martel, French High Commissioner in Syria and Lebanon, to the French Minister for Foreign Affairs, dated in Beirut, 26 Feb. 1937. Ibid., vol. 694, fos. 32–4, a report by the French Consul-General in Jerusalem, to the French Minister for Foreign Affairs, dated in Jerusalem, Feb. 1937.

about the return of Pan-Islam and calculated its political effect on their considerable Muslim populations, the Axis Powers, too, were evaluating Pan-Islam and relating it to their own global interests.

Italy had been involved in the Geneva convention of European Muslims. Arslān had had at least one interview with Benito Mussolini before the convention, in which Italian scholars participated as guests and where meticulous care was taken not to offend Italy by unseemly verbal attacks.[190] The interest of Italy in its Muslim population in Libya and Ethiopia was even more evident shortly afterwards. On 10 March 1937, just prior to Mussolini's visit to Libya (to inaugurate the coastal highway to the Egyptian frontier), Marshal Balbo issued a proclamation stating that 'The *Duce* is the Protector of Islam and as such exalts the Muslim people'.[191] When Mussolini himself visited Tripolitania, he was girded with a so-called 'sword of Islam' and he again promised 'protection and friendship to Muslims everywhere'.[192]

While Muslim reactions were mixed (in Egypt it was argued that a Christian cannot be the 'Protector' of Muslims),[193] Mussolini's intentions to exploit Pan-Islam politically were transparent. Two years later, the Damascene newspaper *al-Ayyām*[194] told its readers that Mussolini had declared that he was ready to lead all Muslims in a second Crusade against Great Britain. This information was based on the Parisian *Marianne.*[195] What the editors of *al-Ayyām* were probably unaware of was that *Marianne* was a weekly committed to a French–British *rapprochement*, hence antagonistic to Hitler's Germany and Mussolini's Italy. Although this piece of information was probably fictitious (one could label it 'disinformation'), it was none the less indicative of the mood of the times

190 Arslān denied, however, any 'deal' with Italy. See Vacca, 'L'Emiro Arslan dichiara che il congresso musulmano Europeo non ha fatto propaganda per la Spagna e per l'Italia', 565–7.

191 *Corriere della Sera* (Rome), 11 Mar. 1937 and other Italian newspapers.

192 Kramer, *Islam Assembled*, 152–3.

193 FO 371/75121, file 13113/1781, note of Foreign Office's Research Dept., 'Signor Mussolini's Claim to the Designation of "Protector of Islam"', dated 27 Oct. 1949.

194 Of 7 Feb. 1939. See also FO 371/23190, file 1172/102.

195 Cf. FO 371/23190, file 1509/102, note by the chancery of the British Embassy in Rome, no. 10/142/39, to the Foreign Office in London, dated in Rome, 24 Feb. 1939.

that it was accepted in good faith by Pan-Islamic circles in Damascus, at least. For the historically minded, Mussolini's stand was like a repeat performance of Kaiser Wilhelm's visit to their city.

The story of Nazi Germany's involvement with Islamic leaders and Pan-Islamists is too complex to be discussed here in detail. Amīn al-Ḥusaynī and several other Muslim religious leaders spent the war years in Berlin, trying to recruit German support for their cause—in which they failed. The Nazi leadership could not commit itself fully to be supportive of Pan-Islam, as it was reluctant to act against the wishes and interests of its major ally, Italy.

No such reservation influenced the global strategy of Japan, which repeatedly strived for a *rapprochement* with Muslims. A part of this policy had a Pan-Islamic orientation, as evidenced in the propaganda activities of Abdürreşid İbrahim, the dedicated Pan-Islamist in Tsarist Russia and the Ottoman Empire. Imitating his own activity in World War I, Abdürreşid İbrahim spent the years of World War II (until his death in 1944) officiating in the Tokyo mosque and providing Pan-Islamic propaganda for his hosts.[196] A pamphlet of his was distributed in Mecca and Medina, appealing to all Muslims, particularly those in India, for union and religious fraternity, as Islam was a religion of unity and the Koran the leader of all Muslims. It warned all Muslims not to put their trust in foreign states which had always tried to drive a wedge among Muslims. Lastly, it called for a *Jihad*, presenting Hitler as a model statesman who had defeated the British in Norway and who had proclaimed that the East was for the Orientals; hence, the people of India were urged to rebel against the British.[197]

Great Britain and France, for their part, demonstrated respect for Islam during World War II, but were wary of encouraging Pan-Islam politically, lest this led to a united Islamic front during—or after—the war. It was only after World War II that Pan-Islam came again into its own, by the efforts of its own protagonists.

[196] FO 371/27043, file 353/53, copy of a letter from Count W. de Bylandt, Acting Secretary-General of the Dutch Ministry for Foreign Affairs, dated in London, 30 Jan. 1941, enclosed in British Minister to the Netherlands Sir Neville Bland's no. 13, to the Foreign Office, dated 31 Jan. 1941.

[197] Summary ibid. Unfortunately, the pamphlet itself is not enclosed.

VI. Pan-Islam in Recent Years: New Ideologies and Formal Organization

1. Rethinking Pan-Islam

In World War I, Pan-Islam, as represented by the Ottomans, had been defeated. In World War II, Islam as such was not involved, though Muslim individuals and groups fought on both sides. The years immediately following 1945 presented an entirely different situation for the evolution of Pan-Islam. One after another, independent Muslim states were established in South and South-east Asia, the Middle East, and North Africa. From these arose new perspectives and challenges for political Pan-Islam.[1] For the first time, numerous sovereign states could determine their own attitudes towards Islam and Pan-Islam and reconsider the old goal of Pan-Islamic union. On the other hand, their capacity to unite politically under the banner of Pan-Islam was curtailed by other interests, both national and local. Nationalism competed increasingly with Islam, in the altered circumstances of nation states concerned with their own specific problems and interests, frequently incompatible with those pertaining to a universal concept of Islamic unity and union.[2] Islam, rather than Pan-Islam, became an integrative value in some of the Muslim nation states. However, divergent emphases on Islam and its public role often prevented even limited consensus on the issue of establishing an all-Muslim union. Consequently, a rethinking of the feasibility of

[1] On some of these, see Kemal A. Faruki, 'Approaches to Muslim Unity', 32–3.

[2] Cf. J.S. Badeau, 'Islam and the Modern Middle East', 64–9. F. Cataluccio, 'Panislamismo e albori di nazionalismo arabo nel secolo XIX'.

political Pan-Islam gradually led to a search for alternative propositions, more acceptable to Muslim entities. The political goal of a unitary Islamic state was replaced by a goal of unity in Islamic policies.

As far as the nationalist/Pan-Islamist dichotomy[3] is concerned in the newly established states, it is quite clear that the former has not entirely superseded the latter.[4] Clashes of views in those states may well indicate that this dichotomy is still relevant. While many a Western observer has considered patriotism in the Muslim nation states a barrier to Pan-Islam, this was not necessarily the opinion of all Islamic thinkers.[5] True, some of them, such as Maududi,[6] decried the impact of particularist nationalism on Pan-Islam, considering the former as a tool of colonialism. Maududi had founded, in 1941, *Jamaʿat-i Islami*,[7] probably the most important Islamist and Pan-Islamist party of those active, subsequently, in Pakistan.[8]

Local nationalism was attacked by religious circles in Egypt and elsewhere (which considered secular nationalism especially noxious to Islam itself).[9] An exponent of this religious view was Fārūq Ḥamāda, a professor of Islamic studies in Rabat. In a book entitled *Constructing the* Umma *between Islam and Contemporary Thinking*,[10] published in 1986, he discussed Islam in the context of both modernism and nationalism. In his view, the appeal of nationalism had adversely affected inter-Muslim relations, for Muslims had become too con-

[3] On which see also *Islam v sovryemyennoy politikye stran Vostoka*, 44–9.

[4] Cf. Majid Fakhry, 'The Theocratic Idea of the Islamic State in Recent Controversies', 451.

[5] Nor even of all non-Muslim ones. See e.g. N.R. Keddie, 'Pan-Islam as Proto-nationalism', 18.

[6] See above, ch. v. He sometimes spelt his name 'Maudoodi'.

[7] M.M.J. Fischer, 'Islam and the Revolt of the Petit Bourgeoisie', 107.

[8] In his 73-page booklet, *The Process of Islamic Revolution* (originally an address at the Aligarh Muslim University, transl. into English in 1947, repr. in 1955), 6 and 8, he argued as follows, 'The distinguishing mark of the Islamic state is its complete freedom from all traces of nationalism and its influence, direct or indirect. It is a state built exclusively on principles. I should call it an ideological state . . . Islam is the only system in the world which seeks to organize the state on the basis of an ideology free from all traces of nationalism and invites mankind to form a non-national state by accepting its ideological basis'.

[9] Cf. Piscatori, *Islam in a World of Nation-states*, 40.

[10] Fārūq Ḥamāda, *Binā' al-umma bayn al-Islām wa-'l-fikr al-mu'āṣir.*

cerned with assisting their local rulers, whether these were Muslims or not, as in India, Sri Lanka, Bulgaria, Indonesia, the Soviet Union, and Afghanistan.[11] Ḥamāda argued that, none the less, Muslim unity and union were more important than ever, for ideological blocs and political alliances in the contemporary world allowed no place for mini-states. Disparity within their respective populations had not prevented China, the Soviet Union, and the United States from establishing political states. Muslims, too, could unite in one state; even if an all-Muslim state was not envisaged for the immediate future, it was likely to be established at some date and the idea ought to be fostered.[12] This was essentially the approach of yet another book, published at about the same time, almost at the other end of Islam, in Lahore. Written by Miskīn Ḥijāzī in Urdu and entitled *The Unity of the World of Islam*,[13] this maintained that Islam—rather than colour, race, or language—offered the best basis for unity and political union.

Not a few other Muslims, however, strenuously attempted in the period under discussion to devise a formula for co-existence, striving to soften the incompatibility between particularist nationalism and supra-national Pan-Islam.[14] Ḥasan al-Turabī, a professor of law who was a founder of Sudan's Muslim Brethren and later became a leader of its National Islamic Front and an important political figure, worked to establish a theocratic Muslim state there[15] and argued in general terms that 'the state is merely the political dimension of the collective endeavour of Muslims'.[16] al-Turabī had some definite ideas about Pan-Islam as well. In his view, a modern Islamist movement ought to proceed from a first target of founding an Islamic community in its immediate environment to the ultimate one of setting up an all-Islamic union by co-operating with similar movements. He warned, however, against undue haste which could cripple the potential of an

[11] Ḥamāda, *Binā' al-umma bayn al-Islām wa-'l-fikr al-muʿāṣir*, 112–13.
[12] Ibid. 130–1.
[13] Miskīn Ḥijāzī, *ʿĀlam-i Islām kā ittiḥād*.
[14] See Esposito, *Islam and Politics*, 232–4.
[15] For al-Turabī, his career and ideas, see A.A. Osman, 'The Ideological Development of the Sudanese Ikhwan Movement', esp. 395–425.
[16] Hassan al-Turabī, 'The Islamic State', 243.

international Islamist movement for achieving unity.[17] Somewhat more pointedly, Aḥmad Ṣidqī al-Dajānī, a researcher of the Palestine Liberation Organization, maintained in 1981 that nationalism (especially of the Pan-Arab category) was complementary with Pan-Islam (or at least with Islamic solidarity).[18] A similar argument was raised, about the same time, in the monthly *al-ʿArabī*, published in Kuwait.[19] Others took great pains to elaborate in some detail their perception that nationalism and Pan-Islam need not be exclusive. One of these, a generation before, was ʿAbd al-Raḥmān al-Bazzāz, a noted Iraqi statesman who, in a lecture in Baghdad in January 1952, had distinguished between the relationship of Arab nationalism to Islam and to Pan-Islam:

> The national government for which we call does not, in any way, contradict Islam. But this is not to imply a call for Panislamism. To say that Islam does not contradict the Arab national spirit is one thing and to make propaganda for Panislamism is another. Panislamism in its precise and true meaning aims to form a comprehensive political organisation which all the Muslims must obey. This organisation, although it may be desired by all the pious Muslims, is not possible under the present conditions . . . the call to unite the Arabs—and this is the clearest and most important objective of Arab nationalism—is the practical step which must precede the call for Panislamism.[20]

al-Bazzāz was a secular thinker and politician, and this fits well with the views he expressed. What is more unexpected is the fact that, in the religious camp, one finds several exponents who belittle the conceptual difference between nationalism and Islam (or Pan-Islam). It is not surprising that some of these were articulate in ʿAbd al-Nāṣir's time, as he attempted to use Pan-Arabism and Pan-Islam simultaneously in his policies. One example is a lecture held by a professor named ʿUmar Bahāʾ al-Dīn al-Amīrī at al-Azhar, on 1 March 1960, about Arabism and Islam.[21] The publication of his lecture as a 22-

[17] Osman, 419–22.

[18] Aḥmad Ṣidqī al-Dajānī, 'Mustaqbal al-ʿalāqa bayn al-qawmiyya al-ʿArabiyya wa-'l-Islām', esp. 67–9.

[19] Aḥmad Kamāl Abū al-Majd, 'Bal al-Islām wa-'l-ʿUrūba maʿan'.

[20] ʿAbd al-Raḥmān al-Bazzāz, 'Islam and Arab Nationalism', 214–15, transl. by S.G. Haim.

[21] ʿUmar Bahāʾ al-Dīn al-Amīrī, *ʿUrūba wa-Islām*.

page booklet by al-Azhar's printing press lent it an approval of sorts by this institution. Amīrī's main arguments were that, since Arabism had developed under the aegis of Islam throughout history, Arab nationalism and Pan-Islam could and would coexist.

In the case of Saudi Arabians commenting on the same issue, Pan-Islam, not surprisingly, was favoured over Pan-Arabism. An example is provided by Zayd Ibn ʿAbd al-ʿAzīz Ibn Fayyāḍ's Arabic book on *Islamic Unity*,[22] published in 1968. In this collection of his own articles in Islamic newspapers, chiefly in Saudi Arabia, the author dutifully maintained that there was no contradiction between a Pan-Arab and a Pan-Islamic union,[23] but went on to proclaim that the latter continued to be the hope of every single Muslim—to set up a huge power of 600 millions with one banner, the unity of Islam in mutual co-operation. This was the *sine qua non* of the reassertion of the ancient glory of Muslims; a Pan-Islamic union was bound to succeed where the Arab League had failed.[24]

There were several other ways of looking at this. A few years later, a Muslim political scientist and prolific editor and writer on Islamic affaris, Kalim Siddiqui, took the same view of the prevailing situation, but arrived at a different conclusion. In a lecture at an education conference held in Mecca in 1977 (and then published as a 16-page pamphlet),[25] Siddiqui maintained that the nation states were there to stay. He conceded that they had not solved any of the problems which were confronting the *umma*; but he held that students and the general Muslim public could—and should—be educated for the ultimate triumph of the Islamic movement in the following phase of history, that of the *umma* in the post-nation-state era. While Siddiqui does not describe himself explicitly as a Pan-Islamist, there is little doubt that he is willing to work educationally and with propaganda for that cause.

Siddiqui's approach is characteristic of one of the trends of thought among the circles which we have called 'Islamist'. In

[22] Zayd Ibn ʿAbd al-ʿAzīz Ibn Fayyāḍ, *al-Waḥda al-Islāmiyya*.
[23] Ibid. 32–4.
[24] Ibid. 35–8.
[25] Kalim Siddiqui, *Beyond the Muslim Nation-states*.

the last forty years, Islamists have tended to interpret, as formerly, all events—domestic as well as external—in the light of their impact on the *umma* and, by extension, on Pan-Islam. In so doing, however, they have overlooked the fact that, in the politics of international relations, the frontiers between nationalism and Pan-Islam are sometimes difficult to identify.[26] The yearning for a united *umma* has maintained its appeal, indeed; but in the final reckoning national interests have generally determined foreign relations.[27] While Islam has often been an ingredient in both domestic politics and external conflicts (Cyprus, Lebanon, Israel, India, Pakistan, and other cases), policy decisions have generally been made and implemented in the light of national, rather than religious, considerations. Islam and (more rarely) Pan-Islam have *influenced*, rather than determined, policy-making, especially in foreign relations.[28] It is hazardous to assess the degree of such influence and its relevance, but most agreements and alliances among Muslim states tend to display *realpolitik* rather than a commitment to Pan-Islam.

Several states have demonstrated the impact of Pan-Islam on their official policies (or, at least, on their political writings), in the last forty years, notably Pakistan, Saudi Arabia, and Egypt.[29]

Pakistan sought a leading role in Pan-Islamic politics very soon after attaining statehood. After all, Islam was the main reason for its establishment and existence.[30] On 19 February 1949, Muhammad Ali Jinnah (1876–1949), its first president, proclaimed in a broadcast that 'The great majority of us are Muslims. We follow the teachings of the Prophet Muḥammad. We are members of the Brotherhood of Islam, in which all are equal in rights, dignity and self-respect. Consequently we have a special and a very deep sense of unity'.[31] Jinnah's words accurately reflected a popular notion (still influenced by

[26] F.W. Fernau, *Moslems on the March*, 85–6.
[27] D. Pipes, *In the Path of God*, 151.
[28] Cf. Adeed Dawisha, ed., *Islam in Foreign Policy*, 4.
[29] Indonesia, also, is a relevant case; however, as stated in the introduction to my study, South-east Asia could not be included.
[30] W. Caskel, 'Western Impact and Islamic Civilization', 339.
[31] Quoted in *al-Islam* (Karachi semi-monthly), 1 May 1953, 19.

Maududi's intensive campaign of propaganda)[32] about Pakistan's mission in promoting the cause of political Pan-Islam.[33] This was interwoven with an appeal to religious fundamentalism, popular with Pakistan's rural population and urban lower middle class, which supported a comprehensive Islamic restructuring of society (chiefly during Zia-ul Haq's presidency, 1977–88). However, there was more to the policy of adopting Islam and supporting Pan-Islam than merely satisfying public sentiment. The Government of Pakistan felt that it was isolated among potentially dangerous neighbours and that its most likely support would come from Muslim states. India, too, was trying to secure the amity of Muslim governments.[34] Hence Pakistan's *rapprochement* with Iran and Turkey (in the Baghdad Pact), its courting of Arab states (chiefly Saudi Arabia and Egypt), and its leading role in the promotion of all-Muslim international organizations,[35] of which more below. In actual practice, its success was very modest; Pan-Islamic ideas failed to keep the two wings of Pakistan together, with its eastern component breaking away in 1971 and establishing Bangladesh as a separate, definitely secular, state.

In Saudi Arabia, Islam has been the basis of life and government for the whole of the twentieth century. Under King Fayṣal (reigned 1964–75), however, the trend to Pan-Islam became more marked, due to the cold war among the Arab states, the verbal attacks by radical Arab governments on Fayṣal's regime, and the involvement of Saudi Arabia in the Yemen War (1962–7). While Saudi intellectuals interviewed by J.P. Piscatori[36] and some other Saudis acknowledged the existence of separate Muslim nation states as unlikely to change, the government felt bound to strive for a universalist Muslim approach. King Fayṣal suspected the intentions of

[32] A. Hyman, *Muslim Fundamentalism*, 10 ff. Maududi took a strong stand against nationalism, see also Esposito, *Islam and Politics*, 232.

[33] M.G. Weinbaum and Gautam Sen, 'Pakistan Enters the Middle East', 598.

[34] Fernau, 83.

[35] In 1966, Pakistan's President, Ayyub Khan (President, 1960–9), and its then-Minister for Foreign Affairs, Zulfikar Bhutto (later, Pakistan's President, 1970–9), were articulate in supporting an all-Muslim union. See Inʿamullah Khan, 'Thoughts on a Muslim Summit Conference', 3–4.

[36] Piscatori, 85 ff. See ibid. 109–10, for other Saudi views.

radical Arab nationalism and, in the war of ideas, developed his own perception of Pan-Islamic solidarity, instead of immediate union, *faute de mieux*.[37] In his speeches, he appealed for a wholehearted return to Islam, based on co-operation, which would eventually lead to the union enjoined by the Prophet Muḥammad. Further, he emphasized that there was no contradiction between Muslim solidarity and Arab unity, both of which, according to his perceptions, faced the same foes—imperialism, Zionism, and Communism.[38] Hence, Fayṣal's intensive activity in setting up and fostering international Islamic organizations and publications, which has been continued by his successors. Much of this was tied up with the self-perception of the Saudi Arabian kings as leaders of the entire Muslim world,[39] 'a Caliphal role as non-titular heads of an increasingly politically conscious Pan-Islamic quasi-community'.[40]

In Pakistan the Pan-Islamic activities of the government were at least partly brought about by popular sentiment and in Saudi Arabia by *raisons d'état* as perceived by its king; in the Republic of Egypt both seem to have combined. From the 1940s, the religious concept of Pan-Islam acquired a political dimension.[41] Jamāl ʿAbd al-Nāṣir (1918–70) acknowledged an 'Islamic circle' in his *Philosophy of the Revolution*, published in 1954.[42] He soon abandoned, however, any Pan-Islamic plans he might have had, in favour of Pan-Arab and non-aligned policies, fiercely opposing Saudi Arabia's projects to set up international Islamic organizations.[43] ʿAbd al-Nāṣir's use of

[37] For details, cf. Abdullah Mohamed Sindi, 'The Muslim World and its Efforts in Pan-Islamism', ch. 5.

[38] For details, see N.O. Madani, 'The Islamic Content of the Foreign Policy of Saudi Arabia', 79–87. V.L. Bodyanskiy and M.S. Lazaryev, *Saudovskaya Araviya poslye Sauda*, 57–75.

[39] As witness an article by an ex-Minister of Saudi Arabia, ʿAbd Allāh al-Saʿd, repr. by Muḥammad Ḥasan ʿAwwād, *al-Taḍāmun al-Islāmī*, 107–13.

[40] M. Ruthven, *Islam in the World*, 30.

[41] U. Steinbach and R. Robert, eds., *Der nahe und mittlere Osten*, ii. 36. One of the Cairene journals concerned with this aspect was *al-ʿĀlam al-Islāmī* (The Islamic World), a monthly started in June 1949, copies of which may be consulted in the British Library's Oriental dept., London.

[42] The relevant passages were transl. into English by B. Lewis, 'The Return of Islam', 22–3.

[43] For his resistance to Saudi Arabia's attempts to form an all-Muslim alliance, see Abbas Kelidar, 'The Struggle for Arab Unity', esp. 297 ff.

Islam, both in Egypt and abroad, was mostly manipulative: using al-Azhar Academy for Islamic propaganda, founding an Islamic World Centre and a 'Voice of Islam' broadcasting station, and the like.[44] ʿAbd al-Nāṣir reverted to a more meaningful and Pan-Islamic stance only after his defeat by Israel in 1967; his general policies regarding Islam were continued, with variations, by Anwar al Sādāt (president, 1970–81).[45] However, the longing for Islamic unity among the Egyptian masses seems to have given an added impetus to the Islamic activity of the Muslim Brethren,[46] which gradually assumed a political Pan-Islamic character, with militant branches in Egypt and in several Muslim countries. Their activities were emulated by other Islamic groups in Egypt, some of which increasingly became more aggressive, particularly during the 1970s and 1980s; in addition to their fundamentalist approach, these adopted, over time, Pan-Islamic traits in their preaching and activity.[47]

Most, possibly all, other Muslim states have been involved in varying degrees with Islamic and Pan-Islamic politics in the post-World War II period. Muʿammar Qadhāfī, since he seized power in Libya in October 1969, has perceived himself a champion not only of Islam, but also of Pan-Islam; he first spoke of a Pan-Islamic Sahel Federation,[48] then tried to forge a union with other Muslim states and continuously meddled in the affairs of Muslims in other lands.[49] Jordan has had to contend with an illegal Pan-Islamic political party, *al-Taḥrīr* (Liberation);[50] while Syria's rulers have had their troubles with the local branches of the Muslim Brethren,[51] and Tunisia with its own underground Islamic and Pan-Islamic groups. In the officially secular Republic of Turkey, a National Salvation

[44] H.E. Tütsch, *From Ankara to Marrakesh*, 128–9.

[45] R. Israeli, 'Islam in Egypt 65 ff.

[46] See above, ch. v.

[47] Abd el-Monein Said Aly and M.W. Wenner, 'Modern Islamic Reform Movements', 341 ff.

[48] Steinbach and Robert, eds., i. 735.

[49] Adeed Dawisha, ed., 56 ff. Asaf Husain, *Political Perspectives on the Muslim World*, 177–80.

[50] For which see Fathi Yakan, *Islamic Movement*, 114–17. Yakan is a thinker, close to the views of the Muslim Brethren.

[51] O. Carré and G. Michaud, *Les Frères musulmans (1928–1982)*, 207 ff. Asaf Husain, 174–7.

Party tried with great zeal during the 1970s to introduce a theocratic regime and to forge ties between their state and Muslim ones, on Pan-Islamic premisses, while writing about a Pan-Islamic union, inspired by the Koran and Islamic virtues, continued unabated.[52]

In this array of contemporary Muslim attitudes to political Pan-Islam, Iran was manifestly absent. It had been lukewarm, under the last Qajars, to Ottoman Pan-Islamic propaganda and official attitudes did not change under the two secular-minded Shahs who followed, Rezā Pahlavī (reigned 1925–41) and his son Moḥammed Rezā Pahlavī (reigned 1941–79), who tried to transform religious into national loyalty.[53] The lack of enthusiasm for a Pan-Islamic union can be explained, first of all, by the antagonism felt by Iran's Shiites (about 90 per cent of its population today) towards the Turks (following a sequence of wars with them) and the even stronger ones towards the Arabs (inherited from their forefathers, conquered and subjugated by the Arabs in the seventh century), as well as their own particularist culture and separate language. Since many of Iran's Sunnites live near its borders, Shiite–Sunnite relations have increasingly been centre–periphery ones. There were, of course, sporadic appeals for an all-Muslim union. After 1945, Pan-Islam was considered in Iran as an antidote to Soviet expansion.[54] Then, a new Ahwaz weekly, *Qārūn*, came out early in 1948 with an editorial by Mīr Ṣadīqī, urging an Islamic union in the Middle East, against the designs of the United States, Great Britain, and the Soviet Union.[55] The same approach was taken, about two-and-a-half years later, by the Tehran newspaper *Kayhān*, which called for a Pan-Islamic union against foreign intrigues aiming at infringing the political sovereignty of Muslim states.[56] A year later, a public

[52] For this party, see J.M. Landau, 'The National Salvation Party in Turkey', 10, 49 ff. Asaf Husain, 180–4. A good, very recent example of preaching Pan-Islamic union under the Koran is the journalist Abdurrahman Dilipak's *Vahdet ama nasıl?*

[53] D. Menashri, 'The Shah and Khomeini', 59 ff.

[54] Goichon, 36.

[55] *Qārūn* (Ahwaz), 18 Jan. 1948, summarized in FO 371/68722, file 3203/520, British Consulate General's dispatch no. 10, dated in Ahwaz, 12 Feb. 1948.

[56] *Kayhān* (Tehran), 17 Aug. 1950, partly transl. in FO 371/81902, file 1016/3.

demonstration by a group calling itself *Mosalmānān-i Mojāhid* ('fighting Muslims'),[57] was addressed by Ayatollah Kāshānī (in a recorded speech), his son Muṣṭafā, and his son-in-law Shams Qanātābādī—all of whom called for Muslim unity against the Great Powers.[58] Kāshānī was also campaigning then for a Pan-Islamic congress in Tehran.[59]

Suggestively, the comment of the British Foreign Office on the report about the demonstration of the *Mosalmānān-i Mojāhid* had been: 'The sense of identity with other Muslim countries is not of the strongest in Persia'.[60] Indeed, one is unable to determine whether even the infrequent recommendations of Pan-Islamic unity noted were advocated on the political or merely the spiritual level. Thus a book, written in 1963 by Ḥasan Abṭaḥī Khorāsānī and entitled *Ittiḥād va-dūstī dar Islām* (Unity and Friendship in Islam),[61] presented the emphasis laid down by Islam on religious unity and camaraderie, entirely avoiding the discussion of political union. This approach continued into the 1970s, for example in a book published in 1979 by ʿAlī Qā'imī and entitled *Tafriqa mas'ala-i rūz-i mā* (Divisiveness is the Problem of Our Times). This is an attack on divisiveness in Islamic society and a call for Islamic unity, under the slogan, 'Islam is the religion of unity'.[62] Again, most of Qā'imī's arguments are theological, largely based on the Koran, and advocate Pan-Islamic brotherhood; the Prophet Muḥammad's actions are given as examples.[63]

This general trend has not changed visibly under the Islamic

[57] This was not the group led by Kāshānī himself, the name of which was *Mojāhedīn-i Islām*.

[58] The meeting was held on 7 Sept. 1951. See FO 371/91463, file 1015/318, Middleton's dispatch—signing for Sir F. Shepherd, British Ambassador to Iran—to Prime Minister C.R. Attlee, dated in Tehran, 11 Sept. 1951.

[59] French transl. of his proclamation, calling for an all-Islamic meeting, in *Le Monde* (Paris daily), 6 Oct. 1951, 3. See also FO 371/98311, file 1782/6, Quarterly Review of Islamic Affairs for the 3rd quarter of 1952, dated 10 Oct. 1952. At about the same time, another active Persian group, *Fidā'iyān-i Islām*, also held Pan-Islamic views; see B. Lewis, *The Middle East and the West*, 112–13.

[60] FO 371/91463, file 1015/318, as above, n. 58.

[61] Ḥasan Abṭaḥī Khorāsānī, *Ittiḥād va-dūstī dar Islām*. A copy can be consulted in the New York Public Library.

[62] ʿAlī Qā'imī, *Tafriqa mas'ala-i rūz-i mā*, 18: 'Islām dīn-i vaḥdet est'.

[63] Ibid. 26–7.

Republic of Iran. To cite one instance, the religious thinker Maḥmūd Ṭāliqānī's 103-page *Vaḥdet ve āzādī* (Unity and Freedom), published in 1982, comprised a long chapter on 'the Koran as the axis for the unity of Muslims'.[64] Much the same holds true of the official spokesmen of the republic during the 1980s. A collection of articles by the spiritual leaders of the Islamic revolution, published in 1983,[65] preach about the need for the unity of all Muslims, stressing the similarities between Shiites and Sunnites. These works were published in Persian. Special interest attaches, however, to works intended for external consumption. Characteristically entitled *Āfāq al-waḥda al-Islāmiyya* (Horizons of Islamic Unity), an 85-page book in Arabic, published in 1982 by Iran's Ministry for Islamic Guidance, quoted the views of the major revolutionary leaders. Again, these focused on Islamic unity, rather than on a political Pan-Islamic union, striving to minimize the significance of Shiite–Sunnite doctrinal differences and frequently quoting from their common source, the Koran.[66] As for the particular views of the main ideologues, Ayatollah Sharīʿatmadārī's ideology, in particular, is not concerned with Pan-Islam.[67] One cannot doubt Ayatollah Rōḥallāh Khomeynī's rejection of the contemporary international system and his support for Islamic universalism.[68] He appealed for an Islamic *umma* to comprise all the world's nations (not merely the Muslim ones).[69] However, while his writings and preaching have recommended Muslim unity as the way to restore Islam's earlier grandeur,[70] they seem to stop short of explicitly advocating one-state political Pan-Islam.[71] His militancy and that of his close associates do imply a policy for the achievement of this universalism, which has been expressed pithily by one of

[64] Maḥmud Ṭāliqānī, *Vaḥdet ve-āzādī*, 95–103: 'Qor'ān miḥvār-i vaḥdet-i Muslimīn'.

[65] *Risāla-i inqilāb-i Islāmī-i Īrān dar tavḥīd-i kalima.*

[66] *Āfāq al-waḥda al-Islāmiyya, passim.*

[67] Cf. D. Menashri, 'Shiite Leadership', 130–1.

[68] See Adeed Dawisha, ed., 16–17.

[69] Cf. e.g. *Imam Humeyni'nin Islam birliği ile ilgili mesaj ve nutkularından seçmeler*, 31–2.

[70] Menashri, 'The Shah and Khomeini', 63 ff.

[71] Id., 'Shiite Leadership', 129 ff. Cf. however the interpretation of Khalid Duran, *Islam und politischer Extremismus*, 34–8.

his admirers, Ḥoseyn Mōsavī, then leader of the ʿAmal pro-Shiite organization in Lebanon: 'We regard the entire Islamic world as our homeland'.[72] Iran's current rulers are well aware that few Sunnite Muslims would unite politically under Shiite leadership.[73] Whether continuing to blame the Sunnites,[74] or conversely (and more often) attempting to heal the rift (for example, having the Sunnite Egyptian theologian Sayyid Quṭb, executed in 1966, on one of their 'martyr stamps'),[75] their reticence in fully committing themselves politically to Pan-Islam seems evident. In the same manner, a monthly publication in Urdu by Iran's Embassy in Islamabad, entitled *Vaḥdat-i Islāmī* (The Islamic Union—or Unity), quoted liberally from the writings of Khomeynī, Montaẓerī, and others. The main arguments, intended for readers in Pakistan, are the *rapprochement* between Shiites and Sunnites but, again, do not explicitly mention a political union.[76] In this context, the stipulation of the 1979 Constitution of the Islamic Republic of Iran (Article 10 of the 'Principles'), that all Muslims are one *umma* and the government should exert itself continuously to achieve 'the political, economic, and cultural unity of the Islamic world', has not yet been implemented.

2. *The Concept of Islamic Solidarity*

Among this somewhat bewildering array of Pan-Islamic solutions, mainstream thinking about a union continued unabated. Two recent examples should suffice. Muḥammad al-Ṣādiq ʿArjūn, formerly Dean of the Faculty for the Principles of [the Islamic] Religion at al-Azhar, in *The Islamic* Umma *as Prescribed by the Great Koran,*[77] published in 1984, developed his views on the very special character of the *umma*. Although his

[72] Quoted in *Time* magazine, 17 Aug. 1987, 13.

[73] J.L. Esposito, *Islam and Politics*, 233.

[74] As in the *Crescent International*, 16–31 Oct. 1983, repr. in Kalim Siddiqui, ed., *Issues in the Islamic Movement 1983–1984*, 85–7.

[75] P. Chelkowski, 'Stamps of Blood', esp. 565. Cf. R. Badry and J. Niehoff, *Die ideologische Botschaft von Briefmarken—dargestellt am Beispiel Libyens und des Iran*, esp. 23 ff., 37 ff.

[76] *Vaḥdat-i Islāmī* (Islamabad), 24 (Rabīʿ evvel 1407H), 2–3, editorial.

[77] Muḥammad al-Ṣādiq ʿArjūn, *al-Umma al-Islāmiyya kamā yurīduhā al-Qurʾān al-ʿaẓīm*.

arguments were mainly theological and historical, he insisted that Islamic religious unity was best served by each Islamic nation developing its own power and co-operating with others in a sort of union (which he called *jāmiʿa*).[78] More politically minded is a book by Hānī Faḥṣ on *Islamic Unity and Divisiveness*,[79] published in 1986. Its author acknowledged that he had written this work 'in order to deepen the awareness of union' (*taʿmīq al-waʿy al-waḥdawī*) among Muslims and to warn against divisiveness and its dangers (*al-tajzi'a wa-makhāṭiruhā*).[80] Addressing his arguments to Muslim statesmen on the practical level, Faḥṣ explained at some length how divisiveness vanquished the Muslims during the Crusader wars, maintaining that a historical lesson for union ought to be learnt from this.[81] Unity was now the sole means of survival for Muslims in the Soviet Union.[82] Faḥṣ's conclusion was that an overall union of Muslims remained the best way of maintaining the unity inherent in Islam.[83]

However, as well as a rethinking of Pan-Islam, a reorganization was needed.[84] Islamic solidarity, advocated by al-Afghānī,[85] foreshadowed in the conventions which met in the inter-war period and later proclaimed by King Fayṣal of Saudi Arabia and others, was a self-evident solution to the nationalist–Islamist impasse. It was easier, of course, to reconcile fervent nationalism and intra-state rivalries with religious and cultural solidarity (and its derivatives, political and economic co-ordination) than with overall political or economic solidarity, the total achievement of which still remains a major task at the time of writing. None the less, it was easier to adopt common attitudes than a binding decision on a Pan-Islamic state.[86]

The term employed for 'solidarity' in Arabic (and adopted in other Islamic languages), *taḍāmun*, is significant. The term *al-Taḍāmun al-Islāmī* spread swiftly and even gave its name

[78] Ibid., esp. 65 ff.
[79] Hānī Faḥṣ, *Fī al-waḥda al-Islāmiyya wa-'l-tajzi'a.*
[80] Ibid. 7.
[81] Ibid. 79–127.
[82] Ibid. 129–62.
[83] Ibid. 197 ff.
[84] See Q. Ahmed-ur-Rehman Alvi, 'Pan-Islamism or Islamic State?', 38.
[85] M. Gavillet, 'Unité Islamique ou unité nationale?', 83 ff.
[86] Snyder, *Macro-nationalisms*, 134.

to a 96-page Arabic monthly of general interest, published in Mecca from the 1970s (continuing another monthly, *Majallat al-ḥajj*, or Journal of the Pilgrimage). This term conveys the notion of 'mutual guarantee' or 'mutual responsibility', which is soothing. Although many pious Muslims have grasped solidarity as a concept derived from their religion, even dictated by it (as explained by Muḥammad Abū Zahra, Dean of the Law Faculty at Cairo University),[87] the Muslim attitude of solidarity is more widespread.[88] Many feel that such solidarity is advantageous not only for them, but for the entire world.[89] However, rational considerations have also played a part. Political Pan-Islam had failed to achieve its objective of uniting all Muslims, chiefly because of their lack of solidarity.[90] This goal has become, in the current generation, even more unattainable due to the proliferation of independent Islamic states with nationalist particularist ambitions.[91] The situation, although condemned by pious Muslims, was gradually accepted by some of them as a fact of life.[92] Still, accustomed to expressing their identity in terms of Islam and Pan-Islam rather than in those of nationalism,[93] they felt threatened by other religions.[94] Some were suspicious of the political solidarity within Christianity;[95] Muslim solidarity has developed a somewhat antagonistic stand towards non-Muslims, as a result.[96]

Moreover, politically aware Muslims were also conscious of the increasing penetration by the superpowers into their section of the globe. Subtle or not, they regarded this impact, too, as a menace, saying so in as many words.[97] While certain

[87] Muḥammad Abū Zahra, *al-Waḥda al-Islāmiyya* (1976). This is a considerably enlarged version of the book he published in 1958.

[88] Kramer, *Political Islam*, 83. See also R.M. Sharipova, 'Dvidzhyeniye Islamskoy solidarnosti', 60 ff.

[89] Reported by W.M. Watt, in 1956, in his 'Thoughts on Islamic Unity', esp. 194.

[90] Cf. P.J. Vatikiotis, 'Islam as a World Force', 11.

[91] D. Pipes, *In the Path of God*, 155.

[92] Esposito, *Islam and Politics*, 232.

[93] See e.g. Mohamed Sid-Ahmed, 'Shifting Sands of Peace in the Middle East', 68–71.

[94] Maḥmūd Diyāb, *Abṭāl al-kifāḥ al-Islāmī al-muʿāṣir*, 290 ff.

[95] Cf. J.G. Hazam, 'Islam and Nationalism', 161–2.

[96] D. Pipes, 'This World is Political!', 14–15. R.M. Sharipova, *Panislamizm syegodnya*, 52.

[97] e.g. Diyāb, 293 ff. See also G.J.-L. Soulié, 'Le Monde musulman à la recherche de son unité', part 3, 638.

Muslim states have closely co-operated with the United States and others with the Soviet Union, many Muslims have regarded them both with suspicion. A Lebanese Muslim, Ṣalāḥ al-Dīn al-Munajjid, examined in 1967 what he called 'Marxist solidarity', comparing it with Muslim solidarity and favouring the latter.[98] Indeed, some exponents of Muslim solidarity adopted an antagonistic stand towards the superpowers[99] (not dissimilar to that prevailing among other spokesmen of the Non-aligned nations, too). Others, however, saw it more as a challenge directed at the Soviets and Communism, in particular. For instance, Ḥasan Kutubī, apparently a Saudi Arabian, discussed Islamic solidarity at some length, in his 1967 book in Arabic.[100] As he perceived it, among its main merits, side-by-side with the co-ordination of foreign policy by the Islamic states[101] and mutual economic assistance,[102] was a joint struggle against Communism.[103] In such difficult circumstances, solidarity-in-faith, rather than Arab or nationalist slogans, would preserve the Islamic state from perdition.[104]

Since it became increasingly evident that an 'Islamistan',[105] or an area ruled by Islam and capable of challenging the superpowers, would be unattainable in the short-term, a number of Muslim thinkers started intensively to advocate solidarity and attempted to explain their understanding of the term. A few examples follow, chiefly from the mainstream of Muslim thinking.[106]

Najīb al-Kīlānī (born in 1931), a physician in the Ministry for Health, Cairo, and a successful novelist, wrote in 1962 a book

[98] Ṣalāḥ al-Dīn al-Munajjid, *al-Taḍāmun al-Marksī wa-'l-taḍāmun al-Islāmī.*

[99] Examples in L.R. Polonskaya, 'Sovryemyenniye Musul'manskiye idyeyniye tyechyeniya', 19–20. Sharipova, *Panislamizm syegodnya*, 50–1.

[100] Ḥasan Kutubī, *Dawrunā fī zaḥmat al-aḥdāth.*

[101] Ibid. 85 ff.

[102] Ibid. 161 ff.

[103] Ibid. 39 ff.

[104] Ibid. 191–2.

[105] Soulié, part 3, 638–9.

[106] We shall not discuss such bizarre projects as increasing Muslim solidarity by redrawing the globe so that the first meridian of longitude passes through the Kaʿba and redrafting accordingly maps and textbooks. See Shah Syed Munirul Huq, *The Islamic Meridian.* This 56-page booklet is a reprint of an article with the same title in *The Islamic Review* (Woking, Surrey, UK) of Mar. 1953.

in Arabic on *The Road to Islamic Unity*, which he published in Tripoli (Libya).[107] This aimed at all-Muslim solidarity rather than union. He appealed for 'a Muslim public opinion',[108] particularly in the mass media. Solidarity, according to al-Kīlānī, ought to be principled and ideological, rather than utilitarian and strategic,[109] and to be based on a tremendous spiritual revolution and readiness for sacrifice.[110] Somewhat more specific is the approach to Muslim solidarity of Muḥammad al-Mubārak, head of the division of Islamic studies in the college for the *sharīʿa* and Islamic studies at Mecca. In 1971, he wrote a book in Arabic, entitled *Contemporary Islamic Society*, which he published in Beirut.[111] This is based on his lectures, in which he posited the cardinal question of whether the Muslims still formed an *umma*.[112] His reply postulated that this depended on solidarity, which he considered deriving from: (i) co-ordination (*tansīq*) among Muslim peoples in teaching, laying down religious law, economic organization, and social conditions—which, in turn, would lead them to political unity; (ii) lowering the barriers among them by suitable legislation; (iii) popularization of the principles of Islam within both the educated and the general public.[113]

In 1979, Necmettin Erbakan, chairman of the National Salvation Party in Turkey, considered the matter in political terms, when he proclaimed in London that, by raising Muslim co-operation to an advanced level, the world's greatest power could be put together.[114] By 1985, Islamic solidarity had become a frequently employed term, along with unity, among the freedom fighters in Afghanistan. A manifesto put out in Peshawar by the Afghan Students' Islamic Federation declared:

[107] Najīb al-Kīlānī, *al-Ṭarīq ilā ittiḥād Islāmī*.
[108] Ibid., ch. 3.
[109] Ibid. 159–60.
[110] Ibid. 161–3.
[111] Muḥammad al-Mubārak, *al-Mujtamaʿ al-Islāmī al-muʿāṣir*.
[112] Ibid. 33.
[113] Ibid. 35–7.
[114] Quoted by Kramer, *Political Islam*, 83. It is no accident that Turkey's Islamists, ideologically close to Erbakan, still write of a Pan-Islamic union. See e.g. Yusuf Ziya İnan, *Kemalist eyleme göre*, esp. 35 ff., and the even more recent article by a well-known Islamist politician, Hasan Aksay, 'İslâm dünyasının birliği'; Aksay advocated this union against what he considered as Western and Jewish designs.

'The fundamental aim of establishing the students' federation is to expand solidarity and conception [of] unity among the Muslim youths of Afghanistan'.[115] Further objectives are unity among the heroes of the battlefield and solidarity with Islamic organizations;[116] subsequently, solidarity among Muslim students is proclaimed[117] and world-wide solidarity of all Muslims is demanded.[118]

Some of the Pan-Islamic writing in the post-World War II era was devoted to economic solidarity. Economic Pan-Islam is a direct result of the penetration of Westernization into the Islamic environment; it was studied by Muslims and initiated by them as a theory of economic solidarity which sometimes bypassed the laws of the *sharīʿa*. Here a bond could be forged not only between the haves and the have-nots, but also among all Muslims, be they liberals or conservatives, nationalists or Islamists—all could find common ground to assert themselves economically, in their own interests. Basically, this is not very difficult, since, as noted by A. Gh. Ghaussy, a Muslim who teaches economics in Hamburg, 'Islam hardly prescribes any guidelines for a concrete economic order beyond a few ethical–moral principles and religiously determined commandments and prohibitions'.[119]

Despite differing views among proponents of Muslim economic solidarity,[120] they saw some connection between it and other Pan-Islamic aspects. One of the most purposeful discussions of the matter was an Arabic book by Gharīb al-Jammāl, a seemingly Saudi author, *Muslim Solidarity in the Economic Domain*.[121] He defined Islamic solidarity as an entity, whose states and peoples have joined around one creed and one set of morals and values.[122] Solidarity, in this case, was focused on the social and economic development of the Islamic world. It

115 *Manifesto of Afghan Students' Islamic Federation*, 5.
116 Ibid.
117 Ibid. 11.
118 Ibid. 9. See extracts in app. U, below.
119 A. Gh. Ghaussy, 'Attempts at Defining an Islamic Economic Order', 24. See also Ghirelli, *El renacimiento musulmán*, 57–8.
120 A. Ionova, 'Islam i myedzhdunarodnoye ekonomichyeskoye sotrudnichyestvo', 15–17.
121 Gharīb al-Jammāl, *al-Taḍāmun al-Islāmī fī al-majāl al-iqtiṣādī.*
122 Ibid. 6.

was a response to the view that each country could survive solely within larger formations. These enabled each Islamic land, developed or developing, to establish with others a just international economic system—while respecting the independence of each single state.[123] Within the framework of economic solidarity, efforts should be made to free the entire Muslim world from foreign exploitation and facilitate the return of natural resources to their owners, increase inter-Muslim commerce, and emphasize common Muslim interests. All co-operation would be based on Muslim religious unity.[124]

One of the most reasoned presentations of Muslim solidarity was offered in an anonymous pamphlet in Arabic, entitled *Islamic Solidarity in Today's World.*[125] It is undated, but was published, probably in Jedda, in the early or mid-1970s. After having described the prevailing divisions among Muslims, the author argued that Islam provided a vital and dynamic factor for Muslim co-operation and solidarity—the best way to overcome their weakness and defend their rights and interests.[126] He himself considered Islamic solidarity as a practical matter, more important than any political alignment. It was an historical and geographical truth and there existed currently sufficient conditions—geographical, political, and economic—to shape Muslim solidarity into a collective of Islamic peoples (*Majmūʿat al-shuʿūb al-Islāmiyya*). This, however, should be based on civilizational rather than on political premisses, thus setting up new patterns for international co-operation. Such a bloc (*tajammuʿ*) would be more powerful than political ones and could aspire to two goals: (i) to use the power deriving from Muslim solidarity towards promoting Muslim interests—economically, socially, ideologically, educationally, and politically; (ii) to organize all popular and governmental associations to implement all-Muslim co-operation. Solidarity within an Islamic bloc should pave the way to solving the Great Problems of Muslims—economic progress, raising the level of earnings, developing the socio-economic and cultural systems —and of the world.[127]

123 Gharīb al-Jammāl, *al-Taḍāmun al-Islāmī fī al-majāl al-iqtiṣādī*, 6–7.
124 Ibid. 7–13.
125 *al-Taḍāmun al-Islāmī fī ʿālam al-yawm.*
126 Ibid. 11–25.
127 Ibid. 29–47.

The examples given may help the reader to grasp how various people understood and interpreted the concept of Islamic solidarity. However, what perhaps mattered more, in the final reckoning, was how Muslim rulers perceived it in a Pan-Islamic context. Among these, special consideration should be given to the views of King Fayṣal of Saudi Arabia who initiated much of the discussion and action to promote this solidarity. Fayṣal's speeches about Islamic solidarity have been collected and analysed recently by Muḥammad Ḥasan ʿAwwād, in an Arabic book entitled *The Great Islamic Solidarity, Sponsored by the Mighty Leader, Fayṣal Ibn ʿAbd al-ʿAzīz*.[128] ʿAwwād noted three sources of inspiration for Muslim solidarity—the Koran, and the sayings and actions of the Prophet Muḥammad.[129] ʿAwwād considered solidarity (*taḍāmun*) as a more comprehensive term than the more functional and local 'league' (*jāmiʿa, rābiṭa*).[130] These were the views of King Fayṣal, also, when he toured his own state, Pakistan, and Egypt, to solicit support for his religion-inspired appeals for the drawing up of principles and establishment of instruments to make Islamic solidarity feasible and viable.[131] Muslims everywhere were to be enabled and encouraged to join and strive together for the glory of Islam and the benefit of all Muslims—by strengthening the bonds of Islamic brotherhood, raising cultural and economic standards, and co-ordinating all efforts.[132] Fayṣal's aim seems to have been to persuade the Muslims to act as one bloc within the international system.[133]

3. *An Islamic Bloc as the* Modus Operandi

While most Muslim writers on Islamic solidarity agreed that it should be based on religious identity and fraternal sentiment,[134]

[128] Muḥammad Ḥasan ʿAwwād, *al-Taḍāmun al-Islāmī al-kabīr fī ẓilāl daʿwat al-qāʾid al-ʿaẓīm Fayṣal Ibn ʿAbd al-ʿAzīz*. I have been able to consult only the 2nd edn. (1976).

[129] Ibid. 22.

[130] Ibid. 29.

[131] Ibid. 35.

[132] Ibid. 36 ff.

[133] Madani, 96.

[134] More examples in Sharipova, *Panislamizm syegodnya*, 52 ff.

only a few discussed the instruments for its implementation. Of these, practically all favoured the establishment of some sort of Islamic bloc (in Arabic, *tajammuʿ Islāmī*)[135] as the best step towards Muslim unity. After all, politically conscious Muslims could not but feel that many of their aspirations would remain pious hopes, on the international level, unless and until they found a *modus operandi*. Models of association were, indeed, very visible[136] in the 1940s and 1950s, embodied in a variety of organizations: the British Commonwealth of Nations, reshaped to include newly independent states; the French Communauté des Pays d'Expression Française, maintaining France's ties with its former colonies; the Soviet bloc, integrating old and new Communist countries; and the camp of the Non-aligned states, aiming at jointly achieving a degree of independent action in international affairs. The idea of an Islamic bloc soon appealed to Muslim statesmen and writers. While some of the former were naturally partial to limited alliances among several Muslim states only (as more achievable and better serving their own interests), they too realized that belonging to larger formations could lessen the particularist odium which the popular masses attached to such alliances.

Without presuming to present a history of the moves and writing about the Islamic bloc, several landmarks may be pinpointed, introduced chronologically.

Attempts before World War II[137] proved premature; only after this war were there practical results. As early as 1947, there were press reports that Jinnah, Pakistan's leader, was discussing in Egypt projects of setting up a world-wide Islamic league;[138] and that St John Philby, adviser to King ʿAbd al-ʿAzīz Ibn Saʿūd of Saudi Arabia, was in India on the king's behalf, paving the way for the establishment of a Pan-Islamic

135 e.g. ʿAbd al-ʿAẓīm Ibrāhīm Manṣūr, *al-Jāmiʿa al-Islāmiyya naẓra fikriyya ʿanhā*, 80 ff.

136 Cf. T.C. Young, 'Panislamism in the Modern World', 199.

137 Alleged attempts by the Crown Prince of Yemen to set up a Muslim bloc, acc. to *al-Difāʿ* (Baghdad), 3 Aug. 1937, reported in NA, RG 59, 890g.00, General Conditions/108, Summaries from the press of Iraq, transmitted with US Chargé d'Affaires a. i. J.C. Satterwaithe's no. 851, to Secretary of State, 'Diplomatic', dated in Baghdad, 19 Aug. 1937.

138 *al-Muṣawwar* (Cairo), summarized in *Le Progrès Egyptien* (Cairo), 28 June 1947. See also FO 371/61563, file 6464/6464, enc.

league.[139] While these reports may have been premature, a new monthly, *Muslim World Islamic Quarterly in English*, appeared in Singapore, the first proclaimed objective of which was to advocate the establishment of a comprehensive Muslim league. Its honorary manager, Syed Ibrahim Bin Omar Alsagoff, who wrote for this first issue an article entitled 'A Muslim Universal League',[140] appealed for such a bloc to unite the entire Muslim world, laying down the relationships between individual Muslim states and providing a framework for coordinating them in international meetings.

Writing soon gave way to intense activity by a number of Muslims. Of these, probably the most mercurial was Chaudry Khaliquzzaman. After having been involved in Muslim politics in India and then in Pakistan, Khaliquzzaman founded his own Muslim People's Organization, committed to setting up a Muslim bloc of states, which he called 'Islamistan'. For this purpose, he visited in 1949 Baghdad, Damascus, Beirut, Mecca, Cairo, and London—without much success.[141] In 1952, a delegation of this organization visited Iraq and Tehran, with similar results,[142] although the press in Pakistan hailed these activities as significant.[143] Khaliquzzaman's views were expressed more precisely in an interview with R.A.B. Burrows, of the Foreign Office in London, on 3 November 1949.[144] Briefly, he emphasized the importance of establishing a bloc of Muslim states, thus broadening the basis of the Arab League

[139] FO 371/61530, file 12008/754, British Resident in the Persian Gulf W.R. Hay's no. 1992-S, 'Secret', to E.P. Donaldson at the Commonwealth Relations Office, dated in Bahrain, 21 Nov. 1947.

[140] The article covers pp. 3–5 of the issue, a copy of which can be consulted in the British Library, London.

[141] FO 371/75120, file 12174/1781, dispatch no. 984/4/49, 'Confidential', from the British Embassy in Iraq to the Foreign Office in London, dated in Baghdad, 27 Sept. 1949. Cf. FO 371/75120, file 12823/1781. FO 371/75121, file 13524/1781, British Ambassador to Egypt Ronald Campbell's no. 566, 'Confidential', to Ernest Bevin, Secretary for Foreign Affairs, dated in Cairo, 3 Nov. 1949.

[142] FO 371/98265, file 1073/4, 'Restricted', the chancery of the British Embassy in Iran, to Eastern Dept., Foreign Office, dated in Tehran, 28 Apr. 1952.

[143] FO 371/81920, file 1016/1, of 1949. FO 371/98311, file 1782/1, Quarterly Review of Islamic Affairs, 4th quarter 1951, dated 25 Jan. 1952.

[144] FO 371/75121, file 13978/1781, Foreign Office minute, dated 5 Nov. 1949.

(which was to serve as its nucleus). He then requested—and failed to obtain—British support for his plan.

Meanwhile, King ʿAbdallāh of Jordan was trying his hand at initiating a Muslim mini-bloc which would simultaneously help him to achieve his project of a 'Fertile Crescent' headed by himself and end his isolation among the Arab states. During 1949, he had contacts with Moḥammed Rezā Pahlavī, Shah of Iran, about setting up an Islamic bloc,[145] possibly made up of Jordan, Iran, Turkey, and Pakistan.[146] Some of the details of ʿAbdallāh's project were revealed, shortly before his assassination, in an interview with a Pakistani journalist, Wares Ishaq. The king spoke of his plan for a Fertile Crescent union among Jordan, Syria, and Iraq as a first step towards both an Arab bloc and an Islamic one involving Turkey and Pakistan as well. He called the latter 'A greater Commonwealth of Muslim nations—a commonwealth as free and flexible as the present British commonwealth, but with a closer alliance because here the religious and the larger interests would be identical'.[147]

ʿAbdallāh's death put an end to this project. Sir Zafrullah Khan (born 1893), Pakistan's Minister for Foreign Affairs in 1947–54, had no better luck when he visited Turkey and Egypt in February and March 1952 and discussed the merits of establishing a Muslim bloc—a suggestion which met with more approval in Egypt than in Turkey.[148] Apparently, the time for a comprehensive Islamic grouping had not yet come. More limited alliances, such as the Baghdad Pact among Turkey, Iraq, Iran, and Pakistan (1955–8), the United Arab Federation between Jordan and Iraq (1958), and the United Arab Republic joining Egypt, Syria, and Yemen (1958–61), were signed;[149] only after they had disintegrated was the situation ripe for a larger Islamic bloc. Meanwhile, writing on the matter continued unabated.

[145] FO 371/75120, file 11745/1781, Cf. Goichon, 37.

[146] FO 371/75120, file 11791/1781, telegrams from Sept. 1949. See also FO 371/75121, file 13978/1781, Burrow's minute of 15 Nov. 1949.

[147] Reported in *The Egyptian Gazette*, 11 July 1951, enc. in FO 371/91190, file 1029/1.

[148] FO 371/98311, file 1782/2, Quarterly Review of Islamic Affairs, 1st quarter of 1952, dated Apr. 1952. See also *Le Monde*, 22 Mar. 1952, 2.

[149] For details, see Rais A. Khan, 'Religion, Race, and Arab Nationalism', esp. 354 ff.

One of the first detailed discussions was a booklet in Arabic by Rāshid al-Barāwī, a professor in the Faculty of Commerce at Cairo University and author of several works on Middle Eastern politics and economics. Based on a public lecture delivered in 1952, it was entitled, quite appropriately, *The Islamic Bloc.*[150] According to al-Barāwī, the idea for the lecture was suggested to him by Zafrullah Khan's visit and Pan-Islamic canvassing in Egypt.[151] Starting with the assumption that co-operation is essential for world peace, al-Barāwī asked whether: (i) The objective conditions were conducive to setting up an Islamic bloc. (ii) The internal and international situation was suitable. (iii) The bloc, if established, would strive for common goals. (iv) It would remain independent. (v) It would be able to resist foreign intrigue.[152] After considering the history of Pan-Islam and the great interest evinced by Muslims in several international Muslim conventions, the author answered affirmatively. While acknowledging that the Muslim lands hardly formed a geographical or economic unit and that their inhabitants differed in language and other respects, he argued that their co-operating in one Islamic bloc was not only feasible, but imperative, both for their own economic future and for maintaining world peace.[153] An Islamic bloc would be particularly effective if it was composed of independent states.[154]

The less than unqualified support by Rāshid al-Barāwī for the idea of the Islamic bloc derived from his misgivings that there might be foreign intrigue via members subservient to colonial masters. The speedy achievement of independence in an ever-increasing number of Muslim states during the 1950s and the success of Muslim international conventions in the 1960s and 1970s[155] prompted a growing number of Muslim intellectuals to advocate, from the mid-1960s, the feasibility and desirability of a Muslim bloc. One of the most enthusiastic was Abu Aʿla al-Maududi who, in his 1967 booklet, *Unity of*

150 *al-Kutla al-Islāmiyya.*
151 Ibid. 7, 28 ff.
152 Ibid. 10–11.
153 Ibid. 37–62.
154 Ibid. 64–5.
155 For which see below, next section.

the Muslim World, affirmed that there could be no valid argument against a Muslim bloc (he used the term 'Muslim conference'), when there existed a British Commonwealth of Nations, an Organization of African States, and a Warsaw Pact. Moreover, he stated, those groups were homogeneous neither by religion, culture, language, nor (in the case of the British) territorial contiguity—while the Muslim bloc was all of these and more.[156] Then al-Maududi rebutted those critics who opposed the projected Muslim bloc by saying that it was unsound to base it on religion; he, on the contrary, held that, united by Islam, it would solve religious, cultural, economic, and defence problems.[157] Its real strength would lie in remaining outside the orbits of both superpowers.[158]

A year later, another Pakistani, Inamul Haq, a writer and lawyer, published a booklet entitled *Islamic Bloc: The Way to Honour, Power and Peace*. His basic assumption was that 'to catch up with modern times it is of utmost necessity that we the Muslim countries must closely integrate our economic, financial and defence policies'.[159] He developed the argument in a chapter entitled 'Unite or perish',[160] appealing for a 'joint and concerted effort'.[161] Unity of action, concentration of resources, industrialization[162]—all possible in the framework of an Islamic bloc—would improve the economy of all Muslim countries and bestow upon them political, military, and scientific power.[163]

The need for an Islamic bloc and the feasibility of its establishment being more or less accepted during the 1960s, the next decade seems to have been devoted to ways and means. It is noteworthy that most of the available works on its preferred form opined for a commonwealth of Islamic countries (of which the first intimation from high quarters seems to have been the project of King ʿAbdallāh of Jordan). Thus, Muhammad Muslehuddin, writing on the subject in a London

[156] Maududi, *Unity of the Muslim World*, 20–4.
[157] Ibid. 25–6.
[158] Ibid. 29. See also Piscatori, 174 n. 9.
[159] Inamul Haq, *Islamic Bloc*, p. x.
[160] Ibid., ch. 3.
[161] Ibid. 33.
[162] Ibid. 39–40.
[163] Ibid. 40–51.

monthly, in December 1970,[164] held that 'the task to be undertaken by the Muslims is to organize themselves so as to be a living force to combat the evils of nationalism and establish peace on earth'.[165] The best form of association would be a commonwealth, flexible enough not to impinge on the rights of its member states. Such a loosely knit group might be modelled on the British Commonwealth of Nations, at least at first. It would be guided by the Koran and the model of the Prophet Muḥammad into a bond of brotherhood and financial co-operation.[166]

A year later, a different kind of work appeared, *The Idea of an Islamic Commonwealth*, in Arabic.[167] Its author, Mālik Ben Nabī (1905–73), wrote chiefly in French on civilization, capitalism, and Afro-Asianism; he participated actively in the 1955 Bandung meeting of Afro-Asian countries and lived in Cairo from 1956 to the independence of Algeria.[168] The book under discussion had been first published in 1960 during Ben Nabī's Cairo exile.[169] In it, he argued that the Muslim world was so divided within itself that it could not share the revolutionary trends of the Afro-Asian peoples.[170] A Muslim union on the pattern of the British Commonwealth of Nations should resolve Muslims' problems, both doctrinaire and bureaucratic. While the British Commonwealth was made up of states, the Islamic one would be made up of nations, enabling Muslims to catch up with progress.[171]

Mention should also be made of two works by Chaudri Nazir Ahmad Khan, a former Pakistani Minister for Industry and Attorney-General of Pakistan, committed to the ideas of Pan-Islam and a Muslim commonwealth. The first, a 215-page book in English,[172] came out in 1972 and was entitled

[164] Muhammad Muslehuddin, 'Two Needs of Islamic World', esp. 26–30.

[165] Ibid. 26.

[166] Ibid. 28–30.

[167] Mālik Ben Nabī, *Fikrat commonwealth Islāmī*.

[168] Ghāzī al-Tawba, *al-Fikr al-Islāmī al-muʿāṣir*, 81 ff. Z.I. Lyevin, *Islam i natsionalizm v stranakh zarubyezhnogo Vostoka (idyeyniy aspyekt)*, 89–90.

[169] I have been able to consult only the 2nd edn. of this book, issued in 1971.

[170] Ibid. 56 ff.

[171] Ibid. 112 ff. A postscript to the 2nd edn., 125–6, pleaded for a start towards achieving the Muslim commonwealth.

[172] The book has allegedly been published in Arabic and French as well. An

Commonwealth of Muslim States: A Plea for Pan-Islamism. It had been first serialized eight years earlier, under the same title but in a much shorter version, in a Karachi monthly.[173] It was published by al-Ahibba,[174] a non-profit association, set up in Lahore in 1962 and presided by the author, to promote the cause of world Muslim unity. He started this book by stating, 'The future of the Muslim countries lies in their strength and their unity ... a common global organization of all Muslim States is essential for safeguarding and promoting the interests of world Muslims'.[175] He then blamed the Muslims' disunity and disorganization for their having no say in world affairs and being discriminated against in many countries. Since international organizations did not care to attend to Muslim interests, Muslims should establish, on a global basis, their own commonwealth of Muslim states with a 'Council of elders' of all Islamic lands to guide it.[176] Despite the difficulties, Pan-Islamic unity was achievable,[177] by attending to the fundamental needs of Muslim countries, most particularly to those of the poor ones. The resources existed and could be exploited jointly. The author suggested various measures to prepare Muslims for joining a commonwealth of their own: a peoples-to-peoples' programme; better acquaintance with culture and politics; improved means of communication and travel between Muslim countries; good libraries, storing up-to-date information about Muslims; Arabic as a lingua franca; popularization and translation of major Pan-Islamic works; economic co-operation (along with a Muslim common market); harnessing the mass media and founding Pan-Islamic associations; establishing a Muslim world news' service; founding a Muslim world development bank; co-ordinating foreign policy; using the mosque as a launching pad for all the above; selecting a permanent secretariat of Islamic coun-

Urdu transl. was being prepared, but I do not know whether it has been published.

173 *The Voice of Islam*, 12/12 (Sept. 1964) and 13/1 (Oct. 1964).

174 Which means 'dear friends', in Arabic. Its Urdu title was *Muḥibbān-i 'ālam-i Islāmī*, or 'Friends of the Islamic World'.

175 Chaudri Nazir Ahmad Khan, *Commonwealth of Muslim States*, p. i.

176 Ibid. 1–38.

177 Ibid. 39 ff., 62 ff.

tries.[178] The author even appended a draft constitution for the projected commonwealth.[179]

The same writer's second book was even larger. A 330-page work, it was published (again, by al-Ahibba) in English, in 1977, and entitled *Thoughts on Pakistan and Pan-Islamism*. It was dedicated to the memory of King Fayṣal of Saudi Arabia, assassinated in 1975. Although the book discussed various matters of Islamic interest, it did this through a Pan-Islamic prism, returning to advocate a commonwealth of Muslim nations[180] and then one of Muslim states.[181] Essentially, it is a summary of the previous book, along with other matters which had been covered in former lectures and articles of the author. An interesting point is that he claimed to have been the first person in Pakistan to broach the idea of a commonwealth of Muslim nations, as early as 1948.[182]

A more recent work advocated a confederated bloc of Islamic states and offered it a constitution. Written in Urdu and entitled *Trend Towards Unity*,[183] this 78-page book was published in 1983. Its author, Āl-i Aḥmad ʿAbīdī, is a Pakistani professor. The premise of this book was that all Muslims were one nation who had had a common Caliphate. The basic recommendations in the constitution suggested were the following: all rulers of Islamic states to provide facilities for Muslim education; no law, which was not in conformity with Islam, to be permitted; *zakāt*, or almsgiving, to be introduced; mosques and religious schools to be encouraged; the gap between rich and poor to be reduced. Further, a president should be elected for five years and a 31-member parliament, made up of representatives of the confederation's components, should meet in Medina. This confederated bloc would be jointly responsible for defending all Muslim countries and deciding on a joint foreign policy.

Despite speeches and writings advocating an Islamic com-

178 Ibid. 131–45.
179 Ibid. 189–94.
180 Chaudri Nazir Ahmad Khan, *Thoughts on Pakistan and Pan-Islamism*, 15–23.
181 Ibid. 47–66.
182 Ibid., pp. v, 7.
183 Āl-i Aḥmad ʿAbīdī, *Jazb-i bāham*.

monwealth,[184] the idea has not been carried through. Perhaps the decision-makers did not wish to bind themselves to a rigid framework; possibly, some of them did not wish to adopt a non-Muslim model like the commonwealth. Meanwhile, Muslim solidarity, the contemporary expression of Pan-Islam, increasingly manifested itself in all-Muslim organizations and conventions.

4. *All-Muslim Organizations*

Writing about a commonwealth of Muslim nations (or states) was anchored in a desire to promote an appropriate organizational framework within which Muslim solidarity could not only manifest itself, but become operational. The politically minded increasingly realized the power of methodical organization, imperative for the success of joint international efforts by a Pan-Islamic movement.[185] Moreover, as has been said, there were models to choose from; Muslim states were members of several important world organizations after World War II. Consequently, Muslims strived to set up an overall framework which would be perceived as both Islamic and participatory in international affairs.[186]

However, several of the attempts made in the early post-war years were less than successful, being of limited significance and no permanence. Significantly, individual attempts failed or were short-lived. Brief mention has been made of Chaudri Khaliquzzaman's attempts, from 1949, to form in Pakistan and elsewhere a Muslim Peoples' Organization as a nucleus for an international organization with a decidedly Pan-Islamic character. This body was intended to serve as a clearing house for debates about, and solutions to, issues of general relevance, such as the liberation of Muslims governed by others.[187] Attempts were made to recruit support for the Muslim Peoples'

[184] e.g. Diyāb, 295–8 (for 1978).
[185] Yakan, 119 ff. Fatḥī Yakan is a Lebanese Muslim.
[186] Afzal Iqbal, *Contemporary Muslim World*, pp. xxv–xxviii.
[187] FO 371/91209, file 1102/19. FO 371/91250, file 1781/4, Burrows' no. P 466, to Maclennen at the Foreign Office in London, dated in Karachi, 14 Dec. 1950, enclosing a report from *Dawn* (Karachi) of 7 Dec. 1950.

Organization in several countries, but it seems that this association was chiefly active in Pakistan. However, its Lebanese branch published an Arabic pamphlet, whose English subtitle was *Bulletin of the Moslem People's Organisation. Central Office in Karachi, Pakistan. Lebanese branch.*[188] This was a collection of articles by contemporary Muslim figures (such as Sayyid Quṭb)[189] and Arab men of letters (such as the Egyptian journalist Aḥmad Ḥasan al-Zayyāt (1885–1968),[190] in support of promoting an international Muslim association. Khaliquzzaman's personal efforts led to some practical results too. In March 1949 and May 1952, he convened in Karachi conferences attended by delegates from Pakistan and several other Muslim states. But his emphasis on involvement in the political problems of all Muslim nations and his slogan 'Break the boundaries between Muslim countries',[191] all in true Pan-Islamic style, may have frightened some delegates away. His organization was not heard of afterwards.

Other non-governmental international Muslim bodies included *Jamiʿat al-Islām* (more properly, *Jamʿiyyat al-Islām*).[192] This society, originally set up in Turkestan in 1868–9 to resist Russian expansion,[193] has lately been active in the West, collecting funds for Muslim refugees everywhere. A Lahore-based association, al-Ahibba, has already been mentioned. Established in 1962 and headed by Chaudri Nazir Ahmad Khan, it published several books advocating Pan-Islam, as well as a quarterly in English and Arabic, in which Pakistanis and other Muslims preached Muslim unity and argued about ways

[188] In Arabic, *Nashra ṣādira ʿan Munaẓẓamat al-shuʿūb al-Islāmiyya, al-markaz al-raʾīsī fī Karātshī, Pākistān. Farʿ Lubnān.* A copy may be consulted in the Library of the Middle East Centre, St Antony's College, Oxford.

[189] Ibid. 31–2.

[190] Ibid. 24–7.

[191] See NA, RG 59, 880.413/6–2450, US Embassy's Counselor F.W. Wolf's no. 686, to Dept. of State, dated in Karachi, 24 June 1950. Ibid. 880.413/9–1350, US Embassy's Counselor Warwick Perkins' no. 453, to Dept. of State, dated in Karachi, 13 Sept. 1950. Ibid. 890d.413/6–752, US Embassy's First Secretary C.D. Withers' no. 1542, to Dept. of State, dated in Karachi, 7 June 1952.

[192] In Arabic it called itself *Jamʿiyyat al-Islām muʾassasa insāniyya taʿlīmiyya*. See *Jamiʿat al-Islam: History—Policy—Program*. A copy of this publication, by the Vienna branch of this society, may be consulted in the Oriental Library, University of Durham, UK.

[193] Ibid. 3.

to achieve it.[194] Published between 1969 and 1977, this was entitled *al-Ahibba*.[195] Chaudri Nazir Ahmad Khan, the association's founder (and first editor-in-chief of the quarterly), defined its objectives as follows, '*al-Ahibba* would humbly try to organize Muslims not on regional, territorial or racial lines, but strengthen the eternal bond of universal brotherhood that Allah bestowed on them ... It is intended through this unity to move make Muslims invulnerable, free and dignified, so that they can undertake their destined responsibility, viz., the leadership of the world'.[196]

One of the most interesting projects for an international Muslim organization was prepared in Arabic, in 1973 (but never put to the test) by Muḥammad Ḍiyā' al-Dīn al-Rīs, a professor of Islamic history at Cairo University and 'Abd al-'Azīz University in Mecca.[197] He suggested that it be named the Organization of Islamic States,[198] to include all Muslim states, each of which would delegate a representative. These would constitute a general assembly, meeting for three-months' sessions annually; the assemblies might meet at the level of Prime Ministers or Ministers for Foreign Affairs. A smaller executive council, elected by the assembly, would meet monthly or *ad hoc* to deal with urgent intra-Muslim matters. The assembly would be sovereign and its resolutions binding on all member states. It would fulfil the role of the Caliphate in past times (but not interfere in the affairs of each individual state or of such institutions as the Arab League). The secretary general would not be a caliph, but solely represent the organization, which would generally act via committees—for political, legal, military, propaganda, cultural, and social affairs. Its seat would be in Jedda during the preparatory stage, then in whatever Muslim city was selected. Its declared character would not be religious, in a narrow sense, but political and cultural. Its aims would be to work for peace;

194 Among the topics debated were various proposals for a constitution for a projected commonwealth of Muslim states. See e.g. *al-Ahibba*, 1/2 (Oct. 1969), 33–9, repr. below, app. T.

195 Sets of which may be consulted in the Library of Congress, Washington, DC, and the Butler Library, Columbia University, New York City.

196 Chaudri Nazir Ahmad Khan, *Thoughts*, 121–2.

197 al-Rīs, 313–25.

198 *Munaẓẓamat al-duwal al-Islāmiyya*.

co-operate with the United Nations; protect Muslim countries; fight Zionism and imperialism; foster Islam in everyday life; raise the levels of morals and science; inform the entire world of Islam's principles; fight racism, fanaticism, and exploitation; strive for a world system based on justice, peace, and fraternity.

Such projects, and others,[199] drawn up by individuals, stood little chance of turning into fully fledged international Muslim organizations, unless sponsored by the state. Even state sponsorship alone was insufficient, as Egypt's brief experiment in 1954–6 with a General Islamic Congress (*al-Mu'tamar al-Islāmī al-ʿāmm*),[200] conjointly with Saudi Arabia and Pakistan, proved.[201] The last two eventually pulled out[202] and this body remained a strictly Egyptian institution. Similar failure attended a rival association, set up by King Ḥusayn of Jordan and President Ḥabīb Būrgība of Tunisia, named the Islamic Conference, with headquarters in Jerusalem.[203] A much wider co-operation with other Muslim states was imperative for achieving durable results. This was true in 1981, when the King and Government of Jordan founded and funded a Royal Academy for Islamic Civilization research.[204] An institution created for international scholarly research, it co-opted learned Muslims from several other states but remained, none the less, a Jordanian body.

Several Pan-Islamic groups sprang up after World War II, such as an Organization of the Muslim Peoples (in Arabic,

[199] For which see Bashīr al-ʿAriḍī, 'al-Tanẓīm al-Islāmī lam yanbathiq ʿām 1969 wa-lakinnahu yarjiʿ ilā awākhir al-qarn al-māḍī', 11. See also T.Y. Ismael and J.S. Ismael, *Government and Politics in Islam*, 132.

[200] al-ʿAriḍī, ibid. 10. B. Lewis, 'The Return of Islam', 23. For the statutes of this body, cf. 'The Proposed Annual Congress at Mecca', 26–7, and Muḥammad Ḥasan ʿAwwād, 36–40. For its activity, see Mahmud Brelvi, 'The Islamic Congress (*al-Mo'tamar al-Islāmī*), Cairo', 13. It was not revived until 1966, see P. Rondot, 'Perspectives du congrès islamique mondial', 31–2. Shaukat Ali, *Pan-movements*, 229–30. Ismael and Ismael, 132.

[201] For the aims of this congress, as proclaimed by Anwar al-Sādāt, its Secretary General, see NA, RG 59, 880.413/9-1954, Jefferson Caffery's no. 493, to Dept. of State, dated in Cairo, 18 Sept. 1954. Cf. *Le Monde*, 19–20 Sept. 1954, 2.

[202] A.B.A. Haleem, 'The Baghdad World Muslim Conference', 175.

[203] Shaukat Ali, *Pan Movements*, 230.

[204] Royal Academy for Islamic Civilization Research, *Beginnings*. See also *Jordan Times* (Amman), 29–30 Jan. and 1 Feb. 1981.

Munaẓẓamat al-shuʿūb al-Islāmiyya),[205] which held its first convention in 1952.[206] Our discussion will focus, however, on the three most important international Muslim organizations in the post-World War II era: the Muslim World Congress, the Muslim World League, and the Organization of the Islamic Conference.[207]

(*a*) *The Muslim World Congress* (in Arabic, *Mu'tamar al-ʿālam al-Islāmī*; in Urdu, *Mo'tamar-i ʿālam-i Islāmī*).

While allegedly a continuation of the Muslim conventions[208] in Mecca in 1926 and in Jerusalem in 1931,[209] this was essentially a new grouping. Its main claim to continuity was the convenor of the Jerusalem convention, Amīn al-Ḥusaynī, who was the *éminence grise* of the Muslim World Congress until the early 1970s, when he was succeeded in this function by Maʿrūf al-Dawālibī, a Syrian statesman (and erstwhile member of the Muslim Brethren). It was set up (or revived) in Karachi, in 1949, probably with the encouragement of the government,[210] as an international Muslim body,[211] succeeding an organization founded in Cairo before World War II and revived in Pakistan in 1948, the Association of Islamic Brotherhood (*Jamāʿat al-ukhuwwa al-Islāmiyya*).[212]

The Muslim World Congress was thus one of the longest extant associations for Islamic solidarity; in 1983, it comprised 36 member states. Its headquarters were in Karachi and it had

[205] Nehemia Levtzion, *International Islamic Solidarity and its Limitations*, 12, calls it 'Congress of the Muslim peoples' (*Mu'tamar al-shuʿūb al-Islāmiyya*).

[206] For its objectives see below, app. s.

[207] On these three organizations, see *Les Organisations islamiques internationales*. Levtzion, *passim*. al-ʿArīḍī, 10–11. Charis Waddy, *The Muslim Mind*, 174–5. D. Pipes, *In the Path of God*, 303 ff. *Islam kratkiy spravochnik*, 127–33. Ismael and Ismael, 131 ff. J. Reissner, 'Internationale islamische Organisationen', in: W. Ende and U. Steinbach, eds., *Der Islam in der Gegenwart*, 539–47. R. Schulze, 'Regionale Gruppierungen und Organisationen'.

[208] See above, ch. v, section 4.

[209] Waddy, 174. Ismael and Ismael, 132. al-ʿArīḍī, 10. Fernau, 84.

[210] Levtzion, 12.

[211] Cf. FO 371/91209, file 1102/19. FO 371/91250, file 1781/5, confidential report from Karachi, dated 29 Jan. 1951.

[212] FO 371/91209, file 1102, memorandum by the Foreign Office's research department. Cf. *Les Organisations islamiques internationales*, 7. For the Egyptian association, see above, ch. v, section 2. For the Pakistani one, Goichon, 39 ff.

opened branches in 60 countries.[213] Its regional offices were in Beirut for the Middle East; Kuala Lumpur for South-east Asia; Manila for the Far East; Dakar for West Africa; and Mogadishu for East Africa.[214] It published several works, as well as *The Muslim World: A Weekly Review of the Motamar*.[215] This was edited in English, in Karachi, by Inʿamullah Khan, Secretary General of the Muslim World Congress, starting in 10 August 1963. Its editorial policy[216] was to deal with 'day-to-day national and inter-Islamic and international issues, and suggest, from time to time, such means and measures so as to remove misunderstandings between Muslim communities the world over, nay contribute our mite to build bridges ... Our only demand is that we want a square deal for this world of 600 million people'. Special emphasis was laid, however, on a Pan-Islamic approach, as follows, 'A vision—however dim—of the need for unity is found among the Muslim peoples everywhere. It is born basically of their religion, and equally also because of their common past, their common concern and needs of today and also because of their common hopes and aspirations for tomorrow'.

The association convoked periodical conventions. At the first of these, held in Karachi in February 1949,[217] it was set up as a non-governmental organization on the following principles:[218] propagating the teachings of Islam; eliminating the nationalist ties and communal hostilities imposed by geographical conditions; co-operating with all Muslim lands in order to promote Islamic unity. In its second convention, held in Karachi in February 1951,[219] it undertook to strive to persuade Muslim governments and peoples to renounce their differences; to propagate Arabic as the lingua franca of all Muslims; to assist any Islamic people which was attacked; and

[213] *Islam kratkiy spravochnik*, 132. A. Akhmyedov, *Islam v sovryemyennoy idyeyno-politichyeskoy borʾbye*, 22.

[214] Inʿamullah Khan, 'The Moʾtamar al-ʿAlam al-Islami', 27.

[215] A set is available in the Library of Congress, Washington, DC.

[216] *The Muslim World: A Weekly Review of the Motamar* (Karachi), 1/1 (10 Aug. 1963), 1.

[217] Sindi, 'Muslim World', 120–2, for details.

[218] al-Barāwī, 25–8.

[219] 'Sixth World Muslim Conference', 24 ff., and Sindi, 'Muslim World', 122–4, for details.

to unite in defence of Islam.[220] At its third convention, in March 1952, in addition to explicitly supporting the political struggle of several Muslim states, the Muslim World Congress recommended that all Muslim states evolve laws of common citizenship; co-ordinate their trade policies with preferential treatment; and frame constitutions and laws in accordance with the *sharīʿa*.[221] Another convention, under the chairmanship of the organization's president, A.B.A. Haleem, met in Baghdad in June 1962; members from 32 Muslim states and Muslim minorities in non-Muslim lands participated, on an individual basis; resolutions urged stronger cultural ties and closer economic consultation.[222] Other conventions met in Mogadishu with 33 states and Muslim minorities represented, in December 1964–January 1965,[223] and in Amman, in September 1967.[224] Throughout, the Muslim World Congress co-operated with the two other international organizations discussed below, chiefly in the cultural domain.[225] However, the principles adopted above, and the organization statutes approved in 1967, stressed not only the spread of knowledge about Islam and raising the social, cultural, economic, and material levels of Muslims, but support for Muslim wars of liberation as well.[226]

The organization had a truly international Islamic character, at least in its administration. Its five Vice-Presidents originated, at its inception, from Indonesia, Iran, Afghanistan, Turkey, and Algeria, while its twelve-member Executive Committee was equally representative. In practice, however, it

220 Details in FO 371/91208, files 1102/12 and 1102/13. NA, RG 59, 880.413/2-1751, US Chargé d'Affaires Warwick Perkins to Dept. of State, 'Restricted', dated in Karachi, 17 Feb. 1951. Inʿamullah Khan, 3–4. For the text of the resolutions, see below, app. R.

221 FO 371/98265, files 1073/1 and 1073/2.

222 Haleem, 179–82. For the proceedings, see Motamar al-Alam al-Islami, *Proceedings of the World Muslim Conference* (1962). Sindi, 'The Muslim World . . .', 125–9.

223 Inʿamullah Khan, 'The Mo'tamar al-ʿAlam al-Islami', 27–9. Sindi, 'King Faisal and Pan-Islamism', 198 n. 3. Id., 'Muslim World', 129–32. Hans Kruse, 'The Politics of the Islamic Call', 210–11; Kruse's text is based on a lecture he delivered in Delhi, in 1966. For the proceedings, see Motamar al-Alam al-Islami, *Proceedings of the World Muslim Conference* (1965).

224 Sindi, 'Muslim World', 132–3.

225 Schultze, 'Regionale Gruppierungen', 476.

226 See text in *Les Organisations islamiques internationales*, 9.

was managed by a triumvirate: two Pakistanis, its President Haleem and its Secretary General Inʿamullah Khan, and a Palestinian, the former Grand Mufti Amīn al-Ḥusaynī.[227] The organization was fortunate in having the same secretary general from 1950 and up to the time of writing. Inʿamullah Khan, a Burmese-born (in 1914) Muslim who had settled in Pakistan soon after its independence, started his public Pan-Islamic career in October 1950, when he presented to UN Secretary General Trygve Lie a scroll with one million Muslim signatures, which he brought himself to New York, together with a memorandum on the Kashmir case.[228] He also obtained Saudi funding for his organization, edited a book entitled *Studies on the Commonwealth of Muslim Countries*,[229] and wrote the introduction to Inamul Haq's *Islamic Bloc: The way to Honour, Power and Peace*. He was a frequent contributor to *al-Ahibba* and other Pan-Islamic magazines. Inʿamullah Khan was awarded, in March 1988, the prestigious Templeton Prize.[230]

(*b*) *The Muslim World League* (in Arabic, *Rābiṭat al-ʿālam al-Islāmī*).

Like the Muslim World Congress, the Muslim World League is a non-governmental organization, but with greater impact on Islamic and Pan-Islamic affairs.[231] Formally established in 1962 in Mecca (where its headquarters are still located), this

[227] NA, RG 59, 880.413/4-152/XR 398.413Ka, Warwick Perkins's no. 1290. 'Confidential', to Dept. of State, dated in Karachi, 1 Apr. 1952.

[228] United Nations' press release M/689, dated 16 Nov. 1950. See also FO 371/84233, file 1015/557. NA, RG 59, 880.413/10-2850, cable from the US Embassy in Karachi, to Secretary of State, dated 28 Oct. 1950.

[229] Secretariat of the Motamar al-Alam al-Islami (Muslim World Congress), compiler, *Studies on the Commonwealth of Muslim Countries*.

[230] *The New York Times*, 3 Mar. 1988, A 18.

[231] There is one book about it, by Sharipova, *Panislamizm syegodnya*; I understand that Reinhard Schulze, too, is writing a monograph on it, *Islamischer Internationalismus im 20. Jahrhundert*. Meanwhile he has published an article about it, 'Die Islamische Weltliga (Mekka) 1962–1987'. See also *Les Organisations islamiques internationales*, 24–75. Mushirul Haq, 'The Rabitah'. G. de Bouteiller, 'La Ligue Islamique mondiale, une institution tentaculaire'. R.M. Sharipova and T.P. Tikhonova, 'Liga Islamskogo Mira'. Kruse, 203 ff. Waddy, 174. al-ʿArīḍī, 11. Ismael and Ismael, 132. *Islam kratkiy spravochnik*, 132. Sindi, 'Muslim World', 143–7. A. Akhmyedov, 'Vsyemirnaya Islamskaya liga'.

has been an unofficial agency of the Saudis,[232] who have been funding it generously (fifty million US dollars in 1974,[233] much more at present). Launched during the Pilgrimage season[234] by a meeting of 111 Ulema and religious dignitaries, it later co-opted political activists such as al-Maududi.[235] Its main administrative organs were a 60-member council,[236] which has met annually at Mecca, during the Pilgrimage, to determine the general policy of the organization; and a secretariat with executive power to carry on the League's day-to-day work. The secretariat has also co-ordinated the branches founded in, and representatives sent to, all Islamic countries and large Muslim groups in Europe, the United States, Canada, South America, and Australia. Its various departments have been concerned with public relations, documentation, press services, Islamic culture, the *sharīʿa*, the Islamization of law in Muslim lands, responses to Christianization in Muslim countries, Muslim minorities, financial assistance to Muslims afflicted by natural disasters, translations of the Koran, students and grants, distribution of the Koran and other religious texts, audio-visual equipment, computer services, a printing press, library services, budgeting and financial allocations, the construction of mosques and training of officials to serve in them, Muslim propaganda, collaboration with other Islamic bodies, and relations with the United Nations (in which the League has had non-governmental accreditation).[237]

As can be deduced from the above, the Muslim World League's activities have been multivaried and, as a former French Ambassador to Saudi Arabia, Georges de Bouteiller, described them, 'tentacular'.[238] They have certainly bypassed the goals defined in its constituting charter in 1962, which focused on propagating Islam, studying its problems, combating its enemies, and eliminating all obstacles which impeded Islamic

[232] E. Mortimer, *Faith and Power*, 177–8.
[233] Akhmyedov, *Islam v sovryemyennoy*, 17, based on the Muslim press.
[234] Levtzion, 18. Sindi, 'King Faisal', 186.
[235] Sharipova, *Panislamizm syegodnya*, 18.
[236] A 50-member council, according to Akhmyedov, *Islam v sovryemyennoy*, 17.
[237] *Les Organisations islamiques internationales*, 27–30. Sharipova, *Panislamizm syegodnya*, 19 ff.
[238] G. de Bouteiller, 'La Ligue Islamique mondiale', 73 ff.

solidarity. Since it has considered itself an association of all Muslim communities,[239] the Muslim World League, in a spirit characteristic of Islam, has perceived its duty to be involved in anything and everything of concern to Muslims everywhere. Hence, it has maintained representatives even in small Muslim concentrations (e.g. in the Maldive Islands). While subscribing to the current theories of 'Islamic economics',[240] most its work has been religious and cultural. Its educational and proselytizing efforts have largely been channelled through the mosques it has been maintaining, the new ones it has been constructing, and the clerics it has been training to officiate in them (and whose salaries it has often paid);[241] regional offices for the mosques have seen to it that these activities maintained a global perspective.[242]

An international seminar (*Nadwa Islāmiyya ʿālamiyya*), convened by the Muslim World League before each Pilgrimage period, focused on a special area of its activities and recommended future priorities. Subjects have included Jerusalem and Palestine; Muslim minorities; and Islam in South-east Asia. These were then published and distributed (usually free of charge). While the stress was on doctrinaire topics, lectures on Islamic solidarity and other matters have had a place as well.[243] Other publications of the Muslim World League were the Koran and its translations, works on religion, and political tracts, such as one by ʿAlī al-Muntaṣir al-Kattānī on the

[239] Not states, a function reserved for the Organization of the Islamic Conference, to be discussed below.

[240] Details in R. M. Sharipova, 'Ekonomichyeskiye kontsyeptsii idyeologov Ligi Islamskogo Mira'. Id., *Panislamizm syegodnya*, 73 ff. For the theories themselves, see A. R. Shibli, *Muslim World Economic Collaboration*. Shibli styles himself 'President, Forum of Muslim world economists, Lahore', which could be another attempt at an all-Muslim organization.

[241] In March 1987, there was quite a row in the secular Republic of Turkey, when the leftist daily *Cumhuriyet* uncovered and published a serialized report by Uğur Mumcu that the Muslim World League had been paying the salaries of Turkish clerics in the Muslim communities in Federal Germany. See also Sami Kohen, 'Row Erupts in Turkey over Saudi Funding of State Clergy'.

[242] Mushirul Haq, 'The Rabitah', 58–60. Akhmyedov, *Islam v sovryemyennoy*, 18–19, mentions an international Muslim convention in Mecca, convoked by the Muslim World League, to discuss the mosques.

[243] See e.g. ʿAbd Allāh Ibn Nūḥ, 'al-Taḍāmun al-Islāmī'.

conditions of Muslims in the Communist bloc[244]—as part of its self-perceived role as protector of Muslim minorities everywhere.[245] Periodical publications, all issued in Mecca, included an Arabic weekly, *Akhbār al-ʿālam al-Islāmī* (News of the Muslim World); an Arabic monthly, started in 1962, *Majallat rābiṭat al-ʿālam al-Islāmī*[246] (Journal of the Muslim World League), its English equivalent *Muslim World League Monthly Magazine*,[247] later renamed the *Muslim World League Journal*; and an Arabic monthly, *Daʿwat al-ḥaqq* (The Call for Righteousness).

No less significantly, the Muslim World League has served effectively as an umbrella organization for numerous other Islamic associations and groups.[248] In addition to funnelling Saudi grants to some of these,[249] the Muslim World League succeeded in integrating many of them into its own activities, in one way or another; this even included rival organizations, such as the Muslim World Congress, which it reportedly subsidized. An international gathering of delegates from Muslim organizations, which it convened in Mecca, in April 1974,[250] acquiesced in the objectives and policies of the Muslim World League,[251] approved its co-ordinating role, and established

[244] ʿAlī al-Muntaṣir al-Kattānī, *al-Muslimūn fī al-muʿaskar al-shuyūʿī.* The Communists were frequently attacked in the Muslim World League's periodicals (no less than the capitalists), see e.g. 'Statement by Muslim World League' in 1965.

[245] Details in Sharipova, *Panislamizm syegodnya*, 123–30.

[246] Later renamed *al-Rābiṭa*.

[247] First publ. in English and Arabic, in July 1963, it became English-language only from July 1965, in which issue it was stated that the monthly would cover 'the ideology, culture, politics and economics of the Islamic countries'.

[248] For a partial list—but quite a lengthy one—of the recently active organizations and institutions, see D. Pipes, *In the Path of God*, 303–4.

[249] Cf. ibid.

[250] An earlier one had met, also in Mecca, at the headquarters of the Muslim World League, in October 1968. See 'First Conference of World Islamic Organizations Held at Mecca', 27–9.

[251] Mushirul Haq, 'The Rabitah', 62. Akhmyedov, *Islam v sovryemyennoy*, 18. Acc. to Khurshid Ahmad, 'Mecca Conference of World Muslim Organizations', the convention was divided into five committees, discussing: (i) Islamic *daʿwa* (propaganda), means and methods; (ii) contemporary ideologies; (iii) Palestine and other problems of the Muslim world; (iv) Muslim minorities; (v) co-ordination of Islamic work. See also 'Resolutions and Recommendations of the World Conference of Islamic Organizations'.

central offices for all five continents and new branches. One example might enable us to gauge better the method and significance of these and other moves, including international conventions of the Muslims of Australia (1975), Africa (1976), the Americas (1977), and Asia (1978), in addition to several for the Muslims of Europe. The Muslim World League's co-ordinating committee for Africa (an important area of the League's activity) met in Dakar to discuss its strategy.[252] The new Secretary General of the Muslim World League, Muḥammad ʿAlī al-Ḥarakān, presided and delegates from the Muslim World Congress, the Organization of the Islamic Conference, and several other Muslim associations, such as the All-Africa Muslim Congress, participated. Representatives from twelve African states, including Senegal's Prime Minister, were invited. In 1978, the Muslim World League's co-ordinating committee for Asia met in Karachi. Similar meetings were held in North America and elsewhere, continuously improving co-ordination, supervision, and effectiveness in increasing numbers, improving education, raising well-being—and repeating slogans of Muslim solidarity in a Pan-Islamic style. Although the Muslim World League's efforts are too recent for a definitive appraisal, it seems that its attempts to employ Islamic and Pan-Islamic ideals to solve old problems by contemporary means are gradually being rewarded.

(c) *The Organization of the Islamic Conference.*

This is an association of Muslim states, while the Muslim World League is one of Muslim communities, so they complement one another; the Muslim World League welcomed, indeed, the setting up of the new organization.[253] The Organization of the Islamic Conference (in Arabic, *Munaẓẓamat al-muʾtamar al-Islāmī*) was initiated at a summit meeting of Islamic heads of state in Rabat, in 1969, although it was Saudi-inspired and financed and its regular seat is in Jedda.[254]

[252] For details about this meeting, see Sharipova, *Panislamizm syegodnya*, 20 ff.

[253] Rābiṭat al-ʿālam al-Islāmī, *Qarār al-majlis al-taʾsīsī—al-dawra al-ḥādiya ʿashra*, 6 ff.

[254] A recent book on it is Hasan Moinuddin, *The Charter of the Islamic Conference and Legal Framework of Economic Co-operation among its*

Although King Fayṣal had prepared the ground, since 1965, by personally visiting Islamic rulers and persuading them to co-operate,[255] and King Ḥasan II of Morocco had repeatedly called, in 1968–9, for an all-Muslim convention,[256] the meeting in Rabat itself was triggered by a psychotic Australian Christian starting a fire in al-Aqṣā Mosque in Jerusalem.[257] Although the fire was put out before serious damage had been caused, and Israeli authorities condemned the arson and arrested and brought to trial the culprit, suspicions of a dark plot by Israel to destroy the mosque were aroused. Heads-of-state and delegates from twenty-four Muslim states and the PLO gathered to blame Israel, in a demonstration of Islamic solidarity, agreeing to institutionalize their co-operation.[258] The anti-Israel stand and the desire to regain Palestine and Jerusalem for Islam were, indeed, the only *constant* factors agreed upon in the organization which was to be set up; to paraphrase Voltaire, had Israel not existed, someone would have had to invent it—to cement the bonds of Islamic solidarity.[259]

The 1969 Islamic summit in Rabat was followed, between March 1970 and March 1972, by conventions of Islamic Ministers for Foreign Affairs[260] who, with the secretariat they

Member States. See also Abdullahil Ahsan, 'Muslim Society in Crisis'. G. de Bouteiller, 'La "Nation islamique"', part 1, 69; part 2, 101ff. *Les Organisations islamiques internationales*, 76–118. L. B. Borisov, 'Organizatsiya Islamskaya Konfyeryentsiya'. A. Ionova, 'Islam i myedzhdunarodnoye ...'. B. Etienne, *L'Islamisme radical*, 268–75. T.P. Miloslavskaya and G.V. Miloslavskiy, 'Nekotoriye voprosi tyeorii i praktiki finansovo-ekonomichyeskoy dyeyatyel'nosti OIK'. A. Akhmyedov, *Islam v sovryemyennoy*, 10–23. G. Feuer, 'L'Organisation de la conférence islamique'. Muazzam Ali, 'Role of O.I.C. in Promoting World Peace'. Taoufik Bouachba, 'L'Organisation de la conférence islamique'. J.-P. Charnay, *L'Islam et la guerre*, 108 ff. 'Organisation of the Islamic Conference, instrument of Islamic solidarity'.

[255] Sindi, 'King Faisal', 188–9.

[256] See *Le Monde*, 4 Jan. 1968, 3; 11 Apr. 1968, 6; 31 May 1969, 9.

[257] Cf. Ruthven, 359.

[258] For this Rabat Summit meeting see, *inter alia*, Shameem Akhtar, 'The Rabat Summit Conference'. Cf. 'La Conférence islamique au sommet', 28–34.

[259] *Noviye Knigi SSSR* (Moscow), 88/11 (11 Mar. 1988), item 41, announced the forthcoming publication in 1989 of a book, authored by A. Kudryavtsyev, entitled *Islamskiy mir i Palyestinskaya problyema*, focusing on the support of the ideologues of Islamic unity for the Palestinians.

[260] The conventions of the Ministers for Foreign Affairs were held as follows: Jedda, 23–5 Mar. 1970; Karachi, 26–8 Dec. 1970; Jedda, 29 Feb.–4 Mar. 1972; Benghazi, 24–6 Mar. 1973; Lahore, 22–4 Feb. 1974; Kuala Lumpur, 21–5 June 1974; Jedda, 12–15 July 1975; Istanbul, 12–15 May 1976. See also Sindi, 'Muslim World', 174–231.

had set up, succeeded in drawing up a charter[261] for the Organization of the Islamic Conference (ratified in March 1972). The principles underlying the charter and the activities of the Organization of the Islamic Conference in the cause of Islamic solidarity are more pertinent to our discussion than its structure, which is patterned after those of other international bodies of our day (chiefly the United Nations—in which it has observer status—and its agencies).[262] The Organization of the Islamic Conference is made up of forty-five states[263] (Egypt has been suspended and reinstated; Afghanistan was suspended following its Sovietization in 1980). The PLO is a member, too, while the Muslims in Cyprus, Nigeria, and the Philippines, along with the United Nations, the Arab League, and the Organization of African Unity, have observer status. The leading body is made up of heads of states and of government, whose summit meetings ought to take place every three years (in practice, these have occurred in 1974, 1981, 1984, and 1987); an annual gathering of Ministers for Foreign Affairs, which is the main decision-making body;[264] and a General Secretariat, to attend to current business. The secretary general[265] originally had four assistants, each directing a department[266] of, respectively, politics and information (as well as minorities); administration and finance; cultural and social matters; and Palestinian (and Jerusalem's) affairs. Evidently, these are the main foci of the Organization's interests and activities (even if some of its work is carried out by *ad hoc* committees, as on the Iran–Iraq war).

The charter of the Organization of the Islamic Conference is

[261] For a detailed analysis of the charter, see Hasan Moinuddin.

[262] For its structure, see *Les Organisations islamiques internationales*, 80ff. Ahsan, 61–5. Ismael and Ismael, 133–4.

[263] Participation applied, at first, solely to states with a Muslim majority. Some member states however have Muslim minorities, such as Uganda, Cameroon, and Burkina-Faso (adherence was based on their presidents being Muslims, at that time).

[264] For its meetings and resolutions, see Islamic Council of Europe, *The Muslim World and the Future Economic Order*, 353–65.

[265] The first was Tunku Abdur Rahman Putra, who resigned his Premiership in Malaysia to assume this position. See his 'Islamic Unity'. The current one is Muḥammad ʿAlī al-Ḥarakān, formerly Saudi Arabia's Minister of Justice. See A. Akhmyedov. 'Vsyemirnaya Islamskaya liga', 139.

[266] A fifth department was set up later, see R. Schulze, 'Regionale Gruppierungen', 472.

very definitely forward-looking.[267] To a large degree, it was a compromise of the clashing interests of its members as well as between conflicting Islamic and secular ideas, most obvious in the attempt to combine the principles of Islam and the mechanisms of a contemporary international organization. The resulting body, while built up on the United Nations' model and paying lip-service to current maxims of modernity ('international peace and security' and the like), has basically been oriented in an Islamic direction. Both in its charter and in its manifold activities, the guiding principle of the Organization of the Islamic Conference has been the consolidation of Islamic solidarity at state level. The Organization's objectives are, indeed, plainly stated (art. II*a*) as follows:

> 1. To promote Islamic solidarity among member states; 2. To consolidate cooperation among member states in the economic, social, cultural, scientific and other fields of activities ... 5. To co-ordinate efforts for the safeguard of the Holy Places and support the struggle of the people of Palestine ... 6. To strengthen the struggle of all Muslim peoples with a view to safeguarding their dignity, independence, and national rights ...[268]

Islam being the most visible bond between the member states of the Organization of the Islamic Conference, this body cannot be considered a regional association. It is, indeed, *sui generis*, superficially resembling the Catholic Church in certain respects, but differing from it in its essentials (the Church, after all, is not an association of Catholic heads of state). It does not aim at a supra-national *umma* expressed in a union of all its members, all of which remain sovereign, nor even at a federation. Its Pan-Islamic character is expressed in its aiming at greater integration of the Islamic states[269] in the remote future and in its promoting Islamic solidarity in the mean time, by co-operation and co-ordination on all possible levels. This has been visible in three main domains—the political, the

[267] As rightly emphasized by its then secretary-general, Ḥasan al-Tuhamī, in an interview. See ʿAbd al-Ḥalīm ʿUways, 'al-Muʾtamar al-Islāmī', esp. 95 ff.

[268] The preamble of the charter laid down practically the same guide-lines, couched in more general terms. The charter itself has been repr. several times, e.g. by Moinuddin, 186–92, and Ahsan, 188–96.

[269] Cf. the views of the organization's first secretary-general, the Tunku, as quoted by Ahsan, 54–5.

economic, and the cultural—all of which mark the activities of the first organization, since the 1924 abolition of the Caliphate, that has taken global measures to institutionalize Islam and Pan-Islam.

Political issues debated and acted upon by the Organization of the Islamic Conference were chiefly those on which a large measure of consensus could be reached: the 'occupation' of Muslim territories by foreign, non-Muslim forces; in particular the cases of Palestine and Afghanistan, of which the former seemed more persistent. The Organization rejected the UN Security Council's Resolution 242 and devised a strategy for supporting the PLO's claims. Its member states co-ordinated their steps in international bodies, at the Vatican, and at the World Council of Churches, as well as their diplomatic propaganda in many of the world's capitals to turn political attitudes against Israel. These were accompanied by economic measures to weaken Israel's economy and reduce international support for it.[270] There was even some discussion among the member states about military co-ordination which, apparently, has not yet been put into effect. In contrast to the *Jihad* proclaimed by the Organization in this case, it did little to prevent or mitigate the Pakistan–Bangladesh War in 1971. Its attitude to the Soviet penetration in Afghanistan was quite moderate, too—since it was dealing, in this case, with a superpower and since several of its member states were reluctant to support an anti-Soviet stand.[271] Hence, little was done besides protesting verbally,[272] suspending Afghanistan's membership of the Organization in 1980, appealing for assistance for the Afghan refugees, and forming a committee to co-operate with the UN Secretary General in attempting to find a solution. Only after the withdrawal of the Soviet forces, was Afghanistan (actually, the government of the fundamentalist guerrillas) reinstated at the meeting of the Islamic Conference

[270] Examples ibid. 68–84.

[271] For the organization's stand *vis-à-vis* the superpowers, cf. Borisov, 107. For the watered-down resolutions of the Islamabad convention of Islamic Ministers for Foreign Affairs in the matter, see *Islamic Conference of Foreign Ministers: Eleventh Session, Islamabad, 17–22 May, 1980*, 86–8.

[272] J.-P. Filiu, 'L'Afghanistan et le mythe de l'internationalisme islamique', esp. 61.

of Foreign Ministers, in Riyad (in mid-March 1989). A similar attempt at solution-by-committee was tried (and was equally unproductive) to end the Iran–Iraq War (the war stopped for other reasons) and to bridge the hostility between Pakistan and Bangladesh.

The Organization of the Islamic Conference has been somewhat more successful in another political matter that is of great concern to it—the situation of Muslim minorities. In addition to sponsoring statistical and other research about these minorities, and assisting them economically and culturally, the Organization took an interest in their political, civil, and religious rights. The secretary general was delegated to press the case of Muslim minorities in Cyprus, Ethiopia, Cambodia, Thailand, and the Philippines (but not in India, the Soviet Union, or the People's Republic of China). In the cases of the Philippines, at least, it has had some success in initiating talks between the government and the rebels within the Muslim minority there.[273] In other cases, it has encouraged Muslim minorities to organize, has assisted them to become economically viable and better aware of their religion, and has fostered their relations with Muslims elsewhere.[274]

Co-operation in economic affairs was even more successful, as common ground could more easily be found; this was the case especially after 1973, when several of the Organization's members became financial giants. The Organization's charter and subsequent resolutions indicated the significance attached to the use of modern economic policy as a lever for Muslim solidarity—an approach labelled 'Economic Pan-Islam' by a French diplomat.[275] Among the objectives of economic co-operation, such matters were mentioned as the eradication of poverty, disease, and ignorance, the struggle against the exploitation of underdeveloped countries by the developed ones, the

[273] G. Fischer, 'Une tentative de protection internationale d'une minorité', esp. 340. Fischer maintains that the intervention succeeded in persuading the Philippines, not a member state of the Organization of the Islamic Conference, to accept its mediation, but failed to provide a solution acceptable to both sides. See also 'Non-talks in Jeddah: The Inside Story'.

[274] Cf. Etienne, 270. Acc. to *Ma'arīv* daily (Tel Aviv), 24 Feb. 1989, weekend supplement, 11, the Organization of the Islamic Conference was funding Muslim fundamentalists in Israel, too.

[275] G. de Bouteiller, 'La "Nation islamique"', part 2, 105–6.

assumption of sovereignty over natural resources, and the establishment of closer economic relations between all Muslim states. A common factor was the intention to combine the material resources of the member states with the religious bonds of Muslims for needed social development. Groups of experts were appointed, plans drawn up, and affiliated institutions established to carry them out. The general goal was to substitute for dependence on industrialized countries a system of investments among Muslim states, which would go a long way towards their economic integration (for instance, between the wealthy in capital and those rich in manpower).

The following major areas were designated by the Organization of the Islamic Conference, in 1981, for this purpose:[276] food and agriculture; trade and industry; transport, communication, and tourism; financial and monetary matters; energy; science and technology; manpower and social affairs; population and health; and technical co-operation. Due to the magnitude of the task and the absence of total commitment by the more affluent member states to share more generously in the enterprise, results have been modest to date. However, some progress has been made, partly thanks to the new affiliated bodies, of which the most important was the Islamic Development Bank.[277] Modelled on the World Bank, this was established in 1974; situated in Jedda, it has been open only to member states of the Organization of the Islamic Conference, 43 of which have joined the bank. Its main functions have been to grant loans to member states for productive enterprises and provide other financial and technical assistance for economic and social development. Since the bank is enjoined by its statutes to act strictly according to the *sharīʿa*, it has been hampered in some of its activities (such as giving or taking interest); not surprisingly, it has therefore encouraged research into 'Islamic economics'.[278] The bank's particular character has been expressed, too, in its priorities of allocating funds in

[276] Ahsan, 196 ff. See also Charnay, 110 ff.

[277] For whose activities the best sources are its *Annual Reports*. See also Islamic Council of Europe, *The Muslim World*, 161–3, 324–8, 336–42. Cf. 'Islamic Development Bank'.

[278] A. Kh. Nakkash, 'Doktrina "Islamskaya ekonomiki" v izlodzhyenii tyeologov Visshyego Sovyeta po Dyelam Islama'.

Africa[279] and elsewhere;[280] and in assisting financially Muslim groups in countries which are not member states of the Organization, including those in the United States and the Republic of South Africa.

Efforts in religious and cultural matters were equally purposeful. The Organization of the Islamic Conference explicitly undertook 'to propagate Islamic culture and protect it'. A special department was entrusted with collecting cultural information and disseminating religious and cultural works. In addition to publishing, it promoted Islamic studies, subsidized training centres, constructed mosques, schools, and hospitals, and convened seminars and workshops. It invested great efforts in co-ordinating all the above in the various member states and among Muslim minorities elsewhere. Most of these activities could be classified as educational, informational, legal, or humanitarian.[281] Of these, the first seems to have been considered the most important. The Organization has worked out plans—and started implementing them—to found several new universities and Islamic departments in older ones, as well as Islamic centres in different parts of the world, with a common core curriculum in Islamic studies; to develop Arabic as the common language of all Muslims—to foster a sense of solidarity; and to develop an identity of the *umma* by intensive study and research into Muslim history and tradition (specialized institutes for this purpose have been set up in Dakar and Istanbul).[282]

In 1980, an affiliated body was set up to deal with Islamic education and culture. Based in Rabat, it is named the Islamic Educational, Scientific, and Cultural Organization, or ISESCO, reminiscent of the better-known UNESCO, on which it is patterned. ISESCO has been implementing the above objectives

[279] K. Mathews, 'Tanzania, the Middle East and Afro-Arab Co-operation', esp. 26–7.

[280] For this and other economic and financial bodies associated with the Organization of the Islamic Conference, see *Les Organisations islamiques internationales*, 83 ff. Feuer, 167. Moinuddin, 69 ff. Schulze, 'Regionale Gruppierungen', 473. 'Myedzhdunarodniye Musul'manskiye organizatsii', 128 ff.

[281] Ahsan, 144.

[282] Another, based in Ankara, has produced, among other works, a collection of agreements and treaties for the protection of migrant workers. See Organization of the Islamic Conference, *International Coordination of Labour Exchange and Social Security*.

and striving to turn Islamic culture into the pivot of education at all levels in member states and in Muslim communities elsewhere (in order to protect the Islamic identity of the latter).[283] Besides organizing conventions, ISESCO has published, since April 1983, a yearbook in Arabic, French, and English, entitled, respectively, *al-Islām al-yawm*, *L'Islam aujourd'hui*, and *Islam Today*. The first issue proclaimed that ISESCO's 'ultimate aim is to strengthen solidarity and complementarity within the Islamic world'.[284]

An international Islamic News Agency was inaugurated in 1972 by the Organization of the Islamic Conference, to broadcast news in Arabic and English. In the legal domain, besides proposing an Islamic Court of Justice, the Organization set up in 1981 an Islamic Academy for Jurisprudence, whose objectives were to achieve 'the practical unity' of the entire Islamic community by bringing it closer to Islam and by resolving all problems according to the *sharī'a*. Humanitarian and other measures, chiefly involving aid for suffering Muslims, were planned in order to bring all Muslims closer to one another, in a spirit of Islamic solidarity.[285] Among the numerous associations involved,[286] the Islamic Council of Europe deserves at least a brief mention. London-based and financially supported by the Organization of the Islamic Conference, it has been researching and publishing studies in areas of concern to the Organization. Examples are its books *The Muslim World and the Future Economic Order* (1979), *Muslim Communities in Non-Muslim States* (1981), and *Islam and Contemporary Society* (1982). All three are specially commissioned collections of papers by Muslim scholars, generally aiming at fostering Muslim solidarity—and explaining it to non-Muslims.

5. *The Changing Role of the All-Muslim Convention*

The success of the all-Muslim organizations varied. They succeeded in implementing a part of their objectives, thanks to their fervour for Islamic solidarity, their modern organization,

[283] For a critical view of ISESCO's activities, see Etienne, 269–71.
[284] *Islam Today* (Rabat), 1 (Apr. 1983), 11.
[285] Details in Ahsan, 166–9.
[286] Details ibid. and *Les Organisations islamiques internationales*, 84 ff.

and the liberal funding with which they were provided. They failed, in part, due to the unwillingness of some of their members (particularly the member states) to risk their particularist national or individual interests.[287] In one respect, however, they achieved immediate results—an increased visibility through their frequent conventions. European models of international congresses had been available in the inter-war period, but had produced few all-Muslim conventions. In the post-World War II era, by contrast, such conventions proliferated, no doubt due to an increase in the number of independent Muslim states and to the ready availability of funds.

The differences between the all-Muslim conventions preceding World War II and those following it go further, however. The latter were generally sponsored by non-governmental associations at first, then increasingly by government bodies; this usually ensured a degree of continuity absent in the former conventions. They have proliferated, particularly since the late 1960s, with hardly a year passing by without at least one all-Muslim convention. An account of all of them would deserve an entire book and, indeed, some works about them have already been published.[288] We shall limit ourselves to indicating some main characteristics, focusing on the Pan-Islamic apsect.

The period under discussion may be divided into two halves of twenty years each. During 1949–68, most of the international gatherings for Islamic solidarity were convened by non-governmental associations (barring a few sponsored by one government—not by several acting together). Some governmental support and funding were a *sine qua non* condition for continuity, which is why the Organization of Muslim Peoples appears to have met in a convention only once, in 1952; the Muslim World Congress has had a modest record only; while the Muslim World League, amply funded, has sponsored numerous conventions. In the second half of the period, that is, from 1969 to date, intra-state co-operation has led to even more numerous international Muslim conventions. Starting with the summit meeting in Rabat, which paved the

[287] Muazzam Ali, 'Role of O.I.C. in Promoting World Peace', 170–3, exaggerates the failure of these organizations.

[288] Of which one of the best is by Levtzion.

way to the inauguration of the Organization of the Islamic Conference, conventions of Islamic heads of state, Ministers for Foreign Affairs, and of numerous bodies affiliated to this organization, or sponsored by it, met frequently.[289] No less significantly, the Organization of the Islamic Conference had the capability to carry out its resolutions, since they had been agreed to at the highest state level and funds had been earmarked accordingly.

Not more than half-a-dozen Muslim states were actively and continuously involved in promoting Muslim solidarity via international conventions. One of the foremost of these was Pakistan, which hosted numerous conventions in Karachi (beginning with the first international Islamic economic conference in 1949), Rawalpindi (an international conference to mark the 1400th anniversary of the Koran's revelation, 1968),[290] Lahore (the second Islamic summit, 1974),[291] and Islamabad (international gathering on the invasion of Afghanistan, 1980).[292] Pakistanis were motivated by popular sentiment favouring Pan-Islam,[293] the feeling that they were (until the breakaway of Bangladesh) the world's most populous Muslim state, and a perceived need of Muslim support for their own international moves.[294] The Islamic conventions gener-

[289] For details, 'Islamic Conference—From Talk to Take Off'. Cf. Levtzion, 21 ff. al-Jammāl, 12–19.

[290] M.A. Khan, ed., *International Islamic Conference February, 1968. English Papers.*

[291] *Report on Islamic Summit 1974 Pakistan Lahore, February 22–24, 1974*. The emphasis in this convention was on Islamic solidarity, in both speeches and resolutions. One of the banners proclaimed 'Quwwat al-waḥda al-Islāmiyya a'zam min al-quwwa al-dhurriyya' (i.e. the power of Islamic unity is greater than nuclear energy). A special 164-page report of basic data on all Muslim countries was prepared for this convention by the secretariat of the Organization of the Islamic Conference, entitled *Le Monde Islamique: Information basique sue les pays membres du secrétariat islamique*. See also 'From Lahore to Brass Tacks'. Mehrunissa Ali, 'The Second Islamic Summit Conference, 1974'. Vijay Saroop, 'The Islamic Summit'.

[292] B. Maddy-Weitzmann, *Arab Politics and the Islamabad Conference, January 1980*. M.B. Olcott, 'Soviet Islam and World Revolution', esp. 491–3. Cf. Bräker, 'The Islamic Renewal Movement', 519, for the subsequent voting behaviour in the UN and elsewhere of Muslim states.

[293] T.C. Young, 198–9. Weinbaum and Sen, 601 ff.

[294] L. Binder, *Religion and Politics in Pakistan*. Mushtaq Ahmad, *Pakistan's Foreign Policy*, 67–8. Sangat Singh, *Pakistan's Foreign Policy*, 165–72. S.M. Burke, *Pakistan's Foreign Policy*, 62–90.

ated great excitement in Pakistan. Muhammad Anwar Amin, a Pakistani intellectual, reflected this in a book which he wrote just before the 1974 Islamic summit in Lahore. Entitled *Long March to Jeddah*, it emphasized that 'the goal of the Muslim unity is very dear to [the] Muslim masses' and that 'the concept of a Federation of Muslim countries is needed to be materialized'.[295] There was no doubt, in his mind, that significant progress toward Muslim union was being made.[296]

Saudi Arabia was equally active, particularly under King Fayṣal, who was a prime mover in the launching of the Organization of the Islamic Conference and its Jedda-based secretariat and Islamic Development Bank. Not only has it hosted several international conventions, for example in Jedda (first Islamic conference of Foreign Ministers, 1970,[297] and sixth Islamic conference, 1975)[298] and Mecca (third Islamic summit, 1981),[299] but it has funded the Organization of the Islamic Conference, many of the bodies affiliated to it, and other Pan-Islamic groups. This was largely due to Saudi Arabia's being the 'keeper' of the Holy Places of Mecca and Medina and to its wish to minimize the impact of the hostility of so-called 'radical' Arab states towards it by drawing on wider Islamic support. The emergence of Pan-Islamic solidarity provided the Saudi élite with the needed counter-strategy to revolutionary activity in Muslim lands.

Libya, most particularly since Qadhāfī's takeover in 1969, has been increasingly involved, in Benghazi (fourth Islamic conference, 1973) and Tripoli (eighth Islamic conference, 1977), in addition to providing considerable funding. This was probably brought about by Qadhāfī's strong Islamic attachment as well as by his desire to be involved in Muslim affairs everywhere. Morocco was very active, also, not only in hosting

[295] Muhammad Anwar Amin, *Long March to Jeddah*, 20.

[296] Ibid. 77–82.

[297] Chaudri Nazir Ahmad Khan, *Commonwealth*, 157–63. Shameem Akhtar, 'The Jeddah Conference'.

[298] 'Islamic Conference: Realities of Muslim Statesmanship'.

[299] Among whose resolutions ('The Mecca declaration') was a demand from all Muslims for 'a spirit of Islamic solidarity in order to restore the unity of their ranks, work for their prosperity and advancement and achieve, once again, an exalted position in the world community and human civilization'. See *Third Islamic Summit Spotlight on Muslim World Problems*, 23–4.

international meetings, in Rabat (first Islamic summit, 1969) and Fez (tenth Islamic conference, 1979), but in sustained support for the Organization of the Islamic Conference and its insistence on ISESCO's being based in Rabat. This was less due to Islamic fervour than to King Ḥasan II's desire to reduce Morocco's isolation in the Maghrib.

Turkey has been a special case. A self-defined secular republic, it was so reluctant to join Islamic manifestations that it refused to participate in the all-Muslim conventions of the inter-war era. Since the end of World War II, however, despite its leaders' reservations about combining Islam with politics, they unbent enough to draw a line between domestic and international affairs. In the latter domain, Turkey has participated in international Islamic conventions since 1954, when it took part in an international Islamic conference in Karachi. Later, it joined the Organization of the Islamic Conference, as a *de facto* member, since it refused—as a secular state—to ratify its charter, and then, since 1976, as a *de jure* one; in that year, Istanbul hosted the seventh meeting of the Islamic conference. It initiated the founding of Islamic research institutes in Istanbul and Ankara, funded by the Organization of the Islamic Conference. The one in Istanbul, entitled Research Centre for Islamic History, Art and Culture, issues a *Newsletter* in English, French, and Arabic, three times a year. The leaders of the republic have attended the Organization's conventions.[300] The main motives were probably Turkey's wish to foster closer economic ties with other Muslim states and obtain their support in its conflicts with Greece and Bulgaria.

The all-Muslim conventions changed in nature, too. Although the religious factor was still very important, of course, other elements were very much in evidence. The inter-war conventions had been highly political. Those subsequent to World War II were still very concerned with politics, but shied away from controversial matters[301] and were largely focused on economic and cultural affairs, on which compromise and overall agreement could be more easily reached. A first Islamic

[300] As recently as January 1987, President Kenan Evren attended the Fifth Islamic Summit in Kuwait and, on his return, asserted that he was pleased with its debates and resolutions. See *Newspot* (Ankara), 6 Feb. 1987, 1, 7.

[301] See however Akhmyedov, *Islam v sovryemyennoy*, 11 ff.

economic convention met unofficially in Karachi in 1949, but was unable to reach tangible results.[302] A second one met in Tehran, in the following year, attended by ten states, and agreed on setting up an international organizational framework in Karachi, to stimulate the advancement of the economy in Muslim countries and promote collaboration and co-operation by research, planning, and annual conventions.[303] Subsequent international Muslim meetings were held by various bodies, with the tempo increasing in the 1970s. Saudi Arabia again had an important share in this, an international convention on Islamic economics being held in Jedda, in 1976, under the auspices of King ʿAbd al-ʿAzīz University[304] and another, under the same auspices, in 1978, to discuss the monetary and fiscal economics of Islam.[305] Economic issues were central to almost every meeting of the Islamic Conference of Foreign Ministers (the most recent of which met in Amman, in March 1988, and in Riyad, in March 1989). But usually the expectations of the poorer states fell short of the results, although an Islamic Solidarity Fund had been set up.[306] Economic issues were frequently dealt with in other international Islamic conventions too, such as those organized by King ʿAbd al-ʿAzīz University or by the Islamic Development Bank.

On the cultural level, great efforts and considerable sums of money were invested in Islamic education, for instance in an international convention on Muslim education, which met in Mecca, in 1977, and recommended a philosophy of integrated knowledge based on the Koran and Muslim tradition with additional information, provided the latter was consistent with the *sharīʿa*.[307] Most of the educational and cultural work was

[302] FO 371/82335, files 1104/5 and 1104/6.

[303] Ibid., file 1104/12, British Ambassador to Iran Sir F. Shepherd's no. 302, 'Confidential', to Ernest Bevin, dated in Tehran, 16 Oct. 1950.

[304] For the proceedings, see Khurshid Ahmad, ed., *Studies in Islamic Economics*.

[305] For the proceedings, see Mohammad Ariff, ed., *Monetary and Fiscal Economics of Islam*. The university is integrated into the Pan-Islamic activity of Saudi Arabia's royal house. It publishes, among other works, a scholarly quarterly about Muslim minorities (since 1979), entitled *Journal Institute of Muslim Minority Affairs* (*sic*), copies of which may be consulted in the Bayerische Staatsbibliothek, Munich.

[306] Levtzion, 60–3.

[307] Cf. Piscatori, 18–19.

left to such organizations as ISESCO which, in the 1980s, convened frequent international Islamic meetings and published the proceedings in its annual *Islam Today*; or the Islamic Council of Europe, which published the proceedings of its conventions in separate volumes; or research institutions which subsidized and printed various studies on Islam that were presented at the symposia they convened, frequently of particular relevance to promoting Islamic solidarity. Recent action was taken to unify the codes of personal status, when experts of the Ministries for Justice met in Kuwait, in April 1988, to discuss the religious aspects involved.[308]

An assessment of the degree of success of the international Islamic conventions is difficult and necessarily provisional, due to the short span of time involved. Such an evaluation is rendered doubly difficult, indeed, since it means passing judgement on the various organizations involved, many of which are new but have long-range objectives. While an impartial observer would tend, perhaps, to conclude that considerable progress has been made on the road towards greater Muslim solidarity (compare the situation, in this respect, of forty years ago!), he would also concede that much can still be achieved. The opinions of those directly concerned, the Muslims themselves, vary considerably; while those involved are generally pleased with the results of their labours, others are not so sure, as witness an editorial by a Pakistani, Mohsin Ali, in the *Dawn Overseas Weekly* of Karachi,[309] on the fifth Islamic summit of heads of state and government, held under the auspices of the Organization of the Islamic Conference in Kuwait, in February 1987. Mohsin Ali pointed out that, while the convention approved a draft statute for an international Islamic Court of Justice and certain proposals for strengthening scientific and technological ties among the member states, nothing had been achieved in the mediation between Iran and Iraq or Libya and Chad, nor in a national reconciliation in Afghanistan, in the resolution of the Arab–Israeli conflict, in alleviating the situation of the Turkish minority in Bulgaria.[310] An opposite

[308] *Le Monde*, 16 July 1988, 3.
[309] Mohsin Ali, 'Tackling Issues before Muslim World'.
[310] Ibid. 1–4. These failures have not prevented foreign observers, however,

view, however, was expressed by a religious leader of the Tatar community in the Soviet Union, S. Syeutov, who, on 10 December 1987, appealed to the Muslim World League in Jedda for assistance, on Islamic and political grounds.[311]

In other words, buttressing Islamic solidarity on political lines is still a formidable task facing international Islamic conventions and the organizations which sponsor them. Perhaps nobody was more conscious of these difficulties than Abul A'la al-Maududi who, writing shortly before the first Islamic summit in Rabat on 'The Task before the Muslim Summit',[312] offered his advice. Some of his requests have been fulfilled, in the mean time, while others are on the agenda; together they make up a prescriptive programme for Muslim solidarity:

> our attention should be focussed not merely on creating a bloc of Muslim countries, but should also try to come to grips with fundamental problems which face the contemporary Muslim world ... Unless we take care of our cultural identity, the future of the Muslim world is doomed ... What we need is an integration of the healthy elements of ... [education] leading to the emergence of a unified system of education. An ancillary problem is that of organizing and fostering research by pooling the resources of all the Muslim countries ... consider setting up of heavy industry and armament factories in the Muslim countries for the defence of the Muslim world ... popularize our common language—Arabic ... The Muslims should jointly set up a body which could arbitrate, adjudicate and help the Muslim countries solve their mutual problems and disputes ... promote the establishment of a Muslim World News Agency ... closer economic ties between the Muslim countries. The passport and visa restrictions between Muslim countries should in the first instance be eased ... [then] eliminate them altogether ... the propagation of Islam in Africa should be taken up in right earnest. Muslim minorities in several countries of the world are being subjected to intolerable persecution ... If the European countries, which lack a unifying ideology, can gradually move towards the goal of United Europe and some of them can even develop their common market, common planning and some kind

from recently expressing concern over continuing Pan-Islamic propaganda; see e.g. C.P. Kyrres, *Tourkía kai Valkánia*, 104–6.

[311] English transl. in *The Central Asian Newsletter*, 7/5–6 (Dec. 1988), 5–6.

[312] Maududi, *Unity of the Muslim World*, 27–34.

of common parliament and common control, why Muslims cannot become united to solve their common problems and meet the external challenge that threatens them all, although they believe in One God, One Prophet, and One Book.[313]

[313] Ibid. 30–4.

Conclusion

The godfather of research into geopolitics and pan-ideologies, Karl Haushofer (1869–1946), a professor at the University of Munich and a major-general in the German army, devoted several books and articles to them, including a 95-page summing-up of his own understanding, *Geopolitik der Pan-Ideen*, published in 1931. Characteristically, he hardly touched on Pan-Islam, since he considered it an ideology which could not be neatly classified into his own categorization of geopolitics—that is, by applying criteria of territorial data, physical power, and irredentist sentiment.

Haushofer's dilemma seems well grounded. The premisses put forward by ideologues of Pan-Islam differ from those postulated by the proponents of other pan-ideologies. While all are comprehensive in outlook and confrontational in character, for Pan-Islam territorial union was not expressed in irredentist terms, but in those of *Dār al-Islām* and *Dār al-ḥarb*. Language variation as an impeding factor was glossed over, at times with lip-service to Arabic as the potentially unifying language of Islam. What essentially mattered to politically minded Pan-Islamists was the shared background of a long historical and cultural tradition, with focal emphasis on a common religion. In other pan-ideologies, religion could be a factor of varying significance. It was important in Pan-Slavism, Pan-Arabism, and Zionism; and somewhat less so in Pan-Hellenism, Pan-Germanism, Pan-Turkism, and Pan-Iranism. In some of these ideologies, religion became diluted into cultural nationalism upon the establishment of nation states. In Pan-Islam, however, religion has remained, by definition, a central constant. After all, Pan-Islam had a strong basis in Islamic theory and history, systematically structuring its own political doctrine on Islam's tenets about religious and political unity. The Islamic reassertion, via Pan-Islam, has indeed constituted a present and future bridge—actually, the only one—among Muslim peoples and states. Every pan-ideology adopts (or creates) its own myths, which serve also as indicators for its objective; those of Pan-Islam were mostly echoes of the heroic age of Islam.[1] Indeed, sensitive to external challenges, many of Pan-Islam's arguments belong to the realm of cultural warfare, idealizing the pure community of Islam.

Most of the above-mentioned pan-ideologies and pan-movements developed simultaneously from the last third of the nineteenth century,[2] that is, the period of military colonialism and social change. Pan-Islam, as noted by Bernard Lewis, preceded Pan-Arabism and Pan-Turkism.[3] Characterized by religiopolitical activism, it was the only one, amongst all pan-ideologies, which was a religion-cum-politics (others were, at most, politics-cum-religion). This was not solely due to the innate political character of Islam, but to the exigencies of the time. True, some Muslim apologists emphasized the religious non-political character of Pan-Islam, even as late as 1945: 'Pan-Islam is a unity built on principles of faith and spiritual ties. It is basically different from Pan-Germanism and Pan-Slavism ... It has no intention whatsoever of territorial expansion or political hegemony'.[4] The spokesmen of Pan-Islam were generally aware, however, of its political objectives and the implications thereof—even though these were mostly defensive until World War II and insistently militant afterwards (especially since the 1960s). After all, it is the political element of Islam that explains why, among the world's three major universalist religions—Islam, Christianity, Buddhism—it has aspired for a closer union of its components, via Pan-Islam (more strongly than the ecumenical initiatives of Christianity). As to the precise nature of this future union, most ideologues of Pan-Islam have avoided, until recently, too rigid a commitment, except for re-emphasizing the early model of Islam and using it to assert the feasibility of such a union, despite disparities among its components in political regimes, social cultures, and language preferences. Since the *Weltanschauung* of Pan-Islam has been throughout a mixture (in varying dosages) of utopian romanticism and modern pragmatism, its

[1] To use A.J. Toynbee's term, in his *Civilization on Trial*, 212.
[2] Pan-Hellenism started earlier and Pan-Iranism later.
[3] B. Lewis, 'The Return to Islam', 13.
[4] Aga Khan and Zaki Ali, *L'Europe et l'Islam*, 20–1 (English transl. mine).

political ideologues generally shunned detailed specifications which could have caused disarray among supporters of divergent views; but they have often found common ground in expressing some reserve about, or even disillusionment with, political nationalisms.

During its lifespan, over 120 years, Pan-Islam has passed through various stages of differing intensity, creative interpretation, and political strategy, all of which have necessitated rethinking, time and time again. Originally a response to external challenges to *Dār al-Islām*, Pan-Islamic reaction to the penetration of Western civilization and secularist innovations, European physical aggression, and Islamic defeats was twofold in the Ottoman Empire: a readiness to defend Islam intellectually and a political policy based on a Pan-Islamic ideal. In the Hamidian era, Pan-Islam was used for obtaining political—and, to a more limited extent, financial—support, both within the empire and on the international scene. During the much shorter Young Turk period, it was drawn upon for political and military support. In both cases, Pan-Islam was fostered by the rulers of an independent Muslim empire. By contradistinction, in both the Tsarist Empire (and then in the Soviet Union) and in British-ruled India, it was more intensely defensive. In the former, Pan-Islam was considered vital for preserving Islamic identity, while in the latter it was promoted (at least partly) to restore the spiritual and political standing of India's Muslims. In some Muslim countries, in the inter-war years, Pan-Islam assisted the nationalist movements by being an intermediary between the universal *umma* and local nationalism (*waṭaniyya*).[5] Nationalists found in religion an important source for national solidarity. However, a situation soon developed in which many Muslims, while nurturing diffuse Pan-Islamic sentiment, indicated no readiness to change political entities in favour of a world-wide Islamic union. Consequently, the writings and conventions of that generation evoked limited interest only.

Since the end of World War II, Pan-Islam has been expressed in an increase of Muslim self-assurance and a growing feeling that Muslim peoples can together resist foreign aggression

[5] Piscatori, *Islam in a World of Nation-states*, 77.

and penetration and reassert their own values, religiously, culturally, politically, and, later, economically. While the idea of a unified universal state has unavoidably suffered a set-back due to the particularist interests of individual Muslim states and societies, proposals for Islamic solidarity via a Muslim bloc potentially leading to some sort of federation (or confederation) have found many supporters. In the changed circumstances of the existence of more than forty independent Muslim states, steps have been taken by some rulers—chiefly in Pakistan and Saudi Arabia—to shape a political and economic world power. Thus, it seems, 'Pan-Islam gave a new nationalist emphasis to the idea of solidarity . . . found in traditional Islam'.[6] Paradoxically, while in its defensive stages the cause of political Pan-Islam had been taken up by many secularized Muslims (Celal Nuri [İleri] and others in the Ottoman Empire, Muṣṭafā Kāmil in Egypt, Gasprinskiy in Tsarist Russia, and the Ali brothers in India), in its more militant stage, since the 1960s, its ideologues have come almost solely from Islamist circles.

Except for very recent years, one knows much less about the organization of Pan-Islam than about its ideologies, with their variations on the theme of political union. Indeed, it has often been assumed that Pan-Islam never had any organization.[7] This view, without being accurate, is understandable. If one accepts, for instance, the organization typology of William Cameron's *Modern Social Movements: A Sociological Outline*,[8] of visibility, formality, types of meeting, and leadership determination, one is hard put to apply these characteristics to political Pan-Islam. It is particularly striking that its visibility was very low, while the other three criteria varied over time and place, rendering a definitive characterization difficult. Indeed, whenever states were involved, one notices a two-tier organization, formal and informal, complementing one another.

Over the years, there occurred a perceptible change in the support systems employed by Pan-Islamists. These were determined by both ideological preference and political suitability.

[6] Keddie, 'Pan-Islam as Proto-nationalism', 26.
[7] See, for example, Kazemzadeh, 'Pan movements', 369.
[8] New York: Random House, 1966, p. 82.

As has been demonstrated, Abdülhamid employed for Pan-Islamic activities and propaganda both his diplomatic service and official missions as well as special, frequently clandestine, emissaries sent by and reporting to him or to persons in his immediate entourage. The Committee of Union and Progress set up a Pan-Islamic league, entitled the Benevolent Islamic Society, which had many of the trappings of modern organization; state controlled and funded, it was responsible for propaganda (and, probably, subversion too), without the Ottoman Empire having to accept direct responsibility (which rested with the state bureaucracy). The Muslims in Russia and India contributed to Pan-Islam's support systems the concept and practice of large conventions, intended to mobilize followers and funding and provide larger and better instruments for Pan-Islamic activity and response to challenges. In the inter-war period, these conventions, individually convoked in general, were the main external sign of continuing Pan-Islam organization. After World War II, internationalism seems to have become the mark of Pan-Islam; since states started to get involved, the two-tier organization pattern was revived, as in the Muslim World Congress, sponsored by the Government of Pakistan and, even more so, in the parallel and complementary activity of the non-governmental Muslim World League and the official secretariat of the Organization of the Islamic Conference, both funded by Saudi Arabia and largely staffed by Saudis.

Much less is known about participation, except for the fact that it was usually limited to small numbers, largely middle class, literate, and interested in politics (a phenomenon characteristic of most, perhaps all, pan-movements, which have generally been élitist in character). By contrast, their potential supporters were to be found in lower class, Islamist circles. It would seem that at no time were there more than a few thousands involved, barring the Khilafat movement, which claimed more than 20,000 adherents—and even that record figure pales when related to the immense human potential of India's Muslims. Thus the movement failed to mobilize the critical mass needed for success. Lack of funds or communication difficulties do not appear to have been crucial, as the current Pan-Islamic movements, carefully crafted and organizationally sophisticated, are not troubled by either, but still count few activists. In political no less than social life, people like those who agree with them and agree with those they like. Since the politics of Pan-Islam appealed to religious intellectuals, in general, it was a small élite, largely drawn from Islamist circles, that mostly became involved throughout the entire 120 years discussed.

Among these, Arabs have had a central—albeit not an exclusive—role as the main bearers of the historical message of Islam. Without the Arabs, Pan-Islam would have been shallow, so that they were pivotal in pulling the strings of Pan-Islam at Abdülhamid's court, no less than in the inter-war conventions and in such contemporary organizations as the Muslim World League and the Organization of the Islamic Conference. The exceptions in Tsarist Russia and British India were important, but short-lived. Even nowadays, when ceaseless efforts are being made to encourage greater participation of Pakistanis, Indonesians, Malayans, and others, it is the Arabs who are the most visible in Pan-Islamic organizations and activities. This seems applicable to government-sponsored bodies, but it remains by and large true of non-governmental ones also (except in some non-Arab Muslim countries). All the above often appeal to the masses but, although they generate an emotive response, the cadres which remain continuously involved are generally recruited from the religious intellectual élite. It is however plausible that, if the strenuous efforts invested by Pan-Islamic bodies in promoting Islamic education are kept up, a wider constituency of Pan-Islamic activists may be formed in the future.

In evaluating the failure and success of Pan-Islam, one should remember that most pan-movements have failed to achieve their ambitions on the political level. Pan-Italianism was a rare instance of success, while such movements as Pan-Romanianism achieved 'a Greater Romania' only for a brief period between the two world wars. The reasons for the failure of pan-movements are not far to seek and are related to the almost insurmountable difficulties of changing the status quo in peace-time. In this respect, Pan-Islam has faced less formidable problems, as it did not aim at changing political borders by force, but merely strived towards Islamic union (and, later,

Islamic solidarity) on the premiss that Islam and Pan-Islam would not acknowledge the legitimacy of frontiers between Muslim lands.

None the less, Pan-Islam has faced gigantic difficulties. Of these, the most evident has been that, for the last 65 years—or more than half of the timespan considered—the *umma* has had no one leader.[9] The size and diversity of the Muslim lands and the Sunnite–Shiite divergence have prevented an Islamic union, while bilateral tension and conflict between Muslim states have marred co-operation and co-ordination. Pan-Islamic solidarity has very rarely been expressed in concerted action; even in international relations, this has made a limited impact only. More than once, the powers intervened, officially or otherwise, to prevent an all-Muslim move. However, the picture is not all dark. A Pan-Islamic ethos has increasingly permeated education and historiography in many Islamic states; this process has been fostered, in recent years, by all-Muslim bodies. Moreover, all Muslim states are independent nowadays and the great majority have joined common administrative structures, of which the most important is the Organization of the Islamic Conference. These provide a forum for discussion and decision-making in the interest of Islam—provided the member states can agree.

Such concerted resolutions and their effective implementation are, of course, the test of the future success of Pan-Islam, as both the Pan-Islamists themselves and others see it. After all, Islam is still the most effective form of consensus within Muslim communities and could serve as the most available mode of commonality among their states, increasingly shaping all-Muslim solidarity. In other words, with Islamism having become a significant political factor in several major Islamic states, Pan-Islam may be seen as a potentially integrative force among these and other states. The numerous agencies which promote this solidarity, as well as the organizational and financial means at their disposal, may cope over time with particularist nationalisms, the secularist challenges to inherited values, and conflicting interests. It is difficult to predict

[9] A point made by Teresa Rakowska-Harmstone, 'Soviet Muslim Nationalism in Comparative Perspective', 479.

whether some acceptable synthesis of Islamism and nationalism will eventually be worked out and, if so, what impact it may have on the future community of Muslims advocated by Pan-Islamists. However, promoting all-Muslim solidarity may well lead, at some future time, to some confederative (rather than federative) form of association, perhaps a new version of a commonwealth of Muslim states (rather than a single Pan-Islamic entity). Much depends on whether and when the current polycentrism changes, with one of the Muslim states acquiring an acknowledged position of political and economic leadership. Then, as large parts of the world are moving towards more concrete forms of association, Pan-Islamists too may well turn a 120-years' old dream from what seemed a utopia into a political reality.

Appendices

A. Islamic Union, 1871 (Extracts) 315
B. Islamic Union, 1884 (Extracts) by Jamāl al-Dīn al-Afghānī 318
C. Caffarel's Report, 1888 (Extracts) 321
D. al-Afghānī's Pan-Islamic Project, 1892 326
E. Lacau's Report, 1902 (Extracts) 328
F. Letter from the Committee of Union and Progress to Muslims in the Caucasus, 22 September 1906 (Extracts) 330
G. A Pan-Islamic Letter in an Egyptian Newspaper, to the Sultan of Muscat, December 1906 331
H. Report of the French Legation in the Hague, 1907 333
I. Pamphlet by Ahmed Hilmi, 1910 (Extracts) 335
J. Reports of the Russian Secret Police on Pan-Islam, 1911–1912 342
K. Shakīb Arslān on the Establishment of the Benevolent Islamic Society, 1913 344
L. Rules for Anjuman Khuddam-i Kaʿba, 1913 346
M. A Universal Proclamation to all the People of Islam, 1914 (Extracts) 351
N. North Africa's Muslims and the *Jihad*, 1915 358
O. Resolutions of the First All-Russian Muslim Convention, Moscow, May 1917 (Extracts) 360
P. Resolutions of the Second All-Russian Muslim Convention, Kazan, July 1917 (Extracts) 362
Q. Perceptions of Pan-Islam at the British Foreign Office, 1919 (Extracts) 364
R. Resolutions passed on 12 February 1951 by the Muslim World Congress (Second Annual Session) 371
S. Objectives of the Organization of Muslim Peoples, May 1952 373
T. Project of a Constitution for a Commonwealth of Muslim States, 1969 375
U. Manifesto of the Afghan Students' Islamic Federation, 1985 (Extracts) 380

APPENDIX A
Islamic Union, 1871 (Extracts)[1]

... Ottoman unity was succeeded by Islamic unity. The latter, like the former, left the reader[2] unconvinced. A smile of pity, at most, was accorded those who took seriously one dream or the other. One might have hoped that Islamic unity would soon recognize its nonentity and would disappear when confronted by commonsense. However, matters turned out differently: indifference by public opinion emboldened the too courageous innovators. Lacking contradiction, they took themselves seriously and paraded their utopias from Constantinople to Vienna and other European capitals. Everywhere, their theories were well received. The more enlightened or the more attentive, perceiving the chimera of this unitary Islamic dream, may have thought that it was an innocent dream, thus leaving the dreamers their sweet illusions. Doubtless, they failed to perceive that these dreamers were, rather, agitators who were preparing public opinion in a premeditated and not incapable manner, taking the silence of contempt to be the silence of esteem and of adherence. In effect, Islamic unity, praised everywhere and criticized nowhere, proving itself in its initial activity, has already become a school [of thought], preaching, and taking action. The proof may be found in the silence of the European press and additional proof in the impressions of the Muslims of Constantinople, in the style of the newspapers of the capital, even in the activities of the Sublime Porte itself. Islamic unity is not—as might have been conjectured at its inception—the fruit of a delirious imagination, nor an ephemeral fantasy of some daily newspaper; it is, rather, a political school, a flag. We may be allowed, then, to start our discussions with this theory which now has a name and a body, being no more [merely] a theory, but a menace. We think that it is timely and useful to denounce this philosophers' stone which some wish to introduce as a panacea for the

[1] From the pamphlet *Unité islamique*, 6–16 (available at the library of the Institut National des Langues et Civilisations Orientales, Paris). The motto of the pamphlet, anonymously published in Paris, was 'Au feu!', i.e. 'Fire!'. English transl. mine.

[2] Of the Paris newspaper *La Turquie*.

wounds of the empire, while, in reality, it threatens to bring about the most deadly results.

Undoubtedly, there must be a significant reason for 'Islamic' replacing 'Ottoman' [unity]. One would have noticed that Ottoman unity, directly clashing with international treaties, was not easy to espouse, and even more so to implement: an attempt to unite the tribute-paying provinces with others within the empire would clash with the articles of the 1856 treaty as well as with the population threatened by annihilation. [Ottoman] unity was less presentable than the declaration of Prince Gorchakov. Hence, a new approach has been promoted—to respect Serbia's and Romania's autonomy and to proclaim the maintaining of Christian rights. One claims less to unite the empire of the Ottoman sovereign than to unite Islam. Thanks to this sleight of hand one believes to have pre-empted all reproach, without renouncing [however] any of the planned projects or abandoning any of the original objectives. One must however acknowledge ill luck: Ottoman unity faces insurmountable difficulties at the moment; but it might become achievable, should the empire have an administration and a government whose population would support the centre spontaneously, with the Cabinet merely ratifying its wishes. To hope for such Ottoman unity definitely means the wooing of a chimera; it would be utopia for a good patriot to wish the empire such a flourishing future that the Serbs, Valachs and others should find it advantageous to attach themselves to it as soon as possible. However, in so far as Islamic unity is concerned, there is no way whatsoever in which a political school might be established in its name; the day when such a school is effectively dominant will be the Ottoman Empire's last. Our innovators and the newspapers which host them[3] have not sensed this, when they have created and repeated the term 'Islamic unity'.

What is Islam unity, indeed? Its creators have not defined it; we shall do it in their stead. If they are interested in keeping it vague, our advantage is to enlighten it; our best assistant is clarity. Islamic unity may be defined in two ways: either the unification of all laws which govern the diverse Muslim sects; or this may mean the Sultan's domination over all those calling themselves Muslims. There is no other Islamic unification. We would let the admirers of this chimera, via the endless phrases of their organs, select the definition which they find preferable—with the option of keeping both of them. What do they desire? To unite the Muslims? No doubt on the religious level? Well, without mentioning the seventy-three sects which share Islam among them, without referring to the Shiites,

[3] i.e. publish their views.

these irreconcilable enemies of Sunnism, Turkey [alone] comprises four main sects: Hanafites, Shafiites, Malikites, and Hanbalites who, albeit orthodox, are irreconcilable among themselves . . .

We are told that the Sultan has power, [as] he is the Caliph. How is it not perceived that this is the bitterest satire directed at a Sultan? You say that he is the Caliph and that, as such, has not solely the duty, but also the authority and the power to reform religious traditions. Nobody denies his platonic right; but, if he is powerful enough to enforce his right, why does he not employ it to institute in his dominions the much-demanded reforms for equality of religions and races? The truth is that, for all practical purposes, the Caliph is as powerless before the traditions of each rite as he is regarding the imposition on the Believers of equality for the Unbelievers. Would he have more power to enforce acknowledgement of his authority by Muslims outside the Ottoman Empire and does the second definition afford a future for Islamic unity? Raising the question is to answer it. Is spiritual authority meant? Where this does not exist, it can be imposed only by material sanctions. If one was thinking of making all the world's Muslims render homage to the Sultan, by paying him tribute in people or money—in the name of Islamic unity, the Caliphate's universality, and religious convictions—the much hoped-for unity would have gained nothing by its [mere] change of name. Having become 'Islamic', it would face the same international obstacles. One knows that the new Turkish policies do not take any account of France's sovereignty in Algeria, as the change of their programme attests; but what do they think of Russia and Great Britain and their Muslim subjects!

The creators of this unitary programme were not in error. They changed names, but not tendencies: the term 'Islamic' was preferred by them, since it appeared appropriate to disguise their aim. The change of fronts has been evidence of some ability, from the start. They seemed to renounce Ottoman unity, which was attracting Europe's attention; they raised the flag of Islamic unity, assuming that the latter would not awaken legitimate suspicions but, rather, satisfy Europe. They intended to recruit all the religious, the Sultan, and the people, and [thus] reap the fruit of their endeavours. If necessary, they would content themselves with this single success; but they would not mind at all surprising the world, some day, by proclaiming Ottoman unity, in case Islamic unity prospered, thanks to the stupid connivance of several Western diplomats at it . . .

APPENDIX B
Islamic Union, 1884 (Extracts) by Jamāl al-Dīn al-Afghānī[1]

'Obey Allah and His Messenger and do not quarrel with one another, lest you fail and die'![2]

The dominions of Islam stretched between the furthest point west to Tonkin, on the borders of China, in a breadth between Fezzan in the north and Sarandib south of the Equator. Muslims inhabited [these] continuous and contiguous lands, which they ruled invincibly. Great kings reigned over them and administered most of the globe with their swords. None of their armies was routed, none of their flags was lowered and none of their words was contradicted.... Their cities were well populated and solidly constructed, competing with the world's cities in the industry of the inhabitants and their originality.... [When] their Abbasid Caliph spoke, the Chinese Emperor would obey him and the greatest Kings in Europe would tremble....

The Muslim navies ruled unrivalled in the Mediterranean Sea, the Red Sea, and the Indian Ocean, being predominant there until recently; their opponents had to yield to the power that defeated them. Muslims abound nowadays in those lands inherited from their forefathers; their number is no less than 400 millions. Individually, their hearts are replete with the tenets of their religion; they are more courageous than their neighbours and better prepared to die. In this, they are the strongest people in their contempt for worldly life and the most unconcerned with vain glories, since the Koran reached them.... Every Muslim perceives himself personally involved whenever a group of Muslims is subjugated by foreigners....

This has been the case, [both] formerly and nowadays. However, the Muslims have come to a halt, retarded in knowledge and industry, after having been teachers to the world. Their countries started losing the lands on their periphery, although their religion forbids them to yield authority to their opponents.... Have the Muslims forgotten Allah's promise that they would inherit the earth? ...

[1] This article was written by al-Afghānī, or under his inspiration by Muḥammad ʿAbduh, in *al-ʿUrwa al-wuthqā*, in Paris in 1884. It has been repr. in Arabic several times since. English transl. mine.

[2] Koran 8, 48.

As we have explained, the only characteristic common to Muslims is their religion, so that the proliferation of kings among them resembles the proliferation of chiefs within one single tribe, while their rulers are [actually] at one.... But the conflict among their princes had caused disunity, so that Muslims failed to oppose the aggression of their enemies.... This is what happened to the Muslim princes, along with shameful losses brought about by their disunity in wars in which no nation could have competed with them [otherwise]. However, corruption penetrates the souls of those princes, over time.... This is what brought down the Muslims of Spain and destroyed the pillars of the Timurid Sultanate in India, erasing its remains. The British established their government there on this Sultanate's ruins...

Agreement and co-operation for strengthening Islamic rule[3] are among the pillars of the Islamic religion; belief in them is a basic doctrine for Muslims, requiring neither a teacher to preach it, nor a book to confirm it.... Were it not for their misguided princes, eager for domination, Muslims east and west, north and south, would have joined a common appeal. For preserving their rights, the Muslims need only to turn their thoughts towards their own defence and agree on common action, when necessary; and merge their hearts in a joint feeling about the dangers threatening their nation.

Looking at the Russians, one notices three characteristics. They are lagging in the arts and sciences behind the rest of Europe's nations; their lands have no natural resources (if there are, nobody can exploit them for industry); and they are abjectly poor. None the less, focusing their thoughts on the defence of their nation, agreeing on its development, and joining their hearts have created a state capable of shaking the whole of Europe. The Russians do not own factories required for most instruments of war, which does not prevent them from obtaining them; the arts of war are less developed in Russia, which does not prevent them hiring officers from other nations to instruct their armies—so that their military forces have acquired awesome strength and an aggressive power feared by European states.

So what has prevented us from resembling others[4] in a matter which is simple for us and which we strongly desire—to preserve our nation's honour, to grieve for what hurts it, and to co-operate to defend a total union[5] against whomsoever attacks it? Those responsible for preventing the movement of thoughts and the rise in enthusiasm are [the princes] sunk in luxury, seeking succulent food

[3] In Arabic, *wilāya Islāmiyya*, which could also mean 'Islamic friendship' or 'Islamic loyalty'.

[4] That is, such as the Russians.

[5] Or 'total unity'. The Arabic reads *al-waḥda al-jāmiʿa.*

and soft bedding . . . who have become a yoke on the necks of Muslims, barring these lions from rising to attack—rather, rendering them a prey to foxes. There is no refuge except in Allah.

O people, descendants of the brave and the noble, has the tide turned against you and has the time for despair come? No, no, may Allah forbid the loss of hope! From Edirne to Peshawar, there is an uninterrupted sequence of Islamic states, united in the religion of the Koran, numbering no less than 50 millions, distinguished by courage and bravery. Should these not agree between themselves on defence and attack, as have all the other nations? Were they to agree between themselves, this would be no innovation,[6] for co-operating is one of the pillars of their religion. Has apathy stricken their senses, so that they are insensitive to one another's needs? Should not each consider his brother with the Koranic precept of 'the Muslims are brethren', so that they set up together the union to stem the waves threatening them from all sides?

I am not implying that one person ought to rule all [Muslims]; this is difficult, perhaps. I do hope, however, that the Koran would be their ruler and religion the focal point of their union; that every leader would do his utmost to preserve the others, since his own life and survival depends on them. Not only is this a pillar of religion, but a necessity, also. The time for an agreement has come! The time for an agreement has come!

[6] In Arabic, *bida*ʿ, plural of *bid*ʿ*a*, strictly, 'a reprehensible innovation', from the religious viewpoint.

APPENDIX C

Caffarel's Report, 1888 (Extracts)[1]

... On the accession to the Sultanate of Abdülhamid II, the Sanūsīs were already too powerful for the Turkish authorities to go on fighting them. The new Sultan did not even consider it. He had grand projects, for which the support of the fraternities was necessary. He was dreaming of re-establishing the empire of the Caliphs and thus regaining, on the religious level, the political prestige lost to the victorious Russians.

Through the intermediary of one of his old confidants, Muhammad Zafir, the Sultan joined the order of the Sanūsiyya and subsequently granted them his official protection.

On his own part, Sid al-Mahdi, cannily postponing the execution of his [own] hostile projects against the Turks, used Sultan Abdülhamid's Pan-Islamic policy for extending his own influence in Hijaz and in Tripolitania's south. The agitation provoked by the central government in Tripolitania, in 1880, had no other result.

Nowadays, it is no longer the Sultan who protects the Sanūsiyya, but the Sanūsiyya who tolerate his authority, while waiting for the moment when they believe that they are capable of overthrowing it. Outside the garrisoned towns, there is no security in Hijaz, Yemen, and Tripolitania except by the adherents of the order.

In the short term, His Majesty seems to fear lest his dignity as a Caliph may be hurt ... If the Sultan does not dare to rely on the loyalty of his Arab subjects, it is not for want of trying to appease them and [attract] their sympathy. He annually sends the chiefs of the main families in Mecca, Medina, Yemen, and Tripolitania rich gifts and numerous distinctions of honour.[2] He has brought 154 youths from the families of notables in the same areas and had them educated, at his expense, in Istanbul's various schools. He maintains a fairly active correspondence with the heads of the Muslim orders,[3] chiefly with Sid al-Mahdi.

[1] AAT, Archives de la Guerre, 7N1629, report no. 146 of E. Caffarel, French military attaché in Istanbul, to the French Minister for War, dated in Péra, 17 Feb. 1888. English transl. mine.

[2] Probably medals.

[3] The sheikhs of the fraternities.

Using various pretexts, the Sultan sends, from time to time, political emissaries to visit outlying regions of his empire. Without mentioning the scientific mission which had crisscrossed Kashgar, at the beginning of the current reign, one should note that, during last summer,[4] three high-ranking military physicians came to Tripolitania, allegedly in order to carry out a sanitary inspection, but actually to discuss political issues with the chiefs of the Arab tribes.

Shortly afterwards, on the orders of His Majesty, the Minister for Education sent to Spain Shaykh Muhammad Shinketi, an intelligent literate Arab, charging him to search for Arabic manuscripts in the libraries of those cities which had been a part of the Caliphate of Cordoba. The sheikh is a native of Mecca, but resides in Istanbul. He was accompanied by a dragoman and a servant. He has returned on 4 January,[5] via Marseilles, on the French steamer *Anatolie*. I have had definite information that, upon his leaving Spain, he went to Morocco. He was carrying some letters; he mentioned to a Muslim whom he met on board ship that he would certainly be received by His Majesty . . .

Sultan Abdülhamid deals with Arab affairs with the assistance of several intimate advisers, each of whom is assigned definite tasks, as if they constituted a real council of ministers.

Shaykh Fadl (Fazil) deals with questions concerning the south sea coast of the Red Sea and the coasts of Arabia, from Aden to Muscat, as well as commercial relations with British India. Shaykh Ahmad Asad (Esat), who should be distinguished from his homonym, the Şeyhülislam, busies himself with issues concerning the administration of the Holy Cities and Arab tribes in Yemen. Shaykh Muhammad Zafir's department is Egypt, Tripolitania, Tunisia, Morocco, and the African regions in general. Shaykh Abu al-Huda, without having any particular attribution, frequently intervenes in matters regarding Arab affairs.

These four personalities vie for the favours of the Sultan; their influence varies from day to day, according to [their] Master's whims; they envy one another, take over matters, spy on and denounce one another. They are supervised themselves by the Sultan, who gets a full record of the guests they receive and of their every movement. This division of the Sultan's advisers can be put to use, to a certain degree, by those Christian nations against which the intrigues of Islam are directed, France and England.

Here are some biographical details of the above-mentioned Arabs.

Shaykh Zafir is a native of the coastal area of the Red Sea. Without

[4] i.e. of 1887.
[5] 1888.

belonging to the negro race, he has a very dark complexion. A few years ago, he sought refuge in Istanbul, following an attempt at rebellion in the Harar territory, of which he pretended to be the legitimate ruler, as a descendant of the country's ancient princes. He persuaded the Sultan that he could render him a great service by combating the growing influence of the British. With the aid of the palace, he organized last year a new attempt at an uprising. He sent to the country of Harar about a hundred Arabs, well armed and with weapons and ammunition, on board a ship which he had equipped at his own cost. However, upon their arrival in Port Said, the Arabs were arrested by the British authorities, which impounded the rifles and the rounds of ammunition. Shaykh Fadl has a tempestuous, active, and plucky nature. He hates the British. At this moment, his influence is rather weak. He resides in a *konak*, in the Taksim Quarter [of Istanbul], given to him by the Sultan.

Shaykh Ahmad Asad was born in Mecca, where he owns vast property. Sent to Egypt in the early stages of ʿUrabi Pasha's uprising, he sagaciously predicted ʿUrabi's downfall and the forthcoming moves of the British. As long as Said Pasha served as Grand Vizir, Ahmad Asad continuously supported him at the Sultan's Court. Reciprocally, Said Pasha allowed Shaykh Asad complete freedom of action in the affairs of Mecca. Asad gained immense sums of money, either by granting positions to his *protégés*, against cash payments, or by dealings related to *waqf* properties, which made him the owner of numerous buildings fallen in disrepair—all this without any financial investment on his own part. After Said Pasha's downfall, complaints against, and denunciations of, Asad started arriving. This imperious sheikh had meanwhile been involved in the assassination of Shaykh Musa, whose exile to Mecca he had arranged as a favour to his own friend, Grand Sharīf Avn-ül-Tevfik. The latter managed to have Shaykh Musa strangled, *en route* to Jedda, by the soldiers escorting him. Asad lost favour with the Sultan; if he was not exiled, that was out of fear that, should he be far away from the palace, Asad would commit, once again, dangerous indiscretions. Shaykh Asad is striving strenuously to regain the good graces of his Master. He is intelligent, energetic, and has an austere and savage look. He is in cahoots with the British. His friend Avn-ül-Tevfik, currently Emir of the Holy Cities, hopes that the British, whom he has rendered many a service, will soon make him totally independent of the Caliphate. The relations of Shaykh Asad with the Grand Sharīf are known by the Sultan, who has never blocked them, in the hope of thus better supervising the Emir and of being informed of his doings.

Shaykh Muhammad Zafir was born in Tripolitania, in a family originating from Medina. He belongs to the Madaniyya Fraternity, which has connections with the Sanūsiyya. Being in Istanbul in 1875, upon his return from the Pilgrimage to Mecca, he was presented by the Tunisian General Khayreddin Pasha and by Behram Agha (then Chief Eunuch) to Prince Abdülhamid, to whom he foretold his future accession to the throne. Upon Abdülhamid's mounting the throne, he invited Shaykh Zafir to Istanbul, installed him in one of the wings of Yıldız Palace and made him his intimate adviser. The influence of Shaykh Zafir on Ottoman policy since then is well known. This has been expressed, internally, by the elevation of Khayreddin to the position of Grand Vizir, in 1878; and, externally, by the agitation which Zafir's brother, Hamza, has been instructed to stir against us[6] in Tripolitania, as well as by Sultan Abdülhamid's agreement with the Sanūsiyya for the triumph of the Islamic movement. Several political failures have somewhat decreased the standing of the two Zafirs in Istanbul. [However,] thanks to their influence in Tripolitania, they have remained intermediaries for the Sultan in his relations with the Sanūsiyya in this region. Abdülhamid has ordered that a handsome *konak* for Shaykh Zafir and a *tekke* for the dervishes of his order be built under the ramp leading to Yıldız Palace. Shaykh Zafir has now been entertaining for a month an enemy of France, Muhyi al-Din, son of Emir ʿAbd al-Qadir. Zafir leads a secluded life. The Sultan, who considers him a devoted a servant, consults him willingly. By contrast, his brother, Hamza Zafir, is much in evidence; he lives in a luxurious *konak*, beneath the palace, in the village of Beşiktaş. The Zafirs have recruited numerous adherents for their fraternity from the Ottoman bureaucracy and the regiments garrisoned around Yıldız Palace. These circumstances are liable to provide them with considerable influence in the case of a governmental crisis.

Shaykh Abu al-Huda, a native of Syria, is the head of one of the branches of the Rifāʿiyya Fraternity. Intelligent, energetic, courageous, and devoid of scruples, he has imposed himself on the Sultan, who fears and avoids him. He scares all the office-holders in the palace. Three years ago, the Sultan had trusted him very much. Abu al-Huda used this to enrich himself, selling positions and honours, and to buttress his own influence in Hijaz, loading with gifts—in His Majesty's name—the chiefs of the Rifāʿiyya in Jedda, Medina, and Mecca. Like the Zafirs, he has participated in Pan-Islamic agitation. He is the sworn enemy of the present Grand Vizir, Kamil Pasha; he has never missed an opportunity to meddle in [state] affairs, wherever he saw a way of checkmating Kamil's plans. He continues to see the Sultan and to get a hearing. Early last year, the Grand Vizir had scheduled a meeting between the government and the Ottoman Bank, in

[6] i.e. against the French.

order to settle accounts. At three different times, Abu al-Huda came to entreat His Majesty not to approve Kamil Pasha's proposals—and he succeeded in this. Six months ago, regarding Egyptian affairs, he has displayed even more energy and activity to cause the Sultan's rejection of the Anglo-Turkish Convention. Abu al-Huda loves money and hates Grand Vizir Kamil Pasha implacably. Appealing to either of these two sentiments can get anything from him.

APPENDIX D
al-Afghānī's Pan-Islamic Project, 1892[1]

Je trouve que la question des afghans, de leur Emir et de l'agitation où ils se trouvent, est une question dont on peut tirer profit. Par leur attachement à la religion, leur soumission au siège du Khalifat, les afghans ne sont entre les mains d'un diplomate habile qu'un instrument qu'il manie à sa guise, une image à laquelle il donne la forme qui lui plait. L'orientation de la politique afghane vers la politique ottomane dont elle serait dépendante n'est point chose difficile pour un diplomate de bonne volonté, possédant expérience des affaires afghanes et sachant s'y prendre, qui leur démontrerait par des arguments logiques que la prospérité de l'Islam ne s'obtient que par leur alliance avec le siège du Khalifat soit pour la paix, pour la guerre ou pour la neutralité.

Si nous parvenons à rallier les afghans à la politique ottomane, la Perse sera obligé [*sic*] de prendre part à cette alliance, contre son gré, parcequ'elle se trouvera isolée entre deux Puissances et nous donnera la première place dans le concert politique. L'Angleterre et la Russie compteront alors avec nous et chercheront et à se rapprocher de nous et à nous seconder dans nos entreprises. Il nous sera loisible alors de choisir l'alliance qui conviendra le mieux à nos intérêts sans nous abaisser à prier.

Cette politique est certainement préférable au rapprochement à l'Allemagne prise comme clef de la triple-allliance, parcequ'il n'est pas improbable qu'elle nous trahisse pour aider l'Autriche à acquérir quelque partie de notre territoire en compensation du territoire autrichien convoité par elle.

N'oublions pas que l'Allemagne et même les Puissances de l'alliance appuient la politique anglaise en Egypte, ce qui est en contradiction avec les protestations d'amitié qu'elles font à la Turquie et est fait pour empêcher l'Angleterre de contrecarrer l'action de l'Allemagne en Afrique.

[1] FO 78/4452, enc. in Arthur Henry Hardinge's no. 144, 'Secret', to the Earl of Rosebery at the Foreign Office, dated in Ramleh, 3 Sept. 1892. This is a French transl. of the original, obtained by British Intelligence. As retranslating the document would have further obscured al-Afghānī's phraseology, the French text is rendered here unchanged.

Il en résulte donc que les Puissances ont en vue leurs intérêts propres non les nôtres, et toutes n'ont qu'un désir, celui de faire disparaître de la terre jusqu'à notre dernière trace. Et en cela il n'y a aucune distinction à faire entre la Russie, l'Angleterre, l'Allemagne ou la France, surtout si elles s'aperçoivent de notre faiblesse et de notre impuissance à résister à leur desseins. Si, au contraire, nous sommes unis, si les musulmans ne font plus qu'un seul homme, nous pourrons alors nuire et être utiles et notre voix sera écoutée.

Si nous parvenons à nous allier aux afghans et à y joindre la Perse, il nous sera facile d'étendre l'influence du Khalifat parmi les Musulmans de l'Inde, et alors l'Angleterre sera contrainte de ramper devant nous et de quitter l'Egypte.

Le fait d'attirer les afghans à notre politique et de nous allier à eux n'est point difficile. Leur Emir ne pourra point s'y opposer lorsqu'il verra le peuple disposé à cette alliance; mais le succès de cette entreprise exige la plus grande discrétion dans les mesures à prendre afin d'empêcher que les Puissances prennent connaisance de ce projet et paralysent nos efforts avant d'arriver à le mettre à exécution.

Je garantis à la Turquie l'accomplissement de ce fait si elle adopte mon avis, écoute mes conseils et me charge de cette mission. Si le sort me favorise d'être auprès d'elle je lui indiquerai plusiers choses utiles que mon expérience me met plus à même que tout autre de connaître.

C'est ce que je crois soumettre pour le moment et que je considère comme un devoir, car le conseil est un précepte de religion et l'aide vient de Dieu.

APPENDIX E
Lacau's Report, 1902 (Extracts)[1]

The Shādhiliyya maintain, at little cost, dervish camps ... According to my information, this fraternity has only few adherents in Tripolitania and exercises a very limited impact. I am unaware as to whether its activity is more powerful and extensive in Egypt. By contrast, the Madaniyya—whose spiritual and political leader, Shaykh Zafir, chaplain of Sultan Abdülhamid, resides in Istanbul—are quite active in Tripolitania, where they possess small *zāwiya*s ...

All the fraternities employ the same means of propaganda, their objectives being those which Islam has always championed, that is, the annihilation of foreign invaders. They excite the ignorant populations of Africa solely by tales of miracles and predictions of a final victory over Christianity. However, their charlatanism does not always prevail over the reality of facts. Thus, in 1882–3, when I was acting Consul-General here, many Tunisian chiefs openly came to see me and, upon my advice and assurance that their rights would be respected, interrupted their exodus and returned to Tunisia—either on board our ships or by land, with their tribes, forming a mass of more than 100,000 people—despite the exhortations and threats of Si Hamza, Pan-Islamic agent of the Sultan and brother of Shaykh Zafir, chaplain of Abdülhamid and head of the Madaniyya Fraternity; also, despite subsidies and lavish promises by the Ottoman Government ...

The Muslim religious–political circles in Tripoli have no official organ to serve their ideas and propaganda. All plans and alliances intended to combat the expansion of European powers into Muslim lands and, more particularly, French penetration into Africa, are hatched in the dark among Arabs, Sudanese, and Turks. This does not mean that Tripolitania's Arabs are happy with Ottoman administration or that they profess much respect for its integrity; but they see the Sultan as a Caliph, the head of their religion. They adore and

[1] AE, Direction Politique, Série B, Carton 80, Dossier 3, Consul General Lacau's report, dated Tripoli (Libya), 12 Feb. 1902. The report, in French, was uncovered by İ.S. Sırma and published in his 'Fransa'nın Kuzey Afrika'daki', 166–9. English transl. mine.

obey him under this title and do not meddle in intrigues against him and his government in order to detach themselves and place themselves under the authority of any European government [instead]. Since it reasserted its rule in Tripolitania and Cyrenaica, the Sublime Porte has kept up these sentiments of the Arabs in these two provinces by using the authority of influential native personalities, on whom it bestows titles and pensions. It corresponds through them with the Sultan of Waday, Shaykh Sanusi, and other leaders of African regions; it sends them its secret messengers and gifts. One may say, without exaggeration, that on their side, these personalities dispose of all the Arabs and the Negroes of the country to assist them in their designs. In one word, the Ottoman Government has shaped Muslim fanaticism into an anti-European weapon which it is trying to make invincible. The efforts of this vast political–religious association—functioning under the patronage of the Sultan, the Şeyhülislam, and the Sharīf of Mecca and meant to respond to the ideas of progress and European penetration into Africa—have not yet produced satisfactory results. Rivalry, envy, and vulgar appetite have undermined this association to such an extent that it has as yet been unable to draw up a general plan and prepare the means to implement it continuously, intelligently, and energetically. The various fraternities live apart from each other and their adherents follow the same pattern. These are, basically, self-excluding clubs. Will the Muslim world in Africa be resourceful enough to rise to the level of present circumstances, in the hour of pressing dangers for Islam in Africa, now that the Sultan–Caliph Abdülhamid has seen the failure of his Pan-Islamic illusions, one by one? Nothing is less probable, considering African and Ottoman apathy and incoherence. However, petty intrigues and sporadic attacks—troubling the public order and damaging the confidence which we ought to instil into these primitive, simple populations—will continue as long as we are not vigorously established in these regions and have not demonstrated our will to be there the only masters, respected and uncontested....

APPENDIX F
Letter from the Committee of Union and Progress to Muslims in the Caucasus, 22 September 1906 (Extracts)[1]

Muslims are ill-treated everywhere, not merely in the Caucasus. The Christians, considering that the saying 'might is right' is valid and respecting nothing but it, persecute Muslims—whom they perceive as weak—everywhere.

European governments are afraid even of the term 'the union of Islam',[2] as it is called by Christians. Is this not the best proof of how great our enemies' fear of a union of Muslims is?

The whole of Africa, from its northern tip to the other end, is inhabited by Muslims. The countries from the Adriatic Sea to the frontiers of China, end to end, are not only made up, religion-wise, of one faith but are occupied by the Turkish race, speaking one language. Africa and India excepted, should just those of Turkish race unite, they could establish the world's most majestic state. This being so, what is the reason that the union has not been arrived at? Can the answer to that be but 'ignorance and indifference'? . . .

We cannot consider it objectionable to strive openly for the union of Islam. Since we are certain that the Majestic Ottoman [Empire] is the first principle for the safety of Islam, we wish, first and foremost, to limit ourselves to transforming the administration of the Ottoman Government into an orderly and reasonable one. You have persevered in education, which lays the basis for union. You have adopted the Turkish [language] of Istanbul. This Turkish will serve you well to advance in education; and, in the future, will prepare the ground for uniting individuals of the same nation . . .

[1] The text of this MS letter was first reproduced by A.B. Kuran, *İnkılap tarihimiz*, 209–10. The letter is signed, on behalf of the Committee of Union and Progress, by Dr Nazım and Dr Bahaeddin. English transl. mine.

[2] In the Turkish text, *İttihad-ı İslam*.

APPENDIX G

A Pan-Islamic Letter in an Egyptian Newspaper, to the Sultan of Muscat, December 1906[1]

Your Highness is renowned for perspicuity, knowledge, and good works; you have inherited the distinction of an ancient and honourable lineage; you come of Rulers noted for nobility of race and deeds. You have seen and read of the distresses endured by Moslems in both Eastern and Western countries, the reasons of which are their discord and dissension and lack of mutual assistance, while the foreigners cleverly arrange their tricks and put them in circulation, destroying Moslem authority. You should study carefully the contents of this paper and consider them with the acute perception, sound deliberation, and shrewd intelligence. Your Highness is looked upon by our Supreme Lord Abdul Hamid-bin-Abdul Majid as the most eminent Chief among those of the Peninsula [Arabia]; it is your duty to propagate these views among the people of your province, as well as among your neighbours the Chiefs of the minor States, such as Lahej, Mukalla, Shihar, Riyadh, &c. You may add your own scholarly remarks, so that every chief may see that his independence is a holy right which should be untouched; these Chiefs being unable by themselves [i.e. without assistance] to retain their independence and to keep up their power and dignity before the enemy who is prepared to attack them either at present or in future, except by the means provided by the religion in the manner stated [i.e. joint effort]; that is an old Mohammedan law by which the Islam and its followers stand firm and their light shines clearly on the standard of the crescent. Similarly the reverse [i.e. not banding together] is a clear heresy and an evident deviation [from religion]. God says, 'And God is not [disposed] to lead a people into error after directing them until that which they ought to avoid is become known to them'. God has taken upon

[1] IO, L/P&S/19, FO Confidential Prints, India Office's dispatch to the Foreign Office no. 43 (9799), dated 25 Mar. 1907, enc. 3, being an English transl. from the Cairene *Nibrās al-Mashāriqa wa-'l-Maghāriba*.

himself to explain this matter in the verse quoted; it does not therefore become a man of honour to prefer error and discord to concord, agreement, and [mutual] succour.

APPENDIX H

Report of the French Legation in the Hague, 1907[1]

Your Excellency is aware that a lieutenant-colonel in the Turkish army, named Ismail Hakki, has written to a certain Nur Muhammad Akun,[2] a Muslim notable in Java, in order to obtain detailed information about Islamism in Java and Sumatra.

Ismail Hakki Bey, an officer in the Sultan's own guard, seems to seek information about different peoples who may participate in the Pilgrimage to Mecca, in view of a study he is preparing on the Hijaz Railway. This is not the first time that the activities of Lieut.-Col. Ismail Hakki have been brought to the attention of my government. On 15 May 1906, the General Residency of France in Tunis informed Mr Constans that this officer was requesting statistical data on the population of the *Régence* [of Tunisia], asking that the reply be dispatched to him at the German Post Office in Istanbul. The Embassy,[3] forewarned, requested our General Resident not to reply.

The following August, my government learned that the very same officer had written to the Governor of Dahomey, Haut-Sénégal and Niger, the General Commissary of Mauritania, and various officials of French Congo, to request statistical data about the Muslim population in these countries. The Governor General of West Africa decided that these requests should not be followed up.

Doubtless understanding, upon these repeated failures, that he would get nowhere by addressing himself to the French authorities, Ismail Hakki turned to the Geographical Society of Algiers. The society, after having consulted our Embassy in Istanbul as to whether it ought to accept Ismail Hakki as a member, responded negatively.

It should be known that Lieut.-Col. Ismail Hakki is the son of His Excellency Tevfik Pasha, the Sultan's Minister for Foreign Affairs. The latter willingly confirms that his son is gathering detailed data about Muslim populations in the French, British, and Dutch colonies.

[1] MBZ, dossier 451 (A. 190), A. 190/B 107 (24534), French Legation to Netherlands' Minister for Foreign Affairs, dated in the Hague, 19 Dec. 1907. English transl. mine.

[2] French text has: Nour-Mohammed-Akoun.

[3] i.e. the French Embassy in Istanbul.

Ismail Hakki, although just 23 years old, is *aide-de-camp* to the Sultan; evidently, in order to please his imperial master, he has undertaken a study which would be well regarded there. According to his father, however, he ought not be considered one of those agents whose task is Pan-Islamic propaganda.

But there is another Hakki Bey, who is an active agent in this propaganda; this is Ibrahim Hakki Bey, a legal counsellor at the Sublime Porte. Since 1902, he has been the Istanbul contact-man of a Pan-Islamic agent who was visiting China and Japan at that time. It is this Ibrahim Hakki Bey who maintains intimate relations with Khualdiyya Salah,[4] self-styled 'President of the Central Islamic Committee', a person who was detained at the prison of Oran in 1906.

It is this same legal counsellor, Ibrahim Hakki Bey, who was recently in Paris to represent the Sublime Porte in the arbitration of the Chemins de Fer Orientaux.

[4] Text reads: Khoualdia-Salah.

APPENDIX I

Pamphlet by Ahmed Hilmi,[1] *1910 (Extracts)*[2]

Hear, O Moslems!

O Moslems, O brothers in the Faith!

Listen to me! The words I shall speak to you are the words of truth. Do you not believe in God and in his book? Now I shall speak to you out of the book of God. Just look at the world, and you will see that of 350,000,000 Moslems, 300,000,000 are slaves under the hands of strangers.

It is a thousand pities that the servants [of God] who bear a banner inscribed with the words of witness should be destroyed; it is a thousand pities that those who believe in the unity of God, who recognise that God is one and his Prophet is true, should have remained under the feet of those who rush to cast down the truth; it is a thousand pities that the servants of God should have become slaves to nations other than the people of the beloved Prophet.

O brothers in the Faith!

Look at the Moslem world for a moment with my eyes; cast a glance at the world that you may be warned. What shall we see? Men groaning with hunger, men dressed in coarse rags; mosques in ruins, tumble-down imarets[3] where the cauldron boils no longer, schools with neither teachers nor scholars; villages and hamlets like ruins, inhabited only by the screech-owl.

Wherever we look—at Bokhara, Kashgar, India, Persia, China, Java, Egypt, the Soudan, Algiers, Tunis, Morocco, Kazan, the Crimea, the Caucasus—in fact everywhere except in the Ottoman Empire, the country of the Khalif who sits on the prayer-carpet of the Prophet, we see the head of the Moslem crushed to the earth, crushed beneath the enemy's yoke. Those enemies have taken their countries from them and made them strangers in their own lands.

O brothers in the Faith!

[1] Writing under the pseudonym Shaykh Mihriddin Arusî.

[2] IO, L/P&S/19, FO Confidential Prints, Sir Gerald Lowther's report no. 23, to Sir Edward Grey, dated in Istanbul, 9 Oct. 1910, enc. 1, English transl. of the Turkish pamphlet.

[3] 'Imaret: an institution attached to mosques for the distribution of food to needy scholars and the poor' (Lowther's footnote).

Why is it that these places, which once were as fruitful as Paradise, have fallen into ruin? Why is it that the Moslems, who once ruled the world, have become slaves?

Is it that once we were many, and have become few? No! Is it that our country was once rich and possesses riches no longer? No! is it that we used to be brave and courageous and now have become cowardly? No! What has happened then? Why have we become slaves?

O brothers in the Faith!

Now I will speak to you a word more bitter than poison, bitterer than hell-fruit; I will speak to you a word of great truth. I will tell you the reason why we are despicable, wretched, poor, enslaved. It is because we have not harkened to nor understood the commands of our God and of our Prophet; because we have been shameless and ignorant; because we have tyrannised over ourselves, over our brothers, over all servants of God ...

O brothers in the Faith!

Our God commands us to help the poor, to love and console our fellow-believers. Do we carry out this order? Do we give alms in full? Do we stretch out a hand to those brethren whom we might save from hunger and misery? Do we strive by advice, by prayer, and by joining hands with them to save the evil from their evil ways?

No! So again we have failed to keep our compact. It is a pity that we are all ignorant to-day. Many of us have strayed into evil ideas not based upon Islam. Science and knowledge have passed to other nations, but we have remained ignorant. Others make railways, astounding bridges, tunnels, harbours, ships, guns, and rifles; but we have none of those things. We do not work; we act as though ignorance and poverty were necessary conditions of Islam. We respect only a few of God's commands, such as those relating to prayer and fasting; we act as though the others did not exist.

O brothers in the Faith!

Islam is one nation. All Moslems being brothers, they all make up one nation, which is called the 'Muslim nation', or the 'nation of Islam'. Every Moslem is one of this nation, to whatever race he belong—whether he be Arab, Turk, Persian, Afghan, negro, Indian, Javanese, Albanian, Kurd, Georgian, Lezki, or Chichen. Those who consider them as separate and different from each other are sowers of discord.

O brothers in the Faith!

The head and leader of all Moslems, the representative and Khalif of our Prophet, is the Ottoman Sultan. It is the duty of us all to obey all the commands of the Khalif, for they are lawful and true; and to disobey the sacred fetva of the Khalifate is a great error.

The method of government of all governments subject to the faith of Islam is that of lawful consultation and constitution. In Islam there is not despotism, i.e., rule according to the arbitrary wishes of one man. If any Moslem accepts any principle other than that of consultation, he is guilty of disobedience towards the tradition of our Prophet and the command of God, and of giving assistance to tyranny.

Everything was made known by God to our Prophet through an angel; the Prophet in turn made known everything to his companions, and always acted according to the opinion of them all. After our Prophet, the great Khalifs also were subject to the principle of consultation. It was the Yezidis who began the system of despotism.

O brothers in the Faith!

The hundreds of Moslem countries which formerly existed in the world have fallen one by one into the hands of other powers. Only four Moslem Governments in the whole world have remained independent: the Ottoman, which contains the seat of the Khalifate; the Persian; the Afghan; and the Moroccan. When the Abbassid Khalifs grew weak and fell into a state in which they could not be of any service to the Moslem faith, God had pity on this religion and on Islam, and made the Ottoman Turks soldiers and defenders of the Faith. They took over the Prophet's trust from the last of the Abbassids; they carried the frontiers of Islam to the very centre of Europe; they became brothers of bold races like the Albanians and with the Arabs, the Albanians, and the Kurds, they have shed their blood for seven hundreds years, and protected Islam.

As long as the Ottomans obeyed the command of God, Islam grew; it never diminished. As long as the Ottomans remained just, good and pious, they were strong. As long as they were strong, Moslem countries were safe from the enemy's wickedness. But during the last two centuries many of them have left the straight path. The Moslems have lost the Danube, the Crimea, Caucasia, Algiers, Tunis, Egypt, Cyprus, and many another country, and Bokhara, Khiva and Kashgar are lost to them. But at last God had pity on them again. The Ottomans opened their eyes and saw their faults; they removed from his throne their Khalif and Sultan, who had acted contrary to God's command and the tradition of the Prophet; they did away with the accursed rule called despotism, and in its place established the principle of consultation and constitution, which is the command of Islam and the tradition of the Prophet. Together with the Ottomans, all other Moslems opened their eyes; they all understood that the path in which they had been treading was a blind alley. Now we have seen that if we are not brothers, if we do not unite, we shall be crushed; they will make slaves of us, and leave us hungry and naked; and through the crowd of men that have overspread our land their priests will try to turn us from our faith.

O brothers in the Faith!

The enemies of Islam have seen that our eyes are opened and that we have begun to love one another, and they are afraid, for they know that if we Moslems give up our evil ways and unite we shall become masters of our honour. That is why they do not wish us to open our eyes and unite. In order to leave us poor and ignorant as of old, and to make us envious of each other, they have bought with money such traitors to their religion and their nation as there are amongst us. These accursed ones, who fear not God, try in three ways to leave us Moslems in slavery. Just see what those three ways are:

First: they are opposed to the principle of consultation and constitution, and say that the constitution is a thing which is taken from Europe and from the Christians, and has never been seen in Islamism. God send his curse upon those who utter this calumny! It was by Islam that 'consultation' was proclaimed for the first time in the world; the divine verse 'and counselors . . .'[4] commands us to accept consultation. In all the affairs of the nation the Prophet used to collect his companions and take counsel with them. It is true that there are five or ten Christians and Jews among the members;[5] their vote, however, is never taken on religious matters, but only on temporal affairs. And those who say that it is not right to take their opinion on temporal matters either are telling lies or speaking against Islam.

O brothers in the Faith!

Are those who say these words the greater, or is the Prophet? Of course our Prophet is greater. Our Prophet said: 'Wisdom is a possession which the believer has lost. Wherever he finds it he seizes it'. This means that we can take anybody's opinion in the affairs of the world provided that it is wise, that it is useful, and that it is not contrary to our religion.

Secondly: they try to deceive you with feigned fanaticism and false ideas. They strive to prevent you from working and from making progress. Take care! Do not listen to these devils. Beware of these lazy fellows who praise beggary and poverty! Give them no money; leave them hungry and let us see what they will say and do.

O brothers in the Faith!

In order to guard our country, and to be able to struggle properly against the enemies of our faith and fatherland, we need millions of poundsworth of battleships, guns, rifles, and ammunition. If Moslems remain poor, how shall they buy these things? If they remain ignorant, how shall they make these things? Is not all poverty dis-

[4] So in the original.
[5] i.e. in the Ottoman Parliament.

aster? That is why we have lost so many lands. And is not all ignorance a calamity? That is why we have lost so many wars.

The seat of the Khalifate is compelled to-day to maintain a million soldiers. If the nation is poor, how shall these soldiers be fed?

Thirdly: they try to make a cleavage between the Moslem races—between the Turks and the Albanians and Arabs. Now, the greatest enemies of Islam are these sowers of discord. These accursed ones know that if the Arabs and the Albanians be separated from the Turks they will crush the Turks and leave Islam very, very weak; and that is why they strive with such devilries to separate these brothers.

O brothers in the Faith!

He who writes these words to you is above all a Moslem, one who holds truth above all things. Now I tell you that wherever the Turks become weak, the Moslem Khalifate was not loved; and where the representative of the Prophet was not loved, that part fell beneath the enemy's feet. Algiers, Tunis, and Egypt went in that way. If the Turks there had not become so weak, those places would have not been so easily lost.

Afghanistan, one of the places where the Khalifate is loved, has saved itself. There is a great example in this: there are some 50,000,000 of Arabs in the world to-day, and of these only 15,000,000 live independent in the Ottoman Empire and bear the Ottoman name, while more than 20,000,000 have fallen into French or English hands. Now our enemies see that if the Moslem nation ceases from separatist quarrels and unites, it will become very strong and protect its lands; that is why certain traitors to Islam among the Arabs bring about demands for separation and propose nationalist separatism.

To propose separation and not to desire and love unity is to desire the destruction of Islam.

O brothers in the Faith!

The blessed and revered Mehmed Reshad V sits now in the seat of the Khalifate, a seat given him by God and our Prophet; and his acceptability cannot be damaged by the untruthful words of those who are induced to speak by money or instigations of our enemies.

Then, too, there are base men who want to stir up the Albanians with talk of racialism. The cause of their enmity to the Turks is that they say, 'Why have the Turks made the Albanians their brothers?' Yes; they are the foes of Islamism. They fill their bellies with the money received from Catholic priests, and they hope that, if the Turks and the Albanians are separated, they will be able to turn the Turks, i.e., Islam, out of Europe, and then the Albanians will gradually become Christians.

The Turks are the outposts of the Moslem army. The enemy know that the day they destroy the Turks they destroy Islam, and, therefore, they try to sow dissension among Ottoman Moslems.

Fourthly: they try to split up the Moslems by sectarian differences; they want to bring about quarrels between Sunnis and Shiʿahs.

O brothers in the Faith!

The God we adore is one, and the Prophet is one. No Moslem, whether Sunni or Shiʿah, adores any other god or recognises any other Prophet. So let us not listen to the enemy; let us not tie the rope around our necks with our own hands. It is a shameful thing; it is a sin.

O brothers in the Faith!

Now let us realise our faults and defects. Night and day, at the five prayer-times and especially before the mosque at Mecca and the Prophet's tomb, let us cry and pray, saying: 'There is one God; there is one nation! Union! Union! For the sake of the unity of God, grant us unity'!

Let us cast out from among us the sowers of discord; let us not give ear to their words. They pretend to desire the strengthening of Islam and of the Khalifate! If the Ottoman Empire is strong, the Khalifate will also grow in strength.

Then let us pray night and day for the strengthening of the Khalifate and the exaltation of Islam. Let there be no other wish or aim in the heart of any one of us.

Unity and fraternity! Union and brotherhood! These are the two forces that will save us. Let us unite and work. Let us spare a mouthful for others. Let us make schools, and leave not one child ignorant amongst us. Let us love our nation and our honour. Let us accept European civilisation, *i.e.*, science and industry, and even carry them still further. But let us not abandon the blessed customs of our religion and our nation, *i.e.*, let us not adopt the material civilisation of the Europeans.

Moslems ought to pay more attention to agriculture and commerce, and we ought to run our agriculture not on the old principles but according to the progress of the age. We must strive to make our land fertile; we must increase our stock of animals. Every one of us ought to try to keep a horse or a mare, as we are a military nation, and it would be of great advantage to us. Besides, in the matter of rearing horses there are the Prophet's encouragement and inducements; and the Koranic verses and sayings of the Prophet on the subject too.

A Moslem ought not to think only of those Moslems belonging to his own race. The brotherhood of Islam and its spiritual unity will appear gloriously in the full meaning of the terms when we learn not to forget our co-religionists in the Soudan, in the Sahara, Senegal, Morocco, Algiers, Tunis, Egypt, Afghanistan, Java, China, India,

Persia, Turkestan, Kazan; when we learn to regard it as a moral and religious obligation to grieve at their sorrows and rejoice at their happiness.

O brothers in the Faith!

Whoever loves his religion, his nation, and his race, and who desires to see Islam exalted, ought to serve the cause of union and brotherhood. Whoever wishes the Moslems to be rescued from slavery, ought to spare something from his store and spend it on the schools. Whoever wishes to go into God's presence with unstained forehead, ought to assist schools and colleges to save his Moslem brethren from ignorance.

O brothers in the Faith!

Whoever wishes to win the favour of God and the intercession of the Prophet, should help to strengthen the Khalifate with hand, with voice, and with purse.

By God, by God, by God! The most acceptable form of worship today is to help the Moslem army to get rifles, even if only to the extent of 10 paras, or to assist the army and the fleet.

To give money for schools and to save a fellow-believer from ignorance is better than setting free a thousand slaves.

O brothers in the Faith!

I write you these words out of the promptings of my heart. It is God who makes me speak them. My brethren, I beg you read this prayer with me. The doors of mercy are open; the future is ours. Let us join our hearts, and pray to our God, saying:

O God! one and great and mighty, who didst promise to Islam thy divine help! Help Islam and give it victory.

O our God! For the sake of thy all-knowing name, make the people of Islam lovers of knowledge and learning. For thy rich and generous name make the faithful rich and prosperous through the love of work and acquiring wealth.

O our God! for the sake of thy unity unite our aims and efforts; separate us not from unity in this world or in the world to come.

O our God! Make the Khalif of Islam to live long in health, and to stand in the way of truth. Help the Moslem army, send abundance upon the servants of Islam, and generously and mercifully reward the good intent of all Moslems.

Thy prayers and thy peace, with infinite grace and with the mercy befitting thee, be upon the Prophet and the house of the Prophet.

APPENDIX J
Reports of the Russian Secret Police on Pan-Islam, 1911–1912[1]

The idea of Pan-Islam was born 27 to 28 years ago in Africa, between Abyssinia and Sudan, in the district of Mataam[2] and was called *Ittihat Islamiyye*, or 'The League of the Union of Muslims'. The promoter of this league was Shaykh Omer-Ejel-Rubini-Mehdi,[3] originally a Catholic Frenchman from Algeria, who headed five million people of the Danakil tribe.

The League of Pan-Islam appeared following the conquest of African territories by Europeans. Soon afterwards, another sheikh, Shaykh Ebu-El-Toeme, at the head of 1,800,000 Muslims—Somalis, Sudanese, Nubians, and Gallas—entered the league, making common cause with the former sheikh. Both started an intense activity, dispatching emissaries to the ends of the globe for propaganda among Muslims, as a result of which, zealous adherents to Pan-Islam were gained among the Afghan, Turkmen, Indian, Chinese, and Japanese Muslims, as well as among the Turkish Kurds.

Their goal was to establish a strong, powerful multi-million Muslim state. The league's programme was as follows: 1. To unite all Muslims, wherever they might be. 2. To spread, via secret pamphlets, the teachings of Muḥammad among the Russians and their Christians—to the detriment of the Pravoslav faith. 3. To meddle clandestinely in the politics of the European states, most particularly in Russia, and to harm the latter, as far as possible. 4. To plant secret agents in every Russian town, so as to keep in hand people who would always be ready to act. 5. To provide their agents clandestinely with funds,

[1] This report, written in Jan. 1911, with additions in Jan. 1912, was publ. by Arsharuni and Gabidullin, 101–13. English transl. mine.

[2] This account of Pan-Islam being initiated in Africa and spreading thence is fanciful. The only historical parallel appears to be the Mahdi's movement in the Sudan, at the time indicated (1883–4), but this had limited impact outside the Sudan.

[3] *Sic.*

weapons, and followers. 6. To conduct propaganda among the Russian Muslims who had not yet joined the League of Islam.[4]

... This fanatical movement, supported in every way by the Young Turk and Young Persian committees, has spread significantly in Russia at present—to a great degree in the Crimea, near the Volga, and in other parts of the empire. In Baku, Kazan, Orenburg, and Elizabetpol [it is] entirely in the hands of the Young Turks, and a large part of the intelligentsia [there] talks of nothing but the Union and Progress party.... The main principle on which the current Pan-Islamic movement focuses—which, so to say, is its spirit, expressed by its propagandists who strive to inculcate it into the consciousness of the popular Muslim masses—is the union of the entire Islamic world under Turkey's leadership, up to the creation of a Pan-Turkic republic.

Noted Muslim journalists, [both] Turkish and Russian, lately undertake a strenuous discovery of tribes belonging to their own race, with the objective of uniting them in a general Muslim federation.... The statutes of the international Muslim congress, meant to be inaugurated in Cairo and Istanbul, follow the same concept.

Pan-Islamists demonstrate great energy in their oral propaganda in the inner life of the Muslims in Russia, concentrating their attention, in particular, on the Muslim high schools, the *medreses* (seminars) ...[5]

Lately,[6] the ideas of Pan-Islam have started to spread intensively within the Muslim population of the empire, in the main thanks to these ideas being propagated in the midst of the rising generation by means of the Tatar schools and seminars, which are entirely independent under the authority of the mullahs, fanatic protagonists of the Pan-Islamic idea ...[7]

[4] The above fantastic account may well be characteristic of the Pan-Islamic paranoia of the Tsarist police.

[5] A lengthy report on Muslim education in Russia follows, then details about suspect Pan-Islamic activity of Turkish agents in Turkestan and Bukhara and about press articles favourable to Islam.

[6] This is the beginning of the second report, dated Jan. 1912, of the Tsarist secret police.

[7] There follow details about the educational and religious activities of the mullahs, which seem however to have been chiefly of Islamic—rather than Pan-Islamic—purport.

APPENDIX K
Shakīb Arslān on the Establishment of the Benevolent Islamic Society, 1913[1]

The late al-Sharīʿī Pasha discussed with me in Istanbul the establishment of an Islamic society to unite all factions, by which the [Ottoman] state will be delivered from disunity between Arabs and Turks as well as between İttihadists and İtilafists.[2] I replied that I was not opposed to the idea, but that I was afraid it would lead to no result. However, he insisted on carrying it out, being supported by Yūsuf Bey Shatwān, too. At last, I was persuaded to work in the matter. We went and presented the issue to the Crown Prince, Yusuf İzzeddin, whereupon he promised to be the honorary president of this society. We talked it over with Emir Saʿīd Ḥalīm, who was later to become Grand Vizir; both he and his brother, Emir ʿAbbās Ḥalīm, agreed. Then we spoke with Arab Emirs, the Sharīf ʿAlī Ḥaydar Pasha and his brother, the Sharīf Jaʿfar, as well as with the Sharīf Nāṣir, the brother of the Sharīf Ḥusayn, the Emir of Mecca. They [all] promised to join the society. We then talked with persons from the *İtilaf* party, such as Salih Pasha Khayreddin and Muhtar Bey, son of Jamaleddin, the Şeyhülislam, İsmail Bey Gümülcineli, and others; [then] with a group from the *İttihad* party, such as Talat Bey, İsmail Muştak Bey, and others, whose names I have forgotten. All agreed to come and we arranged a first meeting at the Pera Palas Hotel. The late Hakkı Pasha, formerly the Grand Vizir and at the time Ottoman Envoy to Berlin, joined the society, too. The late Shaykh ʿAbd al-ʿAzīz Shāwīsh wrote the statutes of the society, which I read [aloud] while the late ʿĀrif Bey al-Mardīnī translated them orally into Turkish, so that each member could make his observations. It was Hakkı Pasha, the

[1] From Shakīb Arslān's autobiography, *Sīra dhātiyya*, 100–2. English transl. mine.

[2] The main political parties of the day. The former, *İttihad ve Terakki*, was essentially made up of the supporters of the Committee of Union and Progress, while the latter, *İtilaf ve Hürriyet*, claimed to be more liberal minded.

mercy of Allah upon him, who presented most of the observations, due to the breadth of his knowledge and intelligence. After having completed correcting the statutes of the society, we named it the Benevolent Islamic Society.[3] Every member made a monthly donation to the society and we rented an apartment for it in the Cağaloğlu Quarter in Istanbul. Yūsuf Bey Shatwān was its first president. This society was conducive to much good.[4] Had the Ottoman state continued in existence after the World War, the society would have developed and grown, setting up branches everywhere. However, because the state was so preoccupied with the Balkan War and then the World War, its activity remained limited. We used to meet to discuss general matters and exchange views. The society carried out some activity for the instruction of Arabic in Istanbul, which it was hoped it could promote—but for the events which caused everyone to wonder and depressed the high-placed and the lowly [alike].[5]

[3] In Arabic, *al-Jamʿiyya al-Khayriyya al-Islāmiyya.*

[4] Or great philanthropy.

[5] This probably refers to the Ottoman set-backs in World War I. It is interesting that Shakīb Arslān, usually perspicacious, failed to find out that he and other dignitaries were being used as a cover for a semi-clandestine Pan-Islamic organization.

APPENDIX L
Rules for Anjuman Khuddam-i Kaʿba, 1913[1]

1. As the safety of the *Kaaba* and the respect of our *Haram*[2] appears to be at stake, and as we have not the same assurance for the welfare of our shrines as we had before, it therefore seems desirable to organise a special *Anjuman* of Muhammadans for the safety and upkeep of the *Kaaba*, which will be hereafter known as *Anjuman Khuddam-i-Kaaba*.

2. The chief aim of this *Anjuman*, will be to maintain the honour of the *Kaaba*, and to do every service to the house of God, which was built by the prophet Ibrahim, and is the very centre of unity in the world. To care for and protect other sacred places in the same way will be considered to be equivalent to service rendered to the *Kaaba* itself. In order to effect its aims the *Anjuman* should adopt the following measures:—

(*a*) To form an association of lovers of the *Kaaba* and of unity, the members of which will be at all times quite ready to sacrifice their lives, property, honour, and comfort for the sake of the *Kaaba*.

(*b*) To inspire such feelings among Muhammadans as may result in producing such lovers in every town, and to promote the true faith of Islam in every heart.

(*c*) To spread Islam, which is the true service to the *Kaaba*, and to send Muhammadans to every land to instruct people in the *Kalima*.

(*d*) To establish Muhammadan schools and orphanages at different places to assist in the propagation of Islam and to prepare the orphans for the true service of the *Kaaba*.

(*e*) To improve the existing relations between Muslims and the Bait-Ulah Sharif, and to render every facility for the voyage (*Haj*).

[1] Published in the *Moslem Gazette* and other newspapers in India, in May and June 1913. Repr. in IO, L/P&S/20/H137.

[2] 'Prob. *haramain*, an Arabic dual word meaning "the two sacred places"' (footnote of the IO).

(*f*) To safeguard the *Kaaba* from every marauding Power and to devise and carry into execution the best measures for the upkeep and safety of *Haram*. To devise plans to save the *Kaaba* from any dangers which may arise from time to time.

3. All the professors of the *Kalima* whether they be male or female, can become members of this *Anjuman*, and the procedure of making a member will be celebrated with great solemnity. Every *Khadim* will have to repeat the following oath before two Muhammadans, placing his hand on the Qoran:—

'I, son of . . . being in the presence of God, after repentance for my past sins, with the *Kalima* on my lips and facing the *Kaaba* (pointing towards the *Kaaba* with his finger) solemnly affirm that I shall try with my whole heart to maintain the respect of the *Kaaba*, and shall sacrifice my life and property against non-Muslim aggressors. I shall fully carry out the orders of the *Anjuman Khuddam-i-Kaaba* given to me'.

Every member who thus joins the *Anjuman* will be called *Khadim-i-Kaaba*.

4. Of these associates those who will be ready to sacrifice their lives, property, and honour at the direction of God and in the interests of the *Kaaba*, and would like to become volunteers, will have to take the following solemn affirmation:—

'I, son of . . . with my face turned towards the *Kaaba*, being thereby in the presence of God, hereby solemnly affirm that I have given up my life for the service of God. I now serve the *Kaaba* only and maintain the respect of the *Kaaba*. The orders of the *Anjuman Khuddam-i-Kaaba* will be my most responsible duty, which I shall be always ready to carry out with my heart and soul, and without any objection and delay. I will, without objection or delay, start for any destination to which I may be ordered to go; no difficulty will keep me back. With this solemn promise I enter the Society of *Shaidaian-i-Kaaba*, swearing for a second time by my God and my Prophet, the Qoran, my religion, and my honour to remain faithful to the above promise'.

5. It would be possible for some members of this society to hold office as *Shaidai Kaaba* for a limited time under the same solemn affirmation.

6. The *Anjuman Khuddam-i-Kaaba* will be responsible for meeting the requirements of a *Shaidai* and his family according to his social position, and the *Anjuman* will bear all expense incurred by the *Shaidaian* in performing the duties entrusted to them. Every effort will be made that their status as *Shaidaian* may be an improvement on their position prior to employment. A *Shaidai Kaaba* entering the

association for a limited time will have the right of enjoying all privileges as long as he is a *Shaidai*.

7. It will be required that all *Khuddam* should exhibit in some prominent place on their dress a yellow crescent having *Khadim Kaaba* embroidered in black in the centre. Whenever they join any gathering of the Society, or are employed on the work of the *Anjuman* they must wear this badge.

Shaidaian i Kaaba will always wear a green *aba*, and a similar crescent embroidered with *Shaidai-Kaaba* instead of *Khadim Kaaba*.

8. All the affairs of this *Anjuman* will be under the control of a central *Anjuman* composed of selected *Khuddam* which will be known as the *Asal Anjuman Khuddam-i-Kaaba*. For the present Delhi has been selected to be the head-quarters of this *Anjuman*. The orders of this *Anjuman* will be final and binding on all. Subordinate *Anjumans* will be organised in all provinces, in independent Muhammadan States, in Mecca, in Medina, in Baghdad, and in any Hindu States where the *Asal Anjuman* will deem proper to establish them. The *Anjumans* will be called *Anjuman Alia Khuddam-i-Kaaba* of the ... Province. These *Anjumans* will be authorised to organise other *Anjumans* in each district subordinate to it. These *Anjumans* will be known in each district as district *Anjuman Khuddam-i-Kaaba*.

The district *Anjumans* will be authorised to organise *Anjumans* in towns and villages to be named after the village or town.

9. Each town and village *Anjuman* will nominate a *Shaidai* from amongst them as their representative to the district *Anjuman* (*Anjuman Khuddam-i-Kaaba*). The District *Anjuman* will similarly nominate a representative for the provincial *Anjuman* (*Anjuman-i-Alia*) and the provincial *Anjuman* for the Central *Anjuman* (*Asal Anjuman*). This will ensure the organisation of the *Anjumans*. Thus the district *Anjuman* will be made of the selected *Shaidais* of towns and villages. The district nominees will make up the provincial, and the provincial nominees will help to form the *Central Anjuman*.

The District *Anjumans* shall be subordinate to the provincial *Anjumans*, and the provincial *Anjuman* to the *Central Anjuman*.

10. In every *Anjuman* only such men will be selected as office bearers, as have already been taken into the *Anjuman* as a *Shaidai* according to section 4.

11. At present representatives for the *Anjuman* will be selected from the following places:—

(*a*) *Abroad*.—Mecca, Medina, Baghdad (only that portion which is in connection with *Karbala-i-Mualla, Najaf Ashraf* Kazmen and the Mausoleum of *Hazrat Ghaus-ul-Azam*). From these places only those men will be selected who are really Indians, but have gone to live there. They will also be selected according to the prescribed rules.

(*b*) *According to provinces.*—Eastern and Western Bengal, Bihar, Orissa, Oudh, United Provinces of Agra, the Punjab, North-West Frontier, Sind, Bombay, Madras, Central India, Berar, Rajputana, Northern India and Burma.

(*c*) *Muslim States.*—Hyderabad, Bhopal, Rampur, Junagarh, Bhawalpur, Khaipur, Sind, and Tonk.

(*d*) *Other States.*—Kashmir and Mysore.

In all these there are twenty-six circles, which will send in their representatives to form the Central *Anjuman*.

12. Efforts will be made to organise branches of this *Anjuman* in all towns and villages where there is a sufficiency of Muhammadans. In small places, this mosque will be made the meeting place of the *Anjuman*. Possibly all the *Imams* and *Muazzans* of mosques will be made members of the *Anjuman*.

13. All the *Anjumans*, whether central or subordinate, shall hold a committee for the election of office bearers every two years.

14. Every *Anjuman* will select a member as a President or Secretary. He shall be responsible for the whole work of that *Anjuman* and will be called the *Khadim-ul-Khuddam* (Servant of Servants). He will be helped by two assistants who will be known at *Muatmiddin Khadim-ul-Khuddam*.

15. The ordinary members of the *Anjuman* will have to obey the orders of the *Khadim-ul-Khuddam*, without any objection or delay. Any member violating this section will be regarded as committing a very heinous offence.

16. Every member of the *Anjuman*, whether *Khadim* or *Shaidai* will have to pay an annual subscription of one and only one rupee. However rich a man may be, nothing more than a rupee will be taken from him, so that the equality of Islam may be held supreme.

17. The sum thus subscribed will be divided into three equal shares and will be spent in the following way:—

(*a*) One-third of this sum will be given to that Muhammadan empire which will have the duty of looking after the *Khana Kaaba*. It will be spent in the service of the *Kaaba* in any way that the respect, dignity or the safety of that sacred place demands.

(*b*) One-third of the sum will be given to different orphanages, and schools which are likely to produce *Shaidais* of Islam. The sum subscribed from each province and district will be spent in the cause of that province or district from which the sum has been collected.

(*c*) One-third of the sum will be reserved, so that it may be advantageously spent in time of need. This sum may be devoted to any undertaking which may be for the good of the *Kaaba* and other holy shrines, as, for example, to purchase vessels in which pilgrims may be conveyed easily and comfortably and cheaply to the holy places of Islam; any other works which may seem equally advantageous may be undertaken.

18. Every Provincial *Anjuman* shall nominate a legal adviser (*mushir*) from amongst the ordinary *Khadim* to act as its representative on the Central *Anjuman*, in addition to the *Shaidai* nominated by it. This legal adviser will not be required to reside at head-quarters but must attend meeting of his *Anjuman*.

19. The following gentlemen have been elected office bearers and members for one year. They can increase their number. It shall be their duty to begin the work of the *Anjuman Khuddam-i-Kaaba* in the whole country, and after establishing *Anjumans* (Alia, District and Aslia) in different districts, they may resign:—

(1) Maulana Maulvi Abdul Bari, Faranghi Mahal, Lucknow, *Khadim-ul-Khuddam* (Servant of Servants).
(2) Hakim Abdul Wali Sahib of Lucknow.
(3) Dr. Nazir-uddin Hasan, Barrister-at-Law, Lucknow.
(4) Mr. Mohamed Ali Sahib, editor of the *Comrade* and *Hamdard*, Delhi.
(5) Mr. Mushir Husain Kidwai, Barrister-at-Law, Lucknow.
(6) Mr. Shaukat Ali Sahib, B.A., of Rampur.

The latter are the two Secretaries.

APPENDIX M

A Universal Proclamation to all the People of Islam, 1914 (Extracts)[1]

Islamic peoples should be the same as their sympathies towards their own peoples and families, and it is incumbent upon their individuals and their communities to strive and put forth every effort to deliver any people of the peoples of Islam and any nation of the nations which believe in the Unity of God if they have fallen into the grasp of the infidels who are idolators and of the oppressive enemies.

Yes, it is an important duty of all the Muslims that they should despise all difficulties and exert their utmost power to help those of them who have fallen under the rule of the infidels and to deliver them from oppression with all their powers, and whoever violates this duty is guilty of a great iniquity and whoever denies it deserves from God painful punishment in hell continually. And this is the case if the hostility falls upon a people or a tribe of the Muslim communities, and how will it be if the oppressive infidels should, by their enmities, attack the center of the Caliphate, and if they erect enmity toward the Caliphate the apostle of God (May God be gracious to him and give him peace!), and if they spread from their mouths hatred toward him, although he is the great example to all Muslims spread abroad on the face of the earth, and the one who descends from them like the descent of the spirit from the flesh! There can be no doubt that it is an imperative duty in this case upon all the people of the Faith that they should vie with one another in striving for the victory,

[1] The original is a 30-page Arabic pamphlet, *Balāgh ʿāmm li-jamīʿ ahl al-Islām nasharathu Jamʿiyyat al-Mudāfaʿa al-Milliya* (A general proclamation to all Muslims, published by the Association of National Defence). A copy is located in NA, RG 59, 867.00/762, with US Consul J.B. Jackson's no. 258, to US Ambassador in Istanbul Henry Morgenthau, dated in Aleppo, 8 Apr. 1915. For Morgenthau's comments, cf. ibid., his no. 303, to the Secretary of State, dated in Istanbul, 14 May 1915. Jackson appended an English transl. of selected passages, on which our appendix is based. Another copy of this transl. is located in MBZ, dossier 453–454 (A. 190), A. 190 (213). A shorter transl. into French is available in AAT, Archives de la Guerre, 7N2103, enc., in dispatch no. 1217, from the French Minister for Foreign Affairs to the Minister for War, dated in Paris, 14 Nov. 1914; this is entitled *La Guerre sainte est obligatoire.*

and for defending the whiteness of Islam with all the power that they can put forth. Therefore every Muslim without exception must be considered as a soldier, and therefore it is an imperative necessity that everyone who is able to bear arms should learn the military duties, and be ready for the Holy War in case of need. And these duties are incumbent upon individuals, and communities, and peoples; for they are called to be responsible for it in accordance with the saying of the Most High 'And oppose to them all that you can command of force'. For this speech includes all of them.

And it is the duty of every Muslim whatsoever his race and birthplace that he should be ready to rise up for this purpose and expend the utmost effort to accomplish it, considering that this message is addressed to him personally. For this Muslim world has arrived at a condition in which it is not fitting that anyone should consider his personal advantage or bodily repose or any private consideration, but it is the duty of every Muslim that he should cast all worldly business behind his back and that he should occupy himself completely with the deliverance of his religion and his nation from the wiles of his enemies. It is the duty of the whole Muslim world to-day to gather all its resources at one point and interrupt every work temporarily except the work of the rescue of the religion of God and the Holy War in the path of God. A very holy office is presented to-day to every people of the peoples of Islam, the inspired Holy War.

The Stage of Degradation to which the World of Islam has arrived

O people of the Faith, and O beloved Muslims, and O true brothers, consider, though it be for a little, the condition of the Islamic world. For if you consider this a little, you will weep long. You see before you an important matter. You see a bewildering condition which causes the tear to fall and prolongs the thought and causes the fire of grieve [*sic*] to blaze. You see the great country of India which contains hundreds of millions of Muslims has fallen on account of the divisions and the weakness of religion into the grasp of the enemies of God, the infidel English. You see forty millions of Muslims of Java shackled in the fetters of captivity and of affliction under the rule of the Dutch, although these infidels are much fewer than they in number and they are not much more elevated than them in their civilisation and knowledge. You see Morocco and Algeria and Tunis and Egypt and the Sudan groaning from the extremity of suffering in the grasp of the enemies of God and of His Apostle. You see the vast Siberia and the great Turkestan and Khiva and Bokhara and the Caucasus and the Crimea and Kazan and Esderhan and Kazakstan their Muslim peoples who believe in the unity of God ground under the conquering power of the oppressors who are the enemies of the religion. You behold the land of Dairan[2] in preparation for division and you even see the center of the Caliphate which has not ceased since long ages to combat the enemies of the religion breast to breast has become the target for oppression and violence by means which its enemies renew every day in new form. And, in brief, you see that the enemies of the religion and especially the English, the Russians and the French have gone to great lengths in the oppression of the Islamic world and the invasion of its rights, and in injuring it to a degree that cannot be enumerated and that passes all limits, by which they desire to destroy Islam and the Muslims from the face of the earth. But let them beware! They will not be able to accomplish their purpose whatever tyranny and obstinacy they may put forth and in whatever numbers they may appear, and with whatever implements. The Most High (may His name be praised!) will perfect his light and will manifest His religion ...

And now, O people of Islam, and O beloved brothers, is it not enough what can come upon us of shame, rise up, awake, this weakness and subjection has reached its limit and this humiliation and this belittling has arrived at its end. The bonds of the Islamic world have been cut and his word has been spread abroad. The Muslims have awakened under the feet of shame. Their honor has been reduced and their nobility has been put to shame, and their mosques have been destroyed and their schools and their places of worship are in ruins, and bells and crosses have been placed on their minarets.

Today the Holy War has become a Sacred Duty for Every Muslim

Arise, awake and know that today the Holy War has become a sacred duty for all people of the Faith. And it is enjoined upon all the peoples of Islam who are spread upon the face of the whole earth that they should unite among themselves and hasten to run for the deliverance of their native lands from the hands of the infidels, and that they should hasten to use every means and every plan for this purpose. The host of the Islamic Caliphate is prepared to day for the Holy War, and thousands of Muslims who surge to and fro on the borders and the sides of it delight in praising and in reciting the Tekbir and they are expecting the raising of the flag of the Holy War.

Know then that the Holy War has to day become an especial duty for all Muslims and the time has come when every means must be used for the deliverance of the native land of Islam from the power of the oppressive infidels.

[2] This refers to Iran.

It is Necessary to Form Secret and Public Unions in the Land of Islam

The time has come when every people of the peoples of Islam should form secret and public unions and stand up in the face of the enemies who rule over them, proclaiming the Holy War against them that they may preserve their native lands from the extremity of extermination and that they may obtain the grace of independence and they should know that afterwards there will not come to them an opportunity in which it will be possible to do this.

It is necessary that they should know from to-day that the Holy War has become a sacred duty and that the blood of the infidels in the Islamic lands may be shed with impunity (except those who enjoy the protection of the Muslim power and those to whom it has given security and those who confederate with it).

They must know that the killing of the infidels who rule over the Islamic lands has become a sacred duty, whether it be secretly or openly, as the great Koran declares in its word; 'Take them and kill them whenever you come across them, and we have given you a manifest power over them by revelation'.

To whoever kills even one single infidel of those who rule over Islamic lands, either secretly or openly, there is a reward like a reward from all the living ones of the Islamic world. And let every individual of the Muslims in whatever place they may be, take upon him this oath to kill at least three or four of the ruling infidels, enemies of God, and enemies of the religion. He must take upon him this oath before God Most High, expecting his reward from God alone, and let the Muslim be confident if there be to him no other good deed than this, nevertheless he will prosper in the day of judgment and we ask the Most High to extend the people of the Faith by the favour of the Lord.

O people of the Faith, and O brothers, we are suffering greatly because of your afflictions, we the company of Islam in the world of the Caliphate, Dar-ul-Khelafat, are greatly grieved because of your subjection to a small number of infidels and your obedience to them without opposition. Yes, the enemy is powerful, but that does not form a pretext for you which can be allowed....

You know that the Muslims who live in the boundaries of the world of the Caliphate have fought the Holy War against the enemies in these last years at another time, and that they are still bearing arms expecting the Jihad, all of them from the youth who does not show his beard to the old man whose head is crowned with white hair. You know that our brothers the Sinoussiya in Africa are fighting the Jihad against a nation of the strong nations of Europe, that they have been pushing them back from their native land for years, that these are the times of the arising of Islam. Yes, this generation is the generation of the Jihad. This is the day of the exaltation of the word of God.

. . . The world of infidelity is in a condition of weakness and retrogression, especially the English, the French, and the Russians, who are the most oppressive to the Islamic world and the most covetous of its lands, and the most persecuting to its people. For these infidels are busy now with themselves and thinking about the matter of their life and possessions. Therefore it is the most imperative duty of the Islamic Kingdoms whose geographical position aids them that they should proclaim the war absolutely in the faces of the oppressive rulers. Seeing that the condition greatly aids the Muslims who are in the lands of India and Java and Egypt and Morocco and Tunis and Algeria and Caucasia and Khiva and Bokhara and Turkestan and Iran to drive out the infidels who have obtained rule over their lands and to gain deliverance from captivity. But if this opportunity passes them by or if they neglect it, no one but God knows when there will come to them a like opportunity. Therefore it is the duty of those who truly profess Islam that they should proclaim the war immediately, and that they should drive out those weak ones, or those who are corrupt, who have sucked the life of the Islamic world, and have expended its treasures and dispersed all its good things. And let every one of us be confident that the power of the Islamic people in its Holy War is sufficient for driving out all the infidels and casting them out of all its lands. And let us be firmly confident that if the Islamic peoples unite in taking up weapons in the face of those enemies who have no connection with the countries of Islam, except by the most brittle of causes, you shall scatter them without delay hither and thither and disperse them completely. And let us be confident that if the united war is proclaimed, the Islamic peoples are enough to drive out these infidels who have seized the rule in their countries and to take the trenches and fortresses and weapons and essentials of war which they possess in these countries. They will be driven out by their own power and killed by their own weapons without necessity of any other expense. Believe, O Muslims, and if you do not believe, then the great Lord who is all-powerful has sworn to you that the situation of the English in India and Egypt and the situation of the Dutch in Java and the situation of the Italians in Tripoli and the situation of the French in Morocco and Algeria and the situation of the Russians in Iran and Bokhara and Caucasia; the situation of all these governments in these lands is greatly weakened. It is necessary that we know this truly and believe it, for the evidences which point to it are more than can be mentioned and there is nothing back (above it) of the establishment of the true knowledge except inspiration. And it has been decreed, as you know, and you must believe and resolve and unite and determine, for in every kingdom of the Islamic kingdoms there are five hundred thousand or six hundred thousand at least of native Muslims and it is not a strange thing that God Most High should prosper them to purify their lands from these infidels. Indeed there is no doubt if they are resolved and patient, that He whose name is praised will extend to them an unchangeable good fortune, and it is sufficient for this blessed movement that there should lead in it a company of those whose faith is strong and whose resolves are sound and whose command is intelligent and who have taken the protection of God as their support.

The Rights of the Muslims

From this time on it must be the purpose of the Islamic peoples and their target at which they aim, to release the Islamic kingdoms and the native lands of Islam from the infidels who have usurped the rule over time. There can be absolutely no partnership in the native lands of Islam, for the rule of infidels over Muslims is not lawful, and it is not allowable that the Muslims should be judged by a non-Muslim at any time whatever, and he cannot be patient under the rule of the infidels, and the honor of the Muslims is that they should not be subjects to others, and it is their glory that they should have the lordship, and that they should always be followed by others. This is what Islam requires. And Islam will be completed and perfected in the Muslim if he knows it and practises it. Since we are in a day when war has risen with the infidels who have usurped the rule over the Islamic kingdoms and who do not cease to put to shame Islam in enmity and who wish evil against it, it is the duty of Muslims that they should not comply with the commands issued by them, that they buy nothing of their goods, since compliance with their commands is complacence with their rule, and we have before shown the strictness of the prohibition of this and the abhorrence of it . . . Therefore we have seen fit to direct to all the people of the Faith the following proclamations concerning the following matters:

1. That the buying of anything although it be a trifle from the infidels who have usurped the rule over the Islamic kingdoms and who continually manifest enmity to the Muslims, is absolutely interdicted.

2. Compliance with the commands issued by the infidels who have usurped the rule over the Islamic kingdoms and who are openly hostile to the Muslims, is absolutely interdicted.

3. That the giving of the taxes to the infidels who have usurped the rule over the Islamic countries and who are hostile to the assembly of Muslims, is absolutely interdicted.

4. The duty of sending the lawful amount of pious gifts to the center of the Caliphate is enjoined.

Otherwise the condition of the Muslims which has been spoken of above does not harmonize in any way whatever with the spirit of Islam. And in truth the question of the Islamic kingdoms as the Dar-ul-Islam (World of Islam), since the designation Dar-ul-Islam does not fit it because of its being under the rule of the infidels. Neither is it correct to describe it as Dar-ul-Harb (World of War), for all its inhabitants are Muslims. Therefore the prayers of the assembly are not true prayer in it, according to the traditions which have been delivered concerning them. But whoever of their peoples is satisfied with this condition, he is absolutely impious as may be understood from the verses mentioned above. And it is more fitting that these countries should be called the World of Impiety (Dar-ul-Fesak) than that it should be called Dar-ul-Islam or Dar-ul-Harb.

But let the Muslims wake up. Is it not necessary that they should arouse themselves from the slumber of carelessness, must we not repent to God and return to Him seeking forgiveness for this political crime of which we have become guilty? And we should hasten to the deliverance of all the Islamic kingdoms from the hands of the infidels. Yes, the time for that has come, and it is incumbent upon us, the company of the Islamic peoples, that we should rise up as the rising up of one man, in one of his hands the sword and in the other the gun, and in his pocket balls of fire and annihilating missiles and in his heart the light of the Faith, and that we lift our voices to the utmost, saying—India for the Muslim Indians, Java for the Muslim Javanese, Algeria for the Algerians, among the Muslims, Morocco for the Moroccans, Tunis for the Muslim Tunisians, Egypt for the Muslim Egyptians, Iran for the Muslim Iranians, Bokhara for the Bokharians, Caucasus for the Caucasians, and the Ottoman kingdom for the Muslim Turks and Arabs.

Such must be the aim of all Muslims from now on, and they must strive for this end and fight with their goods and their selves for this end, seeing that the Holy War is a duty laid down for this object. And we hope that the native land of Islam will be saved after this from being called the World of Impiety, and that it may be closely bound to the seat of the Caliphate, and it may rightly be called the Dar-ul-Islam in all meaning of that name; and that we likewise may be of those whose faces are white in the last day between the hands of the All-Knowing and All-Wise, and that we may escape from the reproach of the honored Prophet (May God be gracious to Him and give Him peace) . . .

APPENDIX N
North Africa's Muslims and the Jihad, 1915[1]

The highest representatives of the French authorities in North Africa have doubled their semi-official visits to the grand chiefs of the Muslim religious fraternities. These visits, as can be guessed, aim at flattering the influential *marabuts*, making false promises—like those made during the 1870 war—and pleading for calm among the native circles where there now reigns a bubbling which will sooner or later assume the form of a *Jihad*. Thus, the Governor General of Algeria, riding and accompanied by his official circle, has recently undertaken a pilgrimage into the boundaries of the Algerian south, in order to sing the *shahada* before the *marabuts* and praise the good deeds of Muslim France!

Basically, these high functionaries of the Republic are very upset at seeing their prestige lowered to a degree where they have to go themselves and visit Muslim notables in the sandy *Bled*. They do not like this upturned world; but the situation of France is serious and customs change. Anyway, Muslims do not perceive in these visits anything but an ephemeral politeness; they are well acquainted with this type of platitude-of-circumstances, which does not deceive any Muslim today.

Truly, the special character of the *Jihad*—or, in other words, a movement of insurrection—is nothing new for the Algerian Arab people. In Morocco, it has been proceeding for a long time. In Algeria, the very recent event in Marguerite (in the province of Oran), during 1904 and 1905—not to speak of other events—have demonstrated the character of the natives' dissatisfaction with the French administration. However, the *Jihad* has acquired at the present hour other traits: self-assertion and the prestige of Islam are inextricably tied up.

[1] French original in AAT, Archives de la Guerre, 7N2104, enc. in a dispatch from the French Minister for the Navy to the Minister for War, dated in Paris, 15 Dec. 1915. This is a pamphlet, appealing to North Africa's Muslims to join the *Jihad*. Acc. to French Intelligence sources, it had been printed in Berlin and sent, in thousands of copies, for distribution in French North Africa. Its author signed 'Lieutenant El Hadj Abdallah'; he was an active contributor to Ottoman and German Pan-Islamic propaganda during World War I. English transl. mine.

As far as the French-dominated Muslims are concerned, France's threats to Turkey—against Mecca and Medina—are an aggression against Islam itself in North Africa. The criminal plan of France and its allies is to first destroy the best Muslims and the Holy land;[2] then to annihilate the others!

So this is no more a matter of a two-nation war, but rather a war of races. France, which has always aimed at Constantinople's destruction, has—in attacking Turkey—acted in particular to annihilate the consciousness of the Muslims living under its domination and thus erase for ever the emblem of Islam! But would a Muslim, wherever he may be, agree with a Frenchman, a Neo-Frenchman, an Italian, or some other cosmopolitan enemy of the Arabs, telling him—in the street, at the coffee-house, even in the trenches where the Muslim fights for France—that the finest monuments of Constantinople, Mecca, and Medina would soon be turned into stables for the horses of officers in the French Expeditionary Force? No, a Muslim is not permitted to support such desecration! Nor is a Muslim allowed to remain indifferent to the increasing threats against Turkey; for—we repeat—France and its allies seek, along with Constantinople's destruction, to destroy the moral attachment of Muslims dominated by them, in order to eliminate any desire of Muslim orientation eastward.

[2] i.e. Hijaz.

APPENDIX O
Resolutions of the First All-Russian Muslim Convention, Moscow, May 1917 (Extracts)[1]

The Form of State Administration[2]

(*a*) The All-Russian Muslim convention, after having deliberated the issue concerning the form of state administration in Russia, resolved to acknowledge that the form of government within the Russian structure, guaranteeing best the interests of Muslim communities, is the democratic republic on national–territorial–federative bases, since a nationality group which does not have a defined territory enjoys national–cultural autonomy [only].

(*b*) In order to regulate all the spiritual–cultural questions of the Muslim communities in Russia and to carry them out in solidarity, a central All-Muslim organ for the whole of Russia, with legislative functions in this domain, is being set up.

The structure, composition, and function of this organ will be defined by the constituent assembly of representatives from all autonomous units.

The Central Organ of the Muslims[3]

In accordance with the wishes that all Russian Muslims may unite under a common political flag, the convention adopts the following resolution.

(*a*) The All-Muslim convention considers it imperative to set up a

[1] The resolutions, in Russian, are repr. in *Programs of the Muslim Political Parties*, 10–33. The extracts selected are, respectively, on pp. 11–12 and 30–1. English transl. mine.

[2] This proved one of the thorniest issues in the convention. The decision recommending territorial autonomy for Muslims within a Russian federation was carried by 446 votes to 271. The opponents favoured centralization, i.e. more cultural autonomy for Muslims within a unified Russian republic. The resolution which won was proposed by Mehmet Resulzade, a well-known Pan-Islamist.

[3] This resolution dealt with the character of the above-mentioned organ for all the proposed autonomous Muslim units in Russia. It was proposed by Ayaz İshaki, another active Pan-Islamist.

joint line of conduct and common activity of all political parties and organizations which have a place among Russian Muslims.

(*b*) In order to carry out the above, the convention considers it imperative to elect a central national council, whose make-up should comprise representatives of both communities and classes.

(*c*) In the national council will be placed the leadership for Muslims until the constituent assembly [meets].

(*d*) Until the constituent assembly [meets], all local organizations should obey all the decisions of the national council, which would conform with the resolutions of the [present] convention.

APPENDIX P
Resolutions of the Second All-Russian Muslim Convention, Kazan, July 1917 (Extracts)[1]

The 22 July 1917 Resolution[2]

The united session of the All-Russian All-Muslim convention, as well as the conventions of the military activists[3] and the religious ones, considering the issue of national–cultural autonomy for the Muslims in the interior of Russia and Siberia, resolved

(*a*) Without waiting for the convocation of a constituent assembly, to start bringing about national–cultural autonomy for the Muslims in the interior of Russia;

(*b*) To instruct the second Muslim convention of Russia, meeting now in Kazan, to work out a detailed project of the basic principles for carrying out the above-mentioned establishment of national–cultural autonomy.

The 31 July 1917 Resolutions[4]

1. The people[5] establishes a national assembly[6] to institute laws for all the national, cultural, and religious issues of the Muslims in the interior of Russia and to supervise their execution.

2. The delegates to the national assembly will be selected by pro-

[1] The resolutions, transl. from Tatar into Russian, can be found in Davletshin, 338–53. Our selections are, respectively, from pp. 338, 343–6. English transl. mine.

[2] This was a joint resolution, arrived at by the three different Muslim conventions, meeting in Kazan at the time.

[3] This convention decided on raising a Muslim army and soon afterwards started implementing the resolution.

[4] Taken by the second All-Russian convention in Kazan. The resolutions comprised, *inter alia*, a request for instituting Tatar as the official language of Muslim-inhabited areas. Our selection however focuses on the projected parliament in the town of Ufa.

[5] In Russian, *narod.*

[6] In Tatar, *millet meclise.*

vincial national election meetings at the rate of one delegate for every 50,000 people ...

6. The members of the national assembly represent the whole nation.[7] They should not represent solely the province which delegates them ...

13. The national assembly presents itself as an organ which passes laws, necessary for Muslims in the interior of Russia, in matters of religion and national culture. The national assembly is also the supreme organ to reach final decisions bearing on communities, *imams*, religious schools and *medreses*, languages, education, teaching and national enlightenment in general, the public purse, finances, the *waqfs*, and other national–cultural affairs. Amending the constitution of the national–cultural autonomy is within the competence of the national assembly. In addition, among its tasks is the regulation of the reciprocal relations between state and nation, as well as those between our nation and other peoples, and the elimination of conflicts. The national assembly has approved the annual national budget.

14. The meetings of the national assembly are open. As far as the premises allow, anyone may come and listen. In cases in which divulging the debates prematurely might have undesirable consequences, the meetings may be closed.

15. The members of the national assembly must enjoy the legal privileges of parliamentary membership and immunity in relation to the government and laws of Russia ...

26. In order to manage all the national affairs of the Muslims in the interior of Russia, a national centre is set up, entitled 'national directorate'.[8] This is the executive organ of the national assembly. The national assembly passes laws and resolutions, [while] the national directorate takes practical steps based on these laws and resolutions ...

28. The national directorate is made up of 21 members.

29. The national directorate comprises three departments—for religious affairs, education, and finances. Each of these consists of a chairman and six members ...

[7] In Russian, *natsiya*.

[8] The Russian term is *upravlyeniye*, which can mean 'government', too. The Tatar term used was *nazaratı*, which is equally ambiguous.

APPENDIX Q
Perceptions of Pan-Islam at the British Foreign Office, 1919 (Extracts)[1]

. . . Now, though the Moslem communities are divided by race, language, and sect, and though such terrible dangers as the Crusades and the Mongol invasion were unable to produce co-operation between them, they have a bond in the possession of one Sacred Book—the Qur'an—in the common name Moslem, and in the history of the same period when they were united and swept all before them. This bond is sufficient to make them take pride in each other's successes and sympathize in each other's misfortunes. And the pious at least among them everywhere look back with regret to the time when the Islamic sovereign could aspire to be ruler of the world. It is natural that some should assume that disunion is the cause of their political weakness, and hold that, if unity could be restored, Islam would again become politically formidable.

The Pan-Islamism of Sultan Abd al-Hamid II

The notion of uniting the Islamic communities for the purpose of resisting European aggression and ultimately ousting European rulers from Asia and Africa is expressed in Arabic and Turkish by a phrase meaning 'Islamic union'. Pan-Islamism, a translation of this phrase, first appeared in English (it would seem) in an article published in *The Times* of January 19, 1882, agreeing in most points with, though apparently independent of, one by M.G. Charmes which had appeared a few months before in the *Revue des deux Mondes*,[2] where the word occurs for perhaps the first time in French. The French writer traces the movement to a league originally founded about 1870 in Bokhara, of which Khudayar Khan, ex-ruler of Khokand, served as emissary in Arabia; he preached to various Arab, Kurd, and Indian sheikhs and Mullas the necessity of a Holy War against Russia and England.

[1] *The Pan-Islamic Movement* (= Handbooks prepared under the direction of the Historical Section of the Foreign Office, no. 96 b, London 1919), 53–62, extracts. A copy can be consulted in IO, L/P&S/20/G 77.

[2] 47 (1881), 924 (footnote in the original).

Owing to the humiliating conditions imposed on Turkey by the Treaty of Berlin, the league desired to place at the head of the Islamic confederation, not the Ottoman Sultan, but rather the Sherif of Meccah, Husain, who was appointed in 1877 and assassinated in 1880.

The notion, however, that the religious headship of Islam might be politically utilized was adopted by Abd al-Hamid II, who had not, like his predecessor at the time of the Crimean War, received support from any Christian Power against Russia. The persons supposed to have impressed him with this idea are Si Muhammad Zafir, a Marabout of Tripoli who had foretold his accession, this Marabout's cousin, Sheikh Asad, and a certain Abu'l-Huda Effendi; they persuaded him that his predecessors had been mistaken in cultivating the friendship of European Christian Governments, and that his true course was to attempt to reunite Islam against Christendom. It is asserted that he sent agents and emissaries to different parts of the Islamic world with this object; and we possess detailed accounts of some of these operations. Ibrahim es-Senusi was sent to the Sultan of Morocco with presents and an invitation to join the *Jihad*; with the aid of the British consul at Tangier he obtained an audience, but only an evasive answer, and one which could not be presented, because the Sultan of Morocco, being unable to set his seal to the document without claiming superiority or acknowledging inferiority to the Sultan of Turkey by the place where he set it, disregarded that formality, and an unsealed letter was valueless. Hamzah, the brother of Sheikh Zafir, was sent in 1882 on a mission to Tripoli, which also failed. Since these missions were from their nature secret, it is natural that they should only be known by hearsay.

It appears that throughout his long reign Abd al-Hamid II adhered to the policy of figuring as the head of Islam; and the official French account of the Islamic Orders in North Africa asserts that some of these organizations were employed by the Sultan for the propagation of the Pan-Islamic idea, especially the Rifaʿi Order, of which Abu'l-Huda was the head, and the Madani Order, of which the Sheikh Zafir was the head. These organizations were (1897) aided by the Sherif of Meccah, who appointed as guides to the pilgrims persons willing to circulate the doctrine. Abd al-Hamid's Turkish biographer observes that his construction of the Hijaz railway was undertaken with the idea of conciliating the Moslems throughout the world; the same motive is found in various pointless acts of aggression, though some suppose the Sultan's action in these matters to have been intended to impress his own subjects rather than those of other Governments. To the end of his reign he was in the habit of sending embassies to Moslem communities outside Turkey, such as those in China and Afghanistan.

It is clear that his policy of uniting Islam differed widely from that of his predecessor Selim I, whose idea was to annex to his empire by force of arms all Islamic states that were outside it; the number of Moslems who in Selim's time were subject to non-Moslem rulers was still insignificant. It is not probable that Abd al-Hamid ever contemplated leading a confederation of Moslem states to war against Christendom; his endeavours to enter into friendly relations with European Powers, especially with Germany, render this unlikely. Still, by representing himself as the natural champion of Moslems who were subject to Christian Governments, he hoped to render himself formidable to those Governments; and, as Germany had far fewer such subjects than Great Britain, France or Russia, his policy was regarded as helpful to Germany. Within his empire, however, his claim to be Pontiff of Islam was disputed in many quarters; and, outside it, India appears to be the country where it gained most adherents.

Theory of the Movement current in the East

Oriental writers hold that the inventor of Pan-Islamism—i.e. the person who thought that by the restoration of unity among the Moslem communities European aggression might be stopped, and lost territory ultimately regained—was a certain Sayyid Jamal al-Din, who lived 1838–96. He is known as 'the Afghan'; but according to the most recent account of him he was born in a village of Persia, near Hamadhan, was educated at Najaf, and resided for some years in Afghanistan, whence he took the name Afghani. He was with Dost Muhammad when he took Herat in 1863. In the civil war which broke out shortly afterwards he was in the employ of Azim Khan. But after this prince's defeat by Shir Ali in 1869 he thought it unsafe to remain in Afghanistan, and proceeded to India, where the British Government would not permit him to stay; and it would seem that his belief that his master's downfall had been engineered by Great Britain made him a bitter enemy of this empire. He travelled and taught in Egypt and Constantinople, whence he was expelled in 1871 on the ground of unorthodoxy. Returning to Egypt, he was granted a small monthly allowance by Riyad Pasha, but was expelled in 1879 for starting an anti-British propaganda. He then went to Hyderabad, but, owing to the outbreak of trouble in Egypt, was compelled by the British Government to reside in Calcutta until the Arabi revolt was over, so that he might be under observation.

Jamal al-Din subsequently took up his abode in Paris, where in 1884 he started an Arabic journal called *The Firmest Handle*, containing violent attacks on England as the European Power which had under its sway the largest number of Moslems and was, according to him, the implacable and treacherous enemy of Islam; her political programme was, he asserted, to deprive all Moslems of their independence, seize their possessions, and generally humiliate them. After a time the publication of this journal was stopped; but it had attracted the attention of the Shah Nasir al-Din, who invited Jamal al-Din to Persia, and made him Minister of War. Presently he suspected the intentions of the Shah and left Persia, to travel in Europe; but, meeting the Shah at Munich in 1889, he accepted an invitation to return to Teheran, whence, owing to his endeavour to spread revolutionary ideas, he was expelled in 1890. He then came to London, where, according to his biographer, he tried to influence the Government to obtain the deposition of the Shah—notwithstanding his disapproval of European, and especially British, interference with Islamic concerns.

While in England he received an invitation to Constantinople, which he accepted; and there he lived in the enjoyment of the Sultan's favour till his death in 1896. During these years he is said to have made an attempt to induce the Persian Shiahs to recognize the Caliphate of Abd al-Hamid. Hundreds of letters were written by him and by his colleague, Ruhi Effendi, to persons of eminence and learning, whose acquaintance Jamal al-Din had made during his stay in Persia, urging that, if the Persians would acknowledge the Turkish Sultan's claim to the sovereignty of Islam, Turkey and Persia united would be able to defy European aggression. The negotiations showed signs of leading to a successful result, when they were betrayed or somehow came to the knowledge of the Persian Government; and the execution of Ruhi Effendi presently resulted.

The most celebrated of Jamal al-Din's assistants was the Sheikh Muhammad Abdo, afterwards a Mufti in Egypt, and friendly to the British occupation. The idea which made up the Pan-Islamism of his master are expressed by him in two papers, first published in *The Firmest Handle* (1884), and reprinted in the Mufti's biography. He shows the political subordination of any Islamic community to non-Moslems is a violation of the Islamic religion, which enjoins on its followers not only to shake off foreign yokes, but also to defy all Powers that are not Islamic. He attributes the actual subjection of the great majority of Moslems to non-Moslem Governments to the neglect by Moslem rulers of their duties and to the private quarrels of these potentates, who invoked the help of outsiders to settle them; these outsiders ultimately becoming masters of the situation. These princes and governors, who thus sacrificed the interests of Islam to their petty interests, 'had become chains on the necks of those lions [the Islamic warriors], keeping them from their prey, nay, making them the food of foxes'. Still there was no occasion to despair. Between Adrianople and Peshawar there was a continuous chain of

Moslem states, all united in their belief in the Qur'an, numbering not less than 50,000,000, distinguished above all other races for their bravery; could they not unite in self-defence, and by co-operation divert from themselves the torrents that were rushing upon them from all sides? The writer concluded by expressing the hope that the call to such union and co-operation would come in the first place from the most exalted of the Islamic potentates.

In a second paper he urged that the two essentials for success in a nation were unity and imperialism. The desire for conquest in a people was like that for food in the frame of the individual; but unity was essential before conquest could be contemplated. These principles were not only taught by experience, but were inculcated in the Qur'an. It was, however, necessary to arouse the Moslems, who appeared to have forgotten them: and in the sanctuary to which every true Moslem goes on pilgrimage there was a centre where joint effort could be made the subject of deliberation and practical plans devised.

As has been seen, the idea of utilizing Meccah for this purpose had been anticipated . . . and the organization of the religious orders was already being turned to account. Abd al-Rahman Kawakibi, whom some name with Jamal al-Din among the chief promoters of Pan-Islamism, but who appears to have been no friend of Abd al-Hamid, proposed holding a Pan-Islamic Congress in Meccah, for which apparently the Sultan's consent could not be obtained. He died in the year 1902.

Pan-Islamism as a Religious Revival

Whereas Jamal al-Din's expedients for restoring the political dominance of Islam were decidedly drastic, his followers wished to prepare the way by founding Islamic colleges, newspapers, and societies. In 1903 a Pan-Islamic society was founded in London, of which the objects were thus enumerated:

1. To promote the religious, social, moral, and intellectual advancement of the Mussalman world.
2. To afford a centre of social reunion to Muslims from all parts of the world.
3. To promote brotherly feelings between Muslims and facilitate intercourse between them.
4. To remove misconceptions prevailing among non-Muslims regarding Islam and the Mussalmans.
5. To render legitimate assistance to the best of its ability to any Mussalman requiring it in any part of the world.
6. To provide facilities for conducting religious ceremonies in non-Muslim countries, and to found centres of Muslim thought.

7. To found branches of the Central Pan-Islamic Society in different parts of the world, hold debates and lectures, and read and publish papers likely to further the interest of Islam.

8. To collect subscriptions from all parts of the world in order to build a Mosque in London and endow it.

It was suggested by the Egyptian translator of this document that four more objects should be added:

9. To support the Ottoman Caliphate.

10. To carry on propaganda in this sense till all Islamic potentates and authorities are agreed upon it.

11. In return for this, to obtain the support of the Ottoman Caliph for the Islamic communities, and his intervention with European Governments in control of such communities for the redress of their grievances.

12. To call the attention of the Islamic Governments to cases wherein their measures violated Islamic law, when such measures affected the whole community and gave Islam a bad name.

The London society presently changed its name from Pan-Islamic to Islamic, and seems to have conducted its operations on a modest scale. Other societies, schools, and newspapers were started. One Egyptian of eminence declared, in the year 1904, that Pan-Islamism by no means signified anything like a crusade or union of the Islamic Powers under one banner for the purpose of fighting Europe; what it did was not, however, apparently very different, viz. the establishment of a bond of unity and friendship between the Moslems of the world, to help each other morally, materially, and politically, and to group themselves round the Ottoman Caliphate as representing the greatest Moslem Power. Should such a bond be realized, it would be difficult for the European to continue to appropriate Moslem lands and the produce of Moslem toil.

One English convert to Islam tries harder than many others to avoid the consequence that, if Pan-Islamism is to effect its object, it must enter the service of the Ottoman Empire; but the result is obscure:

'Instead of being a danger to Europe or civilization, Pan-Islamism is a movement that should have the support of every lover of peace and civilization. If the world at large is ever to see that higher and truer civilization of which it is capable, the Powers must abandon that lust of conquest which is but a drag on all true progress; they must cease to look on the interest of each state as a claim to which the interests of all others must yield, and combine to seek the benefit of all. The more nearly that ideal is reached, the more important will it be that

Islam should be prepared to take its fitting place in the grand scheme of regeneration. That it should do so it must follow now and for ever the ideas that are the mainspring of Pan-Islamism . . .'

APPENDIX R

Resolutions passed on 12 February 1951 by the Muslim World Congress (Second Annual Session)[1]

The Congress:

1. supported the stand taken by the Muslims of Palestine to safeguard their rights:
2. supported the people of the Nile Valley in their demand for the unity of the Valley, and the withdrawal of British troops from the Sudan and Suez Canal zone:
3. condemned the illegal occupation by India of Hyderabad, and the massacre and looting of Muslims; and appealed for the immediate initiation, by the Security Council, of an impartial commission to examine the subject, the freeing of all political prisoners, the recall of all the Indian armies, and a plebiscite under the auspices of the United Nations:
4. appealed for effective steps by the Security Council to implement the UNCIP resolution on Kashmir which was economically, culturally, linguistically, geographically and ethnically, part of Pakistan:
5. was concerned over the Eritrean question and appealed to the United Nations to guarantee the legitimate rights of the Eritrean Muslims:
6. condemned the reprehensible and unjustifiable aggression by India against Junadagh, Manavadar and Mangrol which had opted for Pakistan at Partition, and asked the Security Council to instruct India to withdraw from these territories:
7. was deeply concerned over the sufferings of the Yugoslav Muslims and the attacks on their religious freedom:
8. resolved that Arabic should be the *lingua franca* of Muslim countries but without adversely affecting the existence and progress of regional languages:

[1] FO 371/91208, 1102/8.

9. resolved that it was the duty of the Motamer, as Muslim countries had common beliefs, history, problems, aims and ideals, to make every effort to remove ill-will between these countries and inspire brotherly relations in accordance with the injunctions of Islam:
10. resolved that an act of aggression against one Muslim country should be considered as an act of aggression against all, and appealed to all Muslim countries to take steps to save the country attacked in accordance with the tenets of Islam and the United Nations charter: and
11. appealed to all Muslim peoples and governments to unite and cooperate to defend their tenets, their peoples, their Holy Places and their languages, and to let the world know that this unity and cooperation was for their own protection and the peace of the world in general and not for aggression against anyone.

APPENDIX S
Objectives of the Organization of Muslim Peoples, May 1952[1]

We, the representatives of Muslim peoples, have noted the past of the Muslims while they were united, and [appreciated] their contributions to humanity—in justice, philanthropy, and peace (acknowledged by Islam's opponents no less than by its adherents). We have also considered the current ill fortune and adversity of Muslim peoples, caused by their being disunited, which brought about domination by imperialist forces, occasionally supported by [local] despotic governments, trampling all their rights, [all of which] resulted in political slavery, economic subjugation, and social anarchy.

Because of all that—at a time when the destructive imperialist forces are quarrelling among themselves, with the forces of evil and destruction lying in ambush and preparing their instruments in differently named blocs—we, the undersigned representatives of Muslim peoples,[2] have decided to unite the forces of good in the Islamic world and prepare the available means within its peoples and states in an Islamic bloc[3] for justice and philanthropy,[4] in order to get rid of the remaining chains which encumber their progress in other domains. This [we shall do] by establishing an international body, which we have called the Organization of Muslim Peoples,[5] whose goals are as follows.

(*a*) The organization's first objective is to strengthen the Islamic faith among all Muslims and its advancement by good morals and the providing of all services in the spirit of a generous Islam.

(*b*) To liberate Muslim peoples, politically and economically, from foreign rule, and to bring together all their forces and resources for the benefit of their peoples and states.

(*c*) To raise the intellectual and material levels of the individual in Muslim lands, to guarantee his social and political rights, to work for

[1] Munaẓẓamat al-shuʿūb al-Islāmiyya, *Ahdāf*. The manifesto (in my possession) of this little-known organization was probably published in Karachi, but I have seen only its Arabic version. English transl. mine.

[2] The Arabic manifesto does not include the names of the signatories.

[3] The Arabic term is *majmūʿa*.

[4] Or 'doing good'.

[5] In Arabic, *Munaẓẓamat al-shuʿūb al-Islāmiyya*.

establishing and buttressing economic justice among Muslim peoples, in accordance with the principles of Islam, to publicize true culture among them and to protect them from factors of decline preventing their progress—until the Muslim countries take up their appropriate place among nations.

(*d*) Broadening the scope of foreign language teaching in Islamic lands, in order to facilitate reciprocal understanding, and fostering instruction in the language of the Koran among Muslim peoples, as it is the language of their religion, in which every Muslim ought to be proficient.

(*e*) The organization will strive to strengthen the economic and cultural relations among Muslim lands and to spread mutual understanding and affection between Muslim individuals and groups.

APPENDIX T
Project of a Constitution for a Commonwealth of Muslim States, 1969[1]

Introduction

There has been since long demand in most Muslim countries for a Commonwealth or Association or Confederation of Muslim States so that under it they may be able to unite their strength against aggression from all quarters and combine their resources for their economic and material development. International treaties have, by experience, been found to be weak bonds. Draft of a constitution of Commonwealth of Islamic States has for this purpose been prepared which is being placed before the Muslim States and Muslims of the world. Any criticism of it or any suggestions will be most welcome to its author who will reproduce it in the light of such criticism and suggestions.

The Constitution has been divided into two parts—'General' and 'Optional'. Accession to General Part which leaves sovereignty almost unfettered is essential and a beginning can be made with it, without loss of effective sovereignty. It is after accession in respect of the Optional Part that sovereignty is slightly affected for the purpose of joint defence, but gains against losses are enormous. But the Commonwealth can still be constituted without accession in relation to Optional Part.

If God so will and the Muslims of the world unite under a scheme of this tenor, they will be members of one nation—the 'Ummah'—which is the Quranic concept of nationality for the Muslims of the world.

Draft Constitution

General Part

Article 1. There shall be a Commonwealth of Islamic States, herein-

[1] This project was prepared by Abdul Hamid, a retired judge of the High Court of Pakistan and ex-secretary of the Ministry of Law. It was published in *al-Ahibba* (Lahore), 1/2 (Oct. 1969), 33–9.

after called 'the Commonwealth', which shall consist of such Islamic States as may initially constitute the Commonwealth and such other Islamic States as may thereafter accede to the Commonwealth.

Article 2. 'Islamic State' for the purpose of this Constitution shall mean a state where the following conditions exist, namely:

(*a*) if the form of the government of the state is monarchical, the king is Muslim by faith, and not less than 55 per cent of the citizens of the state are also Muslims by faith;

(*b*) and if the form of government of the state is republican, not less than 66 per cent of the citizens of the state are Muslims by faith.

Article 3. (1) Accession of not less than five Muslim States shall be necessary to constitute the Commonwealth by executing the Instrument of Accession to the Constitution in the form prescribed in Schedule 'A'.

(2) Any other Muslim State may accede to the Commonwealth by executing the Instrument of Accession in the form prescribed in Schedule 'A'.

(3) A Muslim State shall, by executing the Instrument of Accession, become a 'Member State' of the Commonwealth.

(4) The Commonwealth shall cease to exist if at any time it has less than five Member States.

Article 4. A Member State may at any time by giving notice to the Commonwealth secede from the Commonwealth.

Article 5. (1) There shall be a Commonwealth Council which shall consist of three representatives from each Member State who shall be either elected or nominated in the manner hereinafter provided.

(2) The Commonwealth Council shall not be subject to dissolution and shall cease to exist if the Commonwealth ceases to exist.

Article 6. (1) Three representatives to the Commonwealth Council from a Member State which has a republican form of government shall be elected by the Legislature of that state in the manner prescribed by that Legislature.

(2) Three representatives to the Commonwealth Council from a Member State which has a monarchical form of government shall be nominated by the king of the state in the manner prescribed by him, but preferably after consultation with the Legislature of the state if it has any, or after consultation with such other body as may be representative of public opinion of the state.

Article 7. (1) A Member State may withdraw any of its representatives from the Commonwealth Council in the same manner in which the representatives had been nominated or elected and substitute another representative in his place.

(2) Vacancies occuring amongst the representatives may be filled in the manner in which the original representative was elected or nominated.

Article 8. The Commonwealth Council shall elect a President and a Deputy President of the Council in the manner provided by the Regulations made by the Council.

Article 9. (1) The Commonwealth Council shall discuss and take decision in relation to matters brought before it.

(2) Any matter of interest to one or more Member States may be brought before the Commonwealth Council for discussion and decision, and the decision taken thereon by the Council shall be binding upon Member States.

Article 10. The Commonwealth Council shall make Regulations for the conduct of its business and the Regulations shall state the manner in which matters shall be brought before the Council and decisions shall be taken thereon.

Article 11. Any decision taken by the Commonwealth Council shall be binding on Member States and it shall be the obligation of Member States to give effect to the decision in the state by enacting laws therefore in such manner as may be most effective.

Article 12. Except in so far as the provision of the Constitution and decisions of the Commonwealth Council would otherwise require, a Member State shall retain its full sovereignty over its external and internal affairs.

Article 13. (1) A citizen of the Member State shall be the citizen of the Commonwealth (hereinafter called 'Commonwealth Citizen'), but he shall not, by acquiring Commonwealth citizenship, forfeit his own citizenship or nationality.

(2) A Commonwealth Citizen holding a Passport issued to him by his own Government shall be entitled to enter and move about in the country of any other Member State without any visa, unless a Member State for reasons of health and security of the state imposes restriction on his entry into or movement in the state.

Article 14. (1) No Member State shall declare war against any other Member State.

(2) If any dispute relating to a matter exists or arises between two or more Member States such dispute shall be brought before the Commonwealth Council and shall be discussed and decided there, and decision taken thereon shall be binding on the disputant states. A Member State refusing to abide by the decision shall cease to be Member State.

Article 15. No Member State shall render any assistance to any foreign state at war with a Member State, in the prosecution of war, but on the other hand shall render such assistance to the Member

State as may be within its means and not inconsistent with its treaty obligations.

Article 16. Every Member State shall give to the other Member States most favourable treatment in relation to entry into its state of any goods produced or manufactured in the other Member States in respect of tariff and other duties.

Article 17. Any two or more Member States may enter into agreement for the execution of a joint venture for the economic and educational development of those states on such terms and conditions as may be agreed upon or as may be determined by the Commonwealth Council.

Article 18. (1) It shall be the obligation of the Member State to promote Islamic Ideology in the state and render assistance to this end to other Member States in the manner determined by the Commonwealth Council.

(2) It shall be the obligation of a Member State to promote feelings of brotherhood amongst the Commonwealth Citizens by such means as may be determined by the Commonwealth Council.

Article 19. There shall be a Commonwealth Fund to which contributions shall be made by Member States in such sums and in such proportions as may be determined by the Commonwealth Council.

Article 20. There shall be at least one meeting of the Commonwealth Council in each year at such time and place as may be determined by the Regulations of the Council.

Article 21. There shall be laid before the Commonwealth Council every year annual statement of the income and contributions to the Commonwealth Fund and of the expenditure to be incurred out of that Fund and no expenditure from the Commonwealth Fund shall be incurred except with the consent of the Commonwealth Council.

Article 22. The Commonwealth Council may amend this Constitution by a majority of not less than two-thirds of its members.

Optional Part

Article (1) A Member State may accede to the Commonwealth on all or any of the following subjects which shall be called 'Commonwealth subjects':

(*a*) Defence.
(*b*) Foreign Relations.
(*c*) Joint Economic Development.
(*d*) Trade and Commerce.

(2) A Member State shall in the Instrument of Accession indicate the subjects in respect of which the state accedes to the Commonwealth.

Article 2. A Member State which has acceded to the Commonwealth on a Commonwealth subject shall be bound by the decisions of the Commonwealth Council in relation to that subject, but the decisions shall be taken only by the votes of the Member States which have acceded on that subject.

Article 3. A Member State may, after accession to the Commonwealth in respect of Foreign Relations, enter into treaty arrangements with any foreign state, but no such treaty arrangement shall be in conflict with or in derogation of the decision of the Commonwealth Council.

Schedule A

Instrument of accession

I . . . the King of . . . (or the President of the Republic of . . .) in exercise of the authority vested in me, do hereby accede to the Commonwealth of Islamic States on the terms and conditions embodied in the General Part of the Constitution of the said Commonwealth.

APPENDIX U
Manifesto of the Afghan Students' Islamic Federation, 1985 (Extracts)[1]

1. The Aim of Establishment: The fundamental aim of establishing the students' federation is to expand solidarity and conception unity among the Muslim youths of Afghanistan and to stabilize against the atrocious enemy and the bloodshed communism.

2. Solidarity Among the Vanguards of the Revolution: The Afghan Students' Islamic Federation will exert all its efforts for stabilizing unity among the Islamic forces and for eliminating the discords and differences among them which is the main wish of the whole Muslim nation of Afghanistan. This is considered as the fundamental obligation of the federation.

3. Unity Among the Heroes of the Battle-field: To create unity among those heroes who are fighting fearlessly against the bloody Russians in different fronts forms the aim of this federation, and request from all militant leaders to hold them in respect and introduce them to the inhabitants of the world as the real general and the real representatives of the heroic nation of Afghanistan.

4. Solidarity with Islamic Organizations: The Afghan Students' Islamic Federation (ASIF) attribute itself to those who are in fighting under the flag of monotheism against communism and vehemently express its hatred and aversion against all un-Islamic organizations . . .

Political Status

8. Enmity with Islam: We consider that number of governments in Islamic countries as the enemies of Islam who have adopted hostile attitude about our Holy Islamic Jihad and back up the Russians' invasion of Afghanistan and their status in the world-politics.

9. Unity of the World of Islam: We demand the termination of conflicts and differences amongst the Islamic nations, rooting out all plots of the world colonialists between Muslims by the name of

[1] *Manifesto of Afghan Students' Islamic Federation* (Peshawar, 1985), 5, 9—extracts.

Sunnite and Shiite and we consider all the Islamic world as one united and brotherly NATION.

10. The Status of the Islamic World: We demand complete and world-wide solidarity among the Muslims throughout the world and complete support for the problems of the Islamic countries.

11. Liberation of the Islamic Nation: The liberation of all Islamic nations under the domination of the world super powers and restoration of self-determination for them ...

Selected Bibliography

Definite articles have not been considered in the alphabetical sequence.

Archives

Affaires Etrangères (Quai d'Orsay, Paris)
Archives de l'Armée de Terre (Vincennes, Paris)
Auswärtiges Amt (Bonn)
Department of State files (National Archives, Washington, DC)
Foreign Office files (Public Record Office, London)
India Office Library and Records (London)
Ministerie van Buitenlandse Zaken (Algemeene Rijksarchief, The Hague)

Other

Abbasi, M.Y., 'Sir Syed Ahmad Khan and the Reawakening of the Muslims', in: Ahmad Hasan Dani, ed., *The Founding Fathers of Pakistan* (Islamabad, 1981), 1–44.

'Syed Ameer Ali', in: Ahmad Hasan Dani, ed., *The Founding Fathers of Pakistan* (Islamabad, 1981), 45–82.

Abbott, Freeland, *Islam and Pakistan* (Ithaca, NY, 1968).

Abbott, G.F., *The Holy War in Tripoli* (London, 1912).

ʿAbd al-ʿĀlim, Muḥammad Ibn ʿAlī, *al-Risāla al-ittiḥādiyya*, in: Abū Shāma al-Shāfiʿī, *Majmūʿat al-rasāʾil* (Cairo, 1328 H), 576–85.

ʿAbd al-Ḥamīd, Muḥammad Ḥarb, ed. and transl., *Mudhakkirāt al-Sulṭān ʿAbd al-Ḥamīd* (Cairo, 1978).

ʿAbd al-Qādir, ʿAbd al-Shāfī Ghunaym and Ra'fat Ghunaymī al-Shaykh, *Qaḍāyā Islāmiyya muʿāṣira* (Cairo, 1980).

ʿAbd al-Rāziq, ʿAlī, *al-Islām wa-uṣūl al-ḥukm* (Cairo, 1925).

Abdallah, El-Hadj, *L'Islam dans l'armée française (guerre de 1914–1915)* (Istanbul, 1915).

—— *Les Musulmans de l'Afrique du Nord et le 'Djéhad'* (n.p., 1915).

ʿAbduh, Muḥammad, 'Tawḥīd al-Muslimīn wa-waḥdat al-sulṭatayn al-dīniyya wa-'l siyāsiyya', *al-Mu'ayyad* (Cairo), May 1900.

—— *al-Muslimūn wa-'l-Islām*, ed. Ṭāhir al-Ṭanāḥī (n.p. [Cairo], n.d. [1963]).

Abdülhamid'in hatıra defteri (belgeler ve resimler), ed. İsmet Bozdağ (Istanbul, 1975).

Abdülhamit (Sultan), *Siyasî hatıratım* (Istanbul, 1974).

Abdullah, Ahmed, 'Syed Jamaluddin Afghani's Ideas Blaze the Trail', *Pakistan Horizon* (Karachi), 34/2 (1981), 35–43.

Abdullah, Fevziye, *see* Tansel, Fevziye Abdullah

ʿAbīdī, Āl-i Aḥmad, *Jazb-i bāham* (Multan, 1983).

Abū al-Majd, Aḥmad Kamāl, 'Bal al-Islām wa-'l-ʿUrūba maʿan', *al-ʿArabī* (Kuwait), 263 (Oct. 1980), 6–11.

Abu Manneh, B. 'Sultan Abdulhamid II and Shaikh Abulhuda al-Sayyadi', *Middle Eastern Studies* (London), 15/2 (May 1979), 131–53.

Abū Rayya, Maḥmūd, *Jamāl al-Dīn al-Afghānī 1838–1897*[2] (Cairo, 1971).

Abū Zahra, Muḥammad, *al-Waḥda al-Islāmiyya* (Cairo, 1958; 2nd edn., Cairo, 1976).

Adamec, L.W., *Afghanistan 1900–1923: A Diplomatic History* (Berkeley, Calif., 1967).

Adams, C.C., *Islam and Modernism in Egypt: A Study of the Modern Reform Movement Inaugurated by Muḥammad ʿAbduh* (London, 1933).

Āfāq al-waḥda al-Islāmiyya (Tehran, 1403 H).

al-Afghānī, Jamāl al-Dīn, *al-Waḥda al-Islāmiyya wa-'l-waḥda wa-'l-siyāda*, introd. by Muṣṭafā ʿAbd al-Rāziq (n.p., 1358/1938).

Africain, Un, *Manuel de politique musulmane*[3] (Paris, 1925).

Aḥmad, ʿAbd al-ʿĀṭī Muḥammad, *al-Fikr al-siyāsī li-'l-imām Muḥammad ʿAbduh* (n.p., 1978).

Ahmad, Aziz, Sayyid Aḥmad Khān, 'Jamāl al-Dīn al-Afghānī and Muslim India', *Studia Islamica* (Paris), 13 (1960), 55–78.

——*Studies in Islamic Culture in the Indian Environment* (Oxford, 1964).

——*Islamic Modernism in India and Pakistan 1857–1964* (London, 1967).

——'Afghani's Indian Contacts', *Journal of the American Oriental Society*, 89 (1969), 476–504.

Ahmad, Feroz, *The Young Turks: The Committee of Union and Progress in Turkish Politics 1908–1914* (Oxford, 1969).

——1914–1915 yıllarında İstanbul'da Hint milliyetçi devrimcileri', *Yapıt* (Ankara), 6 (Aug–Sept. 1984), 5–15.

——'The Kemalist Movement and India', *Cahiers du GETC* [Groupe d'études sur la Turquie contemporaine] (Paris), 3 (Autumn 1987), 112–23.

Ahmad, Jamil-ud-Din, 'Foundations of the All-India Muslim League', in: Mahmud Husain *et al.*, eds., *A History of the Freedom Movement* (Pakistan Historical Society Publications, 19; Karachi, 1961), iii. 29–61.

Ahmad, Khurshid, ed., *Studies in Islamic Economics* (Leicester, UK, and Jedda, 1980).
——'Mecca Conference of World Muslim Organizations: Seeking Unity at the Grassroots', *Impact International Fortnightly* (London), 4/9 (10–23 May 1974), 8.
Ahmad, Mirza Mahmud, 'The Future of Turkey', *The Moslem World* (Hartford, Conn.), 10/3 (July 1920), 274–81.
Ahmad, Mushtaq, *Pakistan's Foreign Policy* (Karachi, 1968).
Ahmad, Rafiüddin, 'A Muslim View of Abdul Hamid and the Powers', *The Nineteenth Century* (London), 38 (July 1895), 156–64).
——'A Moslem's View of the Panislamic Revival', *The Nineteenth Century*, 47 (Oct. 1897), 517–26.
Ahmed, Syed Moizuddin, 'The Political Ideas of Jamal-ud-Din al-Afghani—the Founder of the Pan-Islamic Movement', *Sind University Research Journal, Arts Series, Humanities and Social Sciences* (Jamshoto, Sind), 19 (1980), 51–60.
Ahmed-Bioud, Abdelghani *et al.*, eds., *3200 revues et journaux arabes de 1800 à 1965* (Paris, 1969).
Ahsan, Abdullahil, 'Muslim Society in Crisis: A Case Study of the Organization of the Islamic Conference', Ph.D. dissertation, University of Michigan, 1985.
Akarlı, E. D., 'Abdulhamid's Islamic Policy in the Arab Provinces', *Turk-Arap ilişkileri: geçmişte, bugün ve gelecekte* (Ankara, 1979), 44–60.
Akhmyedov, A., 'Vsyemirnaya Islamskaya liga', *Argumyenti* (Moscow, 1983), 139–43.
——*Islam v sovryemyennoy idyeyno-politichyeskoy bor'bye* (Moscow, 1985).
Akhtar, Shameem, 'The Rabat Summit Conference', *Pakistan Horizon*, 22/4 (1969), 336–40.
——'The Jeddah Conference', *Pakistan Horizon*, 23/2 (1970), 179–84.
Akiner, Shirin, *Islamic Peoples of the Soviet Union* (London, 1983).
Akira, Nagazumi, 'The Abortive Uprisings of the Indonesian Communist Party and its Influence on the Pilgrims to Mecca: 1926–1927', *The Tōyōshi-Kenkyū The Journal of Oriental Researches* (Kyoto), 38/1 (June 1979), 1–2.
Aksay, Hasan, 'Islâm dünyasının birliği', *Dış Politika* (Istanbul), 1/2 (July 1988), 15–33.
Akşın, Sina, *31 mart olayı* (Ankara, 1970).
Ali, Ameer, 'Moslem Feeling', appendix in Thomas Barclay, *The Turco-Italian War and its Problems* (London, 1912), 101–8.
——'The Caliphate: A Historical and Juridical Sketch', *The Contemporary Review* (London), 107/594 (June 1915), 681–94.
Ali, Mehrunissa, 'The Second Islamic Summit Conference, 1974',

Pakistan Horizon, 27/1 (1974), 29–49.

Ali, Mohsin, 'Tackling Issues before Muslim World', *Dawn Overseas Weekly* (Karachi), 11/7 (11 Feb. 1987), 1, 4.

Ali, Moulaví Cherágh, *The Proposed Political, Legal, and Social Reforms in the Ottoman Empire and Other Mohammadan`States* (Bombay, 1883).

Ali, Muazzam, 'Role of O.I.C. in Promoting World Peace', in: *L'Islam dans les relations internationales* (Aix-en-Provence, 1986), 170–3.

Ali, Shaukat, *Pan-movements in the Third World: Pan-Arabism, Pan-Africanism, Pan-Islamism* (Lahore, n.d. [1976]).

Ali, Syed Murtaza, 'Saiyed Jamal al-Din Afgani: His Visits to India', *Journal of the Asiatic Society of Bangladesh* (Dacca), 18/3 (Dec. 1973), 207–19.

Alsagoff, Syed Ibrahim Bin Omar, 'A Muslim Universal League', *Muslim World Islamic Quarterly in English* (Singapore), 1/1 (Jan. 1949), 3–5.

Alvi, Q. Ahmed-ur-Rehman, 'Pan-Islamism or Islamic State?', *The Islamic Literature* (Lahore), 6/2 (Feb. 1954), 37–42.

Aly, Abd al-Monein Said, and M. W. Wenner, 'Modern Islamic Reform Movements: The Muslim Brotherhood in Contemporary Egypt', *The Middle East Journal* (Washington, DC), 36/3 (Summer 1982), 336–61.

ʿAmāra, Muḥammad, *al-Aʿmāl al-kāmila li-Jamāl al-Dīn al-Afghānī maʿa dirāsa ʿan al-Afghānī: al-ḥaqīqa al-kulliyya* (Cairo, n.d. [1968]).

——*al-Jāmiʿa al-Islāmiyya wa-'l-fikra al-qawmiyya ʿind Muṣṭafā Kāmil* (Beirut, 1976).

——*Naẓariyyat al-Khilāfa al-Islāmiyya* (Cairo, 1980).

——*al-Islām wa-'l-ʿUrūba wa-'l-ʿalmāniyya* (Beirut, 1981).

——*Tayyārāt al-yaqẓa al-Islāmiyya al-ḥadītha* (n.p. [Cairo], 1982).

——*Abū al-Aʿlā al-Mawdūdī wa-'l-ṣaḥwa al-Islāmiyya* (Cairo and Beirut, 1987).

Amīn, Aḥmad, *Zuʿamā' al-iṣlāḥ fī al-ʿaṣr al-ḥadīth*^{3} (Cairo, 1971).

Amin, Muhammad Anwar, *The Long March to Jedda* (Lahore, 1974).

al-Amīrī, ʿUmar Bahā' al-Dīn, *ʿUrūba wa-Islām* (n.p. [Cairo], 1960).

Anees, Munawwar Ahmad, 'End of Empire', *Afkār—Enquiry Magazine of Events and Ideas* (London), 3/2 (Feb. 1986), 45–9.

Antic, Zdenko, 'Pan-Islamic Nationalism Condemned by Yugoslav Official', Radio Free Europe Research, *RAD Background Report, 251, Yugoslavia* (Munich, 15 Nov. 1979).

Aperçu sur l'illégitimité du Sultan turc en tant que Khalife: Quelques remarques relatives aux prétendues visées turques sur Constantinople d'après les sources arabes et l'opinion des Islamistes les plus célèbres (Paris, 1919).

Appel des Musulmans opprimés au Congrès de Berlin: Leur situation en Europe et en Asie depuis le traité de San-Stéfano (Istanbul, 1878).

Arafati, *see* Salem, Mahmoud

Arıburnu, Kemal, *Atatürk: Anekdotlar—anılar* (Ankara, 1960).

al-ʿArīḍī, Bashīr, 'al-Tanẓīm al-Islāmī lam yanbathiq ʿām 1969 wa-lakinnahu yarjiʿ ilā awākhir al-qarn al-māḍī', *al-Yaqẓa* (Kuwait), 26 Dec. 1986, 10–11.

Ariff, Mohammad, ed., *Monetary and Fiscal Economics of Islam* (Jedda, 1982).

Armstrong, Harold, *Turkey in Travail: The Birth of a New Nation* (London, 1925).

Arnold, Thomas W., *The Preaching of Islam: A History of the Propagation of the Muslim Faith* (Lahore, 1913).

——*The Caliphate* (Oxford, 1924).

Arsharuni, A. and Kh. Gabidullin, *Ochyerki Panislamizma i Pantyurkizma v Rossii* (Moscow, 1931).

Arslān, Shakīb, *Sīra dhātiyya* (Beirut, 1969).

Aruri, N. H., 'Nationalism and Religion in the Arab World: Allies or Enemies?', *The Muslim World*, 67/4 (Oct. 1977), 266–79.

Arusî, Mihriddin (= Ahmed Hilmi), *Yırmıncı asırda âlem-i İslam ve Avrupa Müslümanlara rehber-i siyaset* (Istanbul, 1327 H).

Asʿad Efendi, Ṣaḥib-zādeh al-Shaykh, *Bayān hāmm li-ʿālam al-Islām: al-maqāla al-thālitha min bayān al-sirr al-muṭawwī* (Damacus, 1333).

Ashraf, Sulaymān, *al-Balāgh* (Aligarh, n.d. [1914?]).

ʿĀshūr, Muḥammad Sāmī, *al-Muslimūn taḥt al-ḥukm al-shuyūʿī* (Cairo, n.d.).

'The Asian Circle: 2. Panislamism', *Asiatic Review* (London), 23/73–6 (1927), 209–15.

'Les Aspirations politiques des Musulmans russes', *Revue du Monde Musulman* (Paris), 56 (Dec. 1923), 136–48.

Assaf, Michael, 'Die muselmanische Konferenz in Jerusalem und der Panislamismus', *Palästina Zeitschrift für den Aufbau Palästina* (Vienna), 15/1–2 (Jan.–Feb. 1932), 34–43.

Attunisi, Schaich Salih Aschscharif, *La Vérité au sujet de la Guerre Sainte* (Berne, 1916).

'Au congrès musulman d'Europe', *Die Welt des Islams* (Berlin), 17/3–4 (Feb. 1936), 104–11.

ʿAwaḍ, Louis, *Ta'rīkh al-fikr al-Miṣrī al-ḥadīth*, i–ii (Cairo, n.d.).

ʿAwwād, Muḥammad Ḥasan, *al-Taḍāmun al-Islāmī al-kabīr fī ẓill Fayṣal ibn ʿAbd al-ʿAzīz*^{2} (Cairo, 1976).

Ayyubi, N.A., *Kemal Ataturk* (Aligarh, 1982).

Ayyuhā al-Hindiyyūn wa-'l-Miṣriyyūn al-bu'asā' alladhīn atʿasahum

al-hazz wa-'l-qadr (n.p., n.d.).
Ayyuhā al-ikhwān fī al-dīn ay khayr tantaẓirūn min al-Inglīz (n.p., n.d.).
Ayyuhā al-ikhwān al-Muslimūn (n.p., n.d.).
Aziz, K.K., compiler, *The Indian Khilafat Movement 1915–1933: A Documentary Record* (Karachi, 1972).
al-ʿAẓm, *see* Azmzadeh
Azmzadeh, Refik, *İttihad-ı İslam ve Avrupa* (Istanbul, 1327 H). Arabic transl.: Rafīq al-ʿAẓm, *al-Jāmiʿa al-Islāmiyya wa-Ūrubbā* (Cairo, 1383/1963).
Azoury, Negib, *Le Réveil de la nation arabe dans l'Asie turque* (Paris, 1905).
B., C.E., 'Notes sur le Panislamisme', *Questions Diplomatiques et Coloniales* (Paris), 28/307 (1 Dec. 1909), 641–56; 28/308 (16 Dec. 1909), 729–42.
——'Le Parti musulman russe', *Revue du Monde Musulman*, 2/7 (May 1907), 388–9.
Badawi, M.A.Z., *The Reformers of Egypt* (London, 1978).
Badeau, J.S., 'Islam and the Modern Middle East', *Foreign Affairs*, 38/1 (Oct. 1959), 61–74.
Badger, G.P., 'The Precedents and Usages Regulating the Muslim Khalîfate', *The Nineteenth Century*, 2 (Sept. 1877), 274–82.
Badry, Roswitha, and Johannes Niehoff, *Die ideologische Botschaft von Briefmarken—dargestellt am Beispiel Libyens und des Iran* (Tübingen, 1988).
Baer, Gabriel, 'Islam and Politics in Modern Middle Eastern History', in: M. Heper and R. Israeli, eds., *Islam and Politics in the Middle East* (London, 1984), 11–28.
Baghdādī, Jīlānīzādeh al-Sayyid Muṣṭafā, *Ilā man fī juyūsh al-Inglīz min ikhwāninā Muslimī al-Hind wa-Miṣr wa-'l-Afghān* (n.p., n.d.).
Baha, Lal, 'Activities of Turkish Agents in Khyber during World War I', *Journal of the Asiatic Society of Pakistan*, 14/2 (Aug. 1969), 185–92.
Bajolle, Général, 'Le Panislamisme et la paix mondiale', *La Revue Mondiale* (Paris), 140/1 (1 Jan. 1921), 28–36.
Balāgh ʿāmm li-jamīʿ ahl al-Islām nasharathu Jamʿiyyat al-Mudāfaʿa al-Milliyya (n.p., 1333).
Ballvora, Shyqyri, 'La "Ligue de Prizrend" et la question de l'autonomie de l'Albanie', *Studia Balcanica* (Tirana), 5/2 (1968), 49–68.
Bamford, P.C., *Histories of the Non-co-operation and Khilafat Movements* (Delhi, 1925; repr. 1974).
Barakatullah, Mohammad (Maulavie), *The Khilafet* (London, 1924).
Barbar, A.M., 'Resistance to the Italian Invasion: 1911–1920', Ph.D. dissertation, Univ. of Wisconsin, Madison, 1980.

Bardin, Pierre, *Algériens et Tunisiens dans l'empire ottoman de 1848 à 1914* (Paris, 1979).

Bareilles, Bertrand, *Les Turcs: Ce que fût leur empire: Leurs comédies politiques* (Paris, 1917).

Barrier, N. G., *Banned: Controversial Literature and Political Control in British India* (Columbia, Miss., 1974).

Barrū, Tawfīq ʿAlī, *al-ʿArab wa-'l-Turk fī al-ʿahd al-dustūrī al-ʿUthmānī 1908–1914* (Cairo, 1960).

Bartol'd, V., 'Khalif i Sultan', *Mir' Islama* (St Petersburg), 1 (1912), 203–26, 345–400.

——'Panislamizm', in: Bartol'd, *Sochinyeniya* (Moscow, 1966), vi. 400–2.

Başak, Hasan Tahsin, *İslâm birliği ittifak ve ittihat nazariyesi* (Kastamonu, 1377/1957). Arabic transl.: *Yalzam ʿalā al-Muslimīn al-ittifāq wa-'l-ittiḥād* (Kastamonu, 1377/1957).

al-Bashīr, al-Ṭāhir Muḥammad ʿAlī, *al-Waḥda al-Islāmiyya wa-'l-ḥarakāt al-dīniyya fī al-qarn al-tāsiʿ ʿashar* (Khartoum, 1975).

Basu, Aparna, 'Mohamed Ali in Delhi: The *Comrade* Phase, 1912–1915', in: Mushirul Hasan, ed., *Communal and Pan-Islamic Trends in Colonial India* (New Delhi, 1981), 109–25.

Bayur, Yusuf Hikmet, *Türk inkilabı tarihi*, ii, part 4 (Ankara, 1952).

al-Bayyūmī, Zakariyya Sulaymān, *al-Ikhwān al-Muslimūn wa-'l-jamāʿāt al-Islāmiyya fī al-ḥayāt al-siyāsiyya al-Miṣriyya 1928–1948* (Cairo, 1979).

al-Bazzāz, ʿAbd al-Raḥmān, 'Islam and Arab Nationalism', transl. by S.G. Haim, *Die Welt des Islams* (Leiden), NS 3 (1954), 201–18.

Becker, C.H., 'Panislamismus', *Archiv fur Religionswissenschaft* (Leipzig), 7 (1904), 169–92.

——*Deutschland und der Islam* (Stuttgart and Berlin, 1914).

Beling, W.A., *King Faisal and the Modernisation of Saudi Arabia* (London, 1980).

Bellotti, Felice, *Arabi contro Ebrei in Terrasanta* (Milan, 1939).

Ben Nabī, Mālik, *Fikrat commonwealth Islāmī*^{2} (Cairo, 1391/1971).

Bennigsen, Alexandre, 'Islamic or Local Consciousness among Soviet Nationalities?', in: Edward Allworth, ed., *Soviet Nationality Problems* (New York, 1971), 168–82.

——'Muslim Conservative Opposition to the Soviet Regime: The Sufi Brotherhoods in North Caucasus', in: J.R. Azrael, ed., *Soviet Nationalities Policies and Practices* (New York, 1978), 334–48.

——'Panturkism and Panislamism in History and Today', *Central Asian Survey*, 3/3 (1985), 39–68.

——*Self Determination in Soviet Central Asia: Problems and Prospects* (Middle East Technical University, Asian–African Research Group, 29; n.p. [Ankara], 1986).

——'Marxism or Pan-Islamism', *Central Asian Survey* (Oxford), 6/2 (1987), 55–66.

——and Chantal Quelquejay, *Les Mouvements nationaux chez les Musulmans de Russie: Le 'Sultangalievisme' au Tatarstan* (Paris and the Hague, 1960).

——and Chantal Lemercier-Quelquejay, *Islam in the Soviet Union* (London, 1967).

——and S. E. Wimbush, *Muslim National Communism in the Soviet Union: A Revolutionary Strategy for the Colonial World* (Chicago, 1979).

——and Chantal Lemercier-Quelquejay, *Les Musulmans oubliés: L'Islam en Union Soviétique* (Paris, 1981).

——and Marie Broxup, *The Islamic Threat to the Soviet State* (London, 1983).

——and S.E. Wimbush, *Muslims of the Soviet Empire: A Guide* (London, 1985).

——and Chantal Lemercier-Quelquejay, *Le Soufi et le commissaire: Les Confréries musulmanes en URSS* (Paris, 1986).

Bérard, Victor, *Le Sultan, l'Islam et les puissances: Constantinople—La Mecque—Bagdad* Paris, 1916).

Berg, L.W.C. van den, 'Het Panislamisme', *De Gids* (Amsterdam), 64 (Nov. 1900), 228–69; (Dec. 1900), 392–431.

Berkes, Niyazi, *The Development of Secularism in Turkey* (Montreal, 1964).

Bernex, Jules, *Pendant la résurrection de l'Égypte* (Paris, 1919).

Bessis, Juliette, 'Chekib Arslan et les mouvements nationalistes au Maghreb', *Revue Historique* (Paris), 526 (Apr.–June 1978), 467–89.

Bianchini, A.L., 'I movimenti nazionalisti nei paesi maometani', *Rivista Marittima* (Rome), 55 (Sept. 1922), 463–518.

Binder, Leonard, *Religion and Politics in Pakistan* (Berkeley, Calif., 1961).

Blank, Stephen, 'Soviet Politics and the Iranian Revolution of 1919–1921', *Cahiers du Monde Russe et Soviétique* (Paris), 21/2 (Apr.–June 1980), 173–94.

Blunt, W.S., *The Future of Islam* (London, 1882).

——*India under Ripon: A Private Diary* (London, 1909).

Bobrovnikoff, Sophy, 'Moslems in Russia', *The Moslem World*, 1/1 (Jan. 1911), 5–31.

Bociurkiw, B.R., 'The Changing Soviet Image of Islam', *The Search: Journal for Arab and Islamic Studies* (Brattleboro, Vt.), 4/3–4 (Winter, 1983), 59–80.

Bodyanskiy, V.L., and M.S. Lazaryev, *Saudovskaya Araviya poslye Sauda: Osnovniye tyendyentsii vnyeshnyey politiki (1964–1966 gg.)* (Moscow, 1967).

'Le Bolchévisme et l'Islam. II. Hors de Russie', *Revue du Monde Musulman*, 52 (1922).

Borisov, L.B., 'Organizatsiya Islamskaya Konfyeryentsiya: Politichyeskiye aspyekti dyeyatyel'nosti', *Narodi Azii i Afriki* (Moscow), 4 (1983), 101–8.

Bouachba, Taoufik, 'L'Organisation de la conférence islamique', *Annuaire Français de Droit International* (Paris), 28 (1982), 256–91.

Bouteiller, Georges de, 'La "Nation islamique": Utopie ou réalité géopolitique de demain?', *Défense Nationale* (Paris), 37 (Jan. 1981), 59–69.

——'La "Nation islamique": Une réalité géopolitique de demain?', *Défense Nationale*, 37 (Feb. 1981), 101–10.

——'La Ligue Islamique mondiale, une institution tentaculaire', *Défense Nationale*, 40 (Jan. 1984), 73–80.

Bouvat, L., 'Les Derniers congrès des Musulmans russes', *Revue du Monde Musulman*, 1/2 (Jan. 1906), 264–6.

——'La Presse anglaise et le Panislamisme', *Revue du Monde Musulman*, 1/3 (Jan. 1907), 404–5.

——'Le Parti musulman russe', *Revue du Monde Musulman*, 2/7 (May 1907), 388–9.

——'Un projet de parlement musulman international', *Revue du Monde Musulman*, 7/3 (Mar. 1909), 321–2.

Braginskiy, I.S., 'O prirodye Sryedniaziatskogo Dzhadidizma v svyetye lityeraturnoy dyeyatyel'nosti Dzhadidov', *Istoriya SSSR* (Moscow), 6 (Nov.–Dec. 1965), 26–38.

Bräker, Hans, *Kommunismus und Weltreligionen Asiens: Zur Religions-und Asienpolitik der Sowjetunion*, i, parts 1–2 (Tübingen, 1969–71).

——'The Islamic Renewal Movement and the Power Shift in the Near/Middle East and Central Asia', in: Chantal Lemercier-Quelquejay *et al.*, eds., *Turco-Tatar Past Soviet Present: Studies Presented to Alexandre Bennigsen* (Louvain and Paris, 1986), 501–24.

Brelvi, Mahmud, 'The Islamic Congress (*al-Mo'tamar al-Islāmī*), Cairo: A Brief Survey of its Work', *The Islamic Review* (Woking, Surrey, UK), 43/10 (Oct. 1955), 13.

Brémond, Général [René], *L'Islam et les questions musulmanes* (Paris, 1924).

Broadley, A.M., *How We Defended Arábi and His Friends: A Story of Egypt and the Egyptians* (London, 1884).

Brown, E.C., *Har Dayal: Hindu Revolutionary and Nationalist* (Tucson, Ariz., 1975).

Browne, E.G., 'Pan-Islamism', in: F.A. Kirkpatrick, ed., *Lectures on*

the History of the Nineteenth Century (Cambridge, UK, 1902).
——*The Persian Revolution of 1905–1909* (Cambridge, UK, 1910).
Browne, L.E., *The Prospects of Islam* (London, 1944).
Broxup, Marie, 'Islam in Central Asia since Gorbachev', *Asian Affairs* (London), 18/3 (= OS 74; Oct. 1987), 283–93.
Buckler, F.W., 'The Historical Antecedents of the *Khilafat* Movement', *The Contemporary Review*, 121/677 (May 1922), 603–11.
Buheiry, M.R., 'Colonial Scholarship and Muslim Revivalism in 1900', *Arab Studies Quarterly* (Belmont, Mass.), 4/1–2 (Spring 1982), 1–16.
Burke, Edmund, 'Pan-Islam and Moroccan Resistance to French Colonial Penetration, 1900–1912', *Journal of African History* (Cambridge, UK), 13/1 (1972), 97–118.
——'Moroccan Resistance, Pan-Islam and German War Strategy, 1914–1918', *Francia: Forschungen zur Westeuropäischen Geschichte* (Paris), 3 (1975), 434–64.
Burke, S.M., *Pakistan's Foreign Policy* (London, 1973).
Bury, G.W., *Pan-Islam* (London, 1919).
Byelyayev, Ye. A., 'Panislamizm', *Kratkiy nauchno-atyeistichyeskiy slovar'* (Moscow, 1964), 419–20.
Cabaton, Antoine, 'Panislamisme ou commerce', *Revue du Monde Musulman*, 8/6 (June 1909), 281–2.
Carré, Olivier, and Gérard Michaud, *Les Frères musulmans (1928–1982)* (Paris, 1983).
——and Paul Dumont, eds., *Radicalismes islamiques*, ii (Paris, 1986).
Carretto, G.E. 'La conferenza islamica e la Turchia', *Oriente Moderno* (Rome), 56/5–6 (May–June 1976), 105–16.
Cash, W.W., *The Moslem World in Revolution* (London, 1925).
Caskel, Werner, 'Western Impact and Islamic Civilization', in: G.E. von Grunebaum, ed., *Unity and Variety in Muslim Civilization* (Chicago, 1955), 335–60.
Castagné, Joseph, 'Le Bolchévisme et l'Islam. I. Les Organisations soviétiques de la Russie musulmane', *Revue du Monde Musulman*, 51 (Oct. 1922), 1–254.
Cataluccio, Francesco, *Storia del nazionalismo arabo* (Milan, 1939).
——'Panislamismo e albori di nazionalismo arabo nel secolo XIX', *Archivio Storico Italiano* (Florence), 124/1 (1966), 31–75.
Çetinsaya, Gökhan, 'II. Abdülhamid döneminin ilk yıllarında "islâm birliği" hareketi (1876–1878)', MA dissertation, Ankara University, 1988.
Chakrabarty, Dipesh, 'Communal Riots and Labour: Bengal's Jute Mill-hands in the 1890s', *Past and Present* (Oxford), 91 (May 1981), 140–69.
Charmes, Gabriel, 'La Situation de la Turquie', *Revue des Deux*

Mondes (Paris), 47/4 (15 Oct. 1881), 721–61; 49/4 (15 Feb. 1882), 833–69.

Charmes, Gabriel, *L'Avenir de la Turquie: Le Panislamisme* (Paris, 1883).

Charnay, Jean-Paul, *L'Islam et la guerre: De la guerre juste à la révolution sainte* (Paris, 1986).

Chatelier, *see* Le Chatelier

Chejne, A.G., 'Pan-Islamism and the Caliphal Controversy', *The Islamic Literature*, 7/12 (Dec. 1955), 5–23.

Chelkowski, Peter, 'Stamps of Blood', *American Philatelist* (State College, Pa.), 101/6 (June 1987), 556–66.

Chirol, Valentine, 'Pan-Islamism', *Proceedings of the Central Asian Society* (London), 14 Nov. 1906, 1–28.

—— 'Turkey in the Grip of Germany', *The Quarterly Review* (London), 223/442, part 2 (Jan. 1915), 231–51.

—— 'Islam and the War', *The Quarterly Review*, 229/455 (Apr. 1918), 489–515.

—— 'The Downfall of the Ottoman Khilafat', *Journal of the Central Asian Society* (London), 11/3 (1924), 229–43.

Cleveland, W.L., *Islam against the West: Shakib Arslan and the Campaign for Islamic Nationalism* (London, 1985).

—— 'The Role of Islam as Political Ideology in the First World War', in: E. Ingram, ed., *National and International Politics in the Middle East* (London, 1985), 84–101.

Colombe, Marcel, 'Islam et nationalisme arabe à la veille de la première guerre mondiale', *Revue Historique*, 223 (1960), 85–98.

Colquhoun, A.R., 'Pan-Islam', *North American Review* (New York), 182/6 (15 June 1906), 906–18.

Comité 'La France et l'Islam', *Recueil de discours en faveur de l'Islam et de la Turquie* (Paris, 1921).

'La Conférence islamique au sommet', *Maghreb: Études et documents* (Paris), 36 (Nov.–Dec. 1969), 28–34.

'Il congresso dei Musulmani d'Europa a Ginevra', *Annali del R. Istituto Superiore Orientale di Napoli* (Naples), 8/1 (June 1935), 132–3.

Cruickshank, A.A., 'The Young Turk Challenge in Postwar Turkey', *The Middle East Journal*, 22/1 (Winter 1968), 17–28.

al-Dahhān, Sāmī, *Muḥāḍarāt ʿan al-amīr Shakīb Arslān* (n.p. [Cairo], 1958].

—— *al-Amīr Shakīb Arslān ḥayātuh wa-āthāruh* (Cairo, 1960).

al-Dajānī, Aḥmad Ṣidqī, 'Mustaqbal al-ʿalāqa bayn al-qawmiyya al-ʿArabiyya wa-'l-Islām, *al-Mustaqbal al-ʿArabī* (Beirut), 3/24 (Feb. 1981), 62–70.

Dani, Ahmad Hasan, 'Allama Dr. Muhammad Iqbal on Nation and Millet', in: Ahmad Hasan Dani, ed., *Founding Fathers of Pakistan*

(Islamabad, 1981), 137–60.
——ed., *The Founding Fathers of Pakistan* (Islamabad, 1981).
Darwīsh, Madīḥa, 'The Caliphate and its Revival in the Twentieth Century', *Addarah* (Riyad), 7/1 (Aug. 1981), 1–25.
Daulet, Shafiga, 'The First All Muslim Congress of Russia: Moscow, 1–11 May 1917', *Central Asian Survey*, 8/11 (1989), 21–47.
Davison, R.H., *Reform in the Ottoman Empire 1856–1876* (Princeton, NJ, 1963).
Davletshin, Tamurbek, *Sovyetskiy Tatarstan: Tyeoriya i praktika Lyeninskoy national'noy politiki* (London, 1974).
Dawisha, Adeed, ed., *Islam in Foreign Policy* (Cambridge, UK, 1983).
Dawn, C.E., *From Ottomanism to Arabism: Essays on the Origins of Arab Nationalism* (Urbana, Ill., 1973).
A Delegate, 'European Muslim Conference at Geneva', *Great Britain and the East* (London), 26 Sept. 1935, 396–7.
Depont, Octave, and Xavier Coppolani, *Les Confréries religieuses* (Paris, 1897).
——and I.T. d'Eckhardt, 'Panislamisme et propagande islamique', *Revue de Paris* (Paris), 15 Nov. 1899, 3–34.
Deringil, Selim, 'The "Residual Imperial Mentality" and the 'Urabi Paşa Uprising in Egypt: Ottoman Reactions to Arab Nationalism', *Studies on Turkish–Arab Relations Annual* (Istanbul), 1 (1986), 31–8.
——'Aspects of Continuity in Turkish Foreign Policy: Abdülhamid II and İsmet İnönü', *International Journal of Turkish Studies* (Madison, Wis.), 4/1 (Summer 1987), 39–54.
Diercks, G., *Hie Allah! Das Erwachen des Islams* (Berlin, 1914).
Difāʿ al-Miṣrī ʿan bilādih: Muṣṭafa Kāmil wa-'l-Inglīz (Cairo, 1906).
Dilipak, Abdurrahman, *Vahdat ama nasıl?* (Istanbul, [1988]).
Dillon, E.J., 'Tripoli and Constantinople', *The Quarterly Review*, 216/430 (Jan. 1912), 248–57.
Dixit, Prabha, 'Political Objectives of the Khilafat Movement in India', in: Mushirul Hasan, ed., *Communal and Pan-Islamic Trends in Colonial India* (New Delhi, 1981), 43–65.
Diyāb, Maḥmūd, *Abṭāl al-kifāḥ al-Islāmī al-muʿāṣir* (n.p., 1978).
el-Djabri, Ihsan, 'Le Congrès islamique d'Europe', *La Nation Arabe* (Geneva), 5/6 (July–Sept. 1935), 369–74.
'Documents de la deuxième Conférence Générale de l'Isesco, Islamabad 17–19 Dhou-l-Hijja 1405/3–5 Septembre 1985', *L'Islam Aujourd'hui; Revue de l'Organisation Islamique pour l'Education, les Sciences et la Culture (ISESCO)* (Rabat), 4 (Apr. 1986), 202–43.
Duboscq, André, 'Égypte—les projets du Panislamisme', *Le Temps* (Paris), 18 Mar. 1913.
Duguid, Stephen, 'The Politics of Unity: Hamidian Policy in Eastern

Anatolia', *Middle Eastern Studies*, 9/2 (May 1973), 139–55.
Dunyā kī jang kā pahlā sāl (n.p., n.d. [1915?]).
Duran, Khalid, *Islam und politischer Extremismus: Einführung und Dokumentation* (Aktueller Informationsdienst moderner Orient, 11; Hamburg, 1985).
Earle, E.M., *Turkey, the Great Powers and the Bagdad Railway: A Study in Imperialism* (New York, 1924).
Eccel, A.C., *Egypt, Islam and Social Change: al-Azhar in Conflict and Accommodation* (Berlin, 1984).
Eckardt, I.T. von, 'Panislamismus und islamitische Mission', *Deutsche Rundschau*, 98 (Jan.–Mar. 1899), 61–81.
Edib, Eşref, 'Meşhur İslâm seyyahı Abdürreşid İbrahim Efendi', *İslâm-Türk Ansiklopedisi Mecmuası* (Istanbul), 2/53–4 (Jan. 1945), 3–4.
Edwards, Albert, 'The Menace of Pan Islamism', *North American Review*, 197/690 (May 1913), 645–57.
Egypt. No. 1 (1907): Reports of His Majesty's Agent and Consul-General on the Finances, Administration, and Condition of Egypt and the Soudan in 1906 (Cd. 3394; London, 1907).
Elī'ad, Nissīm, 'Tĕnū'a Pan-Islāmīt mĕḥūddeshet', *Tĕmūrōt* (Tel Aviv), 5–6 (Mar. 1980), 27–9.
Emin, Ahmed, *Turkey in the World War* (New Haven, Conn., 1930).
Enayat, Hamid, *Modern Islamic Political Thought: The Response of the Shī'ī and Sunnī Muslims to the Twentieth Century* (London, 1982).
Encausse, H.C. d', *Réforme et révolution chez les Musulmans de l'Empire russe: Bukhara 1867–1924* (Paris, 1966).
——'The Stirring of National Feeling', in: Edward Allworth, ed., *Central Asia: A Century of Russian Rule* (New York, 1967), 172–88.
——'Civil War and New Governments', in: Edward Allworth, ed., *Central Asia: A Century of Russian Rule* (New York, 1967), 234–53.
Ende, Werner, 'Sayyid Abū l-Hudā, ein vertrauter Abdülhamid II. Notwendigkeit und Probleme einer kritischer Biographie', *Zeitschrift der Deutschen Morgenländischen Gesellschaft*, Suppl. iii, part 2 to XIX Deutscher Orientalistentag (1977), 1143–55.
——and Udo Steinbach, eds., *Der Islam in der Gegenwart* (Munich, 1984).
Engelhardt, É.P., *La Turquie et le Tanzimat*, i–ii (Paris, 1882).
Engineer, Asghar Ali, *Indian Muslims: A Study of the Minority Problems in India* (New Delhi, 1985).
Ersoy, Mehmed Âkif, 'Hutbe ve mevaiz—Bayezid kürsüsünden vaaz: İslâm birliği ve milliyetçilik', *Sebilür-Reşad*, 9/230 (29 Safer 1331 H).
Esat, *İttihad-ı İslâm* (Istanbul, n.d.).

Esposito, J.L., 'Muhammad Iqbal and the Islamic State', in: J.L. Esposito, ed., *Voices of Resurgent Islam* (New York, 1983), 175–90.
——*Islam and Politics* (Syracuse, NY, 1984).
——ed., *Voices of Resurgent Islam* (New York, 1983).
Etienne, Bruno, *L'Islamisme radical* (Paris, 1987).
Evren, President Kenan, 'The Islamic Conference has been Successful', *Newspot* (Ankara), 6 Feb. 1987.
'Exploiting the Crescent: Pan-Islamic Aims', *The Times* (London), 27 Apr. 1915, 7.
Fadyeyeva, I.L., *Ofitsial'niye doktrini i idyeologii v politikye Osmanskoy Impyerii (Osmanizm—Panislamizm) XIX-nachalo XX v.* (Moscow, 1985).
Faḥṣ, Hānī, *Fī al-waḥda al-Islāmiyya wa-'l-tajzi'a* (Beirut, 1406/1986).
Fakhry, Majid, 'The Theocratic Idea of the Islamic State in Recent Controversies', *International Affairs* (London), 30/4 (Oct. 1954), 450–62.
Farah, Caesar, 'The Islamic Caliphate and the Great Powers: 1904–1914', *Studies on Turkish–Arab Relations Annual* (Istanbul), 2 (1987), 37–48.
Farid, Mohamed, *Étude sur la crise ottomane actuelle 1911–1912—1914–1915* (Geneva, 1915).
——'La Guerre mondiale, la Turquie et l'Islam', *Bulletin de la Société Endjouman Terekki-Islam (Progrès de l'Islam)* (Geneva), 3/5–6 (Oct.–Nov. 1915), 195–219.
——'Intrigues contre le Califat ottoman: Les Lieux Saints et l'Islam', *Bulletin de la Société Endjouman Terekki-Islam (Progrès de l'Islam)*, 4/3 (Oct.–Nov. 1916), 159–69.
Faird, Mohammed, *Les Intrigues anglaises contre l'Islam* (Lausanne, 1917).
Farīd, Muḥammad, *Ta'rīkh al-dawla al-ʿaliyya al-ʿUthmāniyya* (Cairo, 1330/1912).
——*Awrāq Muḥammad Farīd, i. Mudhakkirātī baʿd al-hijra (1904–1919)* (Cairo, 1978).
Farooqi, Naimur Rahman, 'Pan-Islamism in the Nineteenth Century', *Islamic Culture* (Hyderabad). 57/4 (Oct. 1983), 283–96.
Faruki, Kemal A., 'Approaches to Muslim Unity', *The Criterion* (Karachi), 7/9 (Sept. 1972), 21–33.
al-Faẓā'iʿ al-Rūsiyya (n.p. [Istanbul?], n.d. [1916]).
Feduchy, M.M., *Panislamismo: Estudios sobre el Islam y la política imperialista* (Madrid, 1934).
Fernau, F.W., *Moslems on the March: People and Politics in the World of Islam* (New York, 1954), transl. by E.W. Dickes.
Fesch, Paul, *Constantinople aux derniers jours d'Abdul Hamid* (Paris, n.d. [1908]).

Feuer, Guy, 'L'Organisation de la conférence islamique', in: *L'Islam dans les relations internationales* (Aix-en-Provence, 1986), 161–9.

Fevret, A., 'Le Croissant contre la croix', *Revue du Monde Musulman*, 2/7 (May 1907), 421–5.

Filiu, J.-P., 'L'Afghanistan et le mythe de l'internationalisme islamique', *Cosmopolitiques* (Paris), 6 (Mar. 1988), 57–61.

'First Conference of World Islamic Organizations at Mecca', *The Islamic Review*, 56/11–12 (Nov.–Dec. 1968), 27–9.

Fischer, Georges, 'Une tentative de protection internationale d'une minorité: La Conférence islamique et les Musulmans philippins', *Annuaire Français de Droit International*, 23 (1977), 325–41.

Fischer, M.M.J., 'Islam and the Revolt of the Petit Bourgeoisie', *Daedalus* (Cambridge, Mass.), 3/1 (Winter 1982), 101–25.

Fisher, Alan, *The Crimean Tatars* (Stanford, Calif., 1978).

Forbes, A.D.W., *Warlords and Muslims in Chinese Central Asia: A Political History of Republican Sinkiang 1911–1949* (Cambridge, UK, 1986).

Frechtling, L.E., 'Anglo-Russian Rivalry in Eastern Turkistan 1863–1881', *Journal of the Royal Central Asian Society* (London), 26/2 (Apr. 1939), 471–89.

Frémont, P., *Abd Ul-Hamid et son règne* (Paris, 1895).

Froberger, Joseph, *Weltkrieg und Islam*, in: Sekretariat Sozialer Studentenarbeit, *Deutschland und das Mittelmeer: Sechs Abhandlungen aus der Sammlung 'Der Welkrieg'* (Gladdach, 1916), 55–75.

Froelich, J.C., 'Panislamisme en Afrique noire', *Études* (Paris), 331 (Nov. 1969), 514–26.

'From Lahore to brass tacks', *Impact International Fortnightly*, 14/5 (8–21 Mar. 1974), 4.

Galiyev, *see* Sultan-Galiyev

Galli, Gottfried, *Der heilige Krieg des Islams und seine Bedeutung im Weltkriege* (Freiburg, 1915).

—— *Wesen, Wandel und Wirken des heiligen Krieges des Islams* (Halle a S., 1918).

Gardos, Harald, 'Ballhausplatz und Hohe Pforte im Kriegsjahr 1915: Einige Aspekte ihrer Beziehungen', *Mitteilungen des Österreichischen Staatsarchivs* (Vienna), 23 (1970), 250–96.

Gasprinskiy, Ismail, *Russkoye Musul'manstvo: Misli, zamyetki i nablyudyeniya Musul'manina* (Simferopol, 1881; repr. by the Society for Central Asian Studies, repr. ser. 6, Oxford, 1985).

Gavillet, Marguerite, 'Unité Islamique ou unité nationale? La Position de Jamāl al-Dīn al-Afghānī', in: Simon Jargy, ed., *Islam communautaire (al-Umma): Concept et réalités* (Geneva, 1984), 81–91.

Georgeon, François, *Aux origines du nationalisme turc: Yusuf Akçura* (Paris, 1980).

Gershoni, Israel, *Mitsrayim beyn yiḥūd lĕ-aḥdūt* (n.p., 1980).
——'The Emergence of Pan-Nationalism in Egypt: Pan-Islamism and Pan-Arabism in the 1930s', *Asian and African Studies* (Haifa), 16/1 (Mar. 1982), 59–94.
Ghali, Boutros Boutros, 'Le Grand Dessein de al-Kawakibi', *La Nouvelle Revue du Caire*, 2 (1978), 93–102.
Gharīb, Muḥammad 'Alī, *Muḥammad Farīd al-fidā'ī al-awwal* (n.p., 1958).
Ghaussy, A.Gh., 'Attempts at Defining an Islamic Economic Order', *Economics* (Tübingen), 37 (1988), 9–39.
Ghersi, Emanuele, *I movimenti nazionalistici nel mondo musulmano* (Padua, 1932).
Ghirelli, Angello, *El renacimiento musulmán* (Barcelona, 1948).
Ghunaym, 'Ādil Ḥasan, 'al-Mu'tamar al-Islāmī al-'āmm', *Shu'ūn Filasṭīniyya* (Beirut), 25 (Sept. 1973), 119–35.
Gibb, H.A.R., 'The Islamic Congress at Jerusalem in December 1931', in: A.J. Toynbee, ed., *Survey of International Affairs 1934* (London, 1935), 99–109.
——*Modern Trends in Islam* (Chicago, 1947).
——'Unitive and Divisive Factors in Islam', *Civilisations* (Brussels), 7/4 (1957), 507–14.
Goichon, A.-M., 'Le Panislamisme d'hier et d'aujourd'hui', *L'Afrique et l'Asie* (Paris), 9 (1950), 18–44.
Gökalp, Ziya, *Türkleşmek, İslamlaşmak, muasırlaşmak* (Istanbul, 1918).
Goldschmidt, jun., Arthur, 'The Egyptian Nationalist Party: 1892–1919', in: P.M. Holt, ed., *Political and Social Change in Modern Egypt: Historical Studies from the Ottoman Conquest to the United Arab Republic* (London, 1968), 308–33.
——'The National Party from Spotlight to Shadow', *Asian and African Studies* (Haifa), 16/1 (Mar. 1982), 11–30.
Gooch, G.P., and Harold Temperley, eds., *British Documents on the Origins of the War 1898–1914*, 10/1 (London, 1936).
Gopal, Ram, *Indian Muslims: A Political History 1858–1947* (London, 1959).
Gordlyevskiy, Vl., 'Musul'manskoye uchyonoye obschyestvo v'Moskvi', *Musul'manskiy Mir* (Petrograd), 1/1 (1917), 69–72.
Gouilly, Alphonse, *L'Islam devant le monde moderne* (Paris, 1945).
Grande, Julian, 'Germany's Press Propaganda: Her Activities among Mahomedañ Peoples', *The Daily Graphic* (London), 2 Aug. 1916.
Grimme, Hubert, *Islam und Weltkrieg* (Kriegsvorträge der Universität Münster i. W., 7; Münster, 1914).
Grothe, Hugo, *Deutschland, die Türkei und der Islam: Ein Beitrag zu den Grundlinien der deutschen Weltpolitik im islamischen Orient* (Leipzig, 1914).

Gubaydullin, G., 'Krakh Panislamizma vo vtoroy impyerialistichyeskoy voyni', *Kul'tura i Pis'myennost' Vostoka* (Baku), 4 (1919), 103–18.

Guenena, Nemat, *The 'Jihad' as an 'Islamic Alternative' in Egypt* (Cairo Papers in Social Sciences, 9, monogr. 2; Cairo, 1986).

Guidi, Michelangelo, 'Islām e Arabismo', *Conferenze e letture del Centro di Studi per il Vicino Oriente, I. Aspetti i Problemi del Mondo Musulmano* (Rome, 1941), 7–28.

Güngörge, Mustafa Talip, *Islâm birlik ister: Bolcülüğe karşı millî birlik ve İslâm kardeşliği* (Istanbul, 1986).

Habib, Mohammad, 'Recent Political Trends in the Middle East: Pan-Islamism to Nationalism', *India Quarterly* (New Delhi), 1/2 (Apr. 1945), 134–9.

Haim, S.G., 'The Abolition of the Caliphate and its Aftermath', in: T.W. Arnold, *The Caliphate*[2] (New York, 1966), 205–44.

——ed., *Arab Nationalism* (Berkeley, Calif., 1976).

Halaçoğlu, Yusuf, 'Binbaşı Ismail Hakkı Bey'in Kaşgar'a dâir eseri', *Tarih Enstitüsü Dergisi* (Istanbul), 13 (1983–7), 521–49.

Haleem, A.B.A., 'The Baghdad World Muslim Conference', *Pakistan Horizon*, 15/3 (1962), 169–76.

Halid, Halil, *The Diary of a Turk* (London, 1903).

——*The Crescent and the Cross* (London, 1907).

——'La Loyauté forcée des Indiens', *Bulletin de la Sociéte Endjouman Terekki-Islam* (*Progrès de l'Islam*), 3/7 (Dec. 1915), 302–13.

——'Panislamische Gefahr', *Die Neue Rundschau* (Berlin), 27/3 (Mar. 1916), 289–309.

Halid, M. Halil, *Turk hâkimiyeti ve İngiliz cihangirliği* (n.p. [Istanbul], 1341 H).

Halim Pacha, Saïd, 'Notes pour servir à la réforme de la société musulmane', *Orient et Occident* (Paris), 1/1 (Jan. 1922), 18–54.

Halim Paşa, Said, *Buhranlarımız* (Istanbul, 1985).

Ḥamāda, Fārūq, *Binā' al-umma bayn al-Islām wa-'l-fikr al-mu'āṣir* (Casablanca, 1406/1986).

Hamet, Ismaël, 'Le Congrès musulman universel', *Revue du Monde Musulman*, 4/1 (Jan. 1908), 100–7.

Hamsa, Abdul Malik, 'Der Panislamismus: Seine Bedeutung und seine Grenzen', *Die Islamische Welt* (Berlin), 1/1 (Nov. 1916), 18–20.

——'Der Panislamismus: Seine praktische Ziele', *Die Islamische Welt*, 1/7 (1 June 1917), 384–6.

Haq, Inamul, *Islamic bloc: The Way to Honour, Power and Peace* (Karachi, n.d. [1968?]).

Haq, Mushir U., *Muslim Politics in Modern India 1857–1947* (Meerut, 1970).

Haq, Mushirul, 'The Rabitah: A New Trend in Panislamism', *Islam and the Modern Age* (New Delhi), 9/3 (Aug. 1978), 55–65.

Haq, Mushir-ul, 'The Authority of Religion in Indian Muslim Politics', in: Mushirul Hasan, ed., *Communal and Pan-Islamic Trends in Colonial India* (New Delhi, 1981), 358–66.

Haq, S. Moinul, 'Maulana Mohamed Ali', in: Mahmud Husain *et al.*, eds., *A History of the Freedom Movement* (Pakistan Historical Society Publ., 19; Karachi, 1961), iii. 140–74.

——'The Khilafat Movement', in: Mahmud Husain *et al.*, eds., *A History of the Freedom Movement* (Pakistan Historical Society Publ. 19; Karachi, 1961), iii. 205–39.

Harb, Mohamed Talaat, ed., *L'Europe et l'Islam: M. G. Hanotaux et le cheikh Mohammed Abdou* (Cairo, 1905).

Hardy, P., *The Muslims of British India* (Cambridge, UK, 1972).

Hartmann, Richard, *Die Krisis des Islam* (Morgenland Darstellungen aus Geschichte und Kultur des Ostens, 15; Leipzig, n.d.).

Ḥasan, ʿAbd al-Bāsiṭ Muḥammad, *Jamāl al-Dīn al-Afghānī wa-atharuh fī al-ʿālam al-Islāmī al-ḥadīth* (Cairo, 1982).

Hasan, Mushirul, 'Religion and Politics in India: The Ulema and the Khilafat Movement', in: Mushirul Hasan, ed., *Communal and Pan-Islamic Trends in Colonial India* (New Delhi, 1981), 1–26.

——ed., *Communal and Pan-Islamic Trends in Colonial India* (New Delhi, 1981).

Hasbi, Süleyman, 'Risale-yi ittihadiye li-saadet-i millet-i el-İslamiye' (Istanbul, 1298/1881; Istanbul University's Central Library, Türkçe Yazma/Turkish MS 4397).

Haushofer, Karl, *Geopolitik der Pan-Ideen* (Berlin, 1931).

Hayit, Baymirza, *Turkestan zwischen Russland und China* (Amsterdam, 1971).

——*Islam and Turkestan under Russian Rule* (Istanbul, 1987).

——*Sovyetler Birliğinde'ki Türklüğün ve İslamın bazı meseleleri* (Istanbul, 1987).

Hazam, J. G., 'Islam and Nationalism', *The Islamic Literature*, 11/5 (May 1959), 155–62.

Heine, Peter, 'Ṣaliḥ ash-Sharif al-Tunisi. A North African Nationalist in Berlin during the First World War', *Revue de l'Occident Musulman et de la Méditerranée* (Aix-en-Provence), 33/1 (1982), 89–95.

——and Reinhold Stipek, *Ethnizität und Islam: Differenzierung und Integration muslimischer Bevölkerungsgruppen* (Gelsenkirchen, 1984).

Henze, P.B., 'The Central Asian Muslims and their Brethren Abroad: Marxist Solidarity or Muslim Brotherhood', *Central Asian Survey*, 3/3 (1985), 51–68.

Heper, Metin, and Raphael Israeli, eds., *Islam and Politics in the Modern Middle East* (London and Sydney, 1984).
Herly, Robert, 'L'Influence allemande dans le Panislamisme contemporain', *La Nouvelle Revue Française d'Outre-Mer* (Paris), NS, 47/12 (Dec. 1955), 591–602; 48/1 (Jan. 1956), 31–6; 48/2 (Feb. 1956), 82–91; 48/3 (Mar. 1956), 129–38.
Heyworth-Dunne, J., *Religious and Political Trends in Modern Egypt* (Washington, DC, 1950).
Ḥijazi, Miskīn, *'Ālam-i Islām kā ittiḥād* (Lahore, 1984).
Hoetzch, Otto, ed., *Die internazionale Beziehungen im Zeitalter des Imperialismus: Dokumente aus den Archiven der Zarischen und Provisorischen Regierung*, ser. 2, 6/1 and 6/2 (Berlin, 1934).
Hostler, C.W., *Turkism and the Soviets: The Turks of the World and Their Political Objectives* (London, 1957).
Hourani, Albert, *Arabic Thought in the Liberal Age 1798–1939* (London, 1970).
Howard, C.H.D., ed., *The Diary of Edward Goschen 1900–1914* (Royal Historical Society, Camden Ser., 25; London, 1980).
Hubert, Lucien, *Avec ou contre l'Islam* (n.p. [Paris], 1913)
Huq, Shah Syed Munirul, *The Islamic Meridian: A Way to Muslim Solidarity* (Dacca, 1979).
Hurgronje, C.S., 'Les Confréries religieuses, la Mecque et le Panislamisme', *Revue de l'Histoire des Religions* (Paris), 44 (1901), 262–81.
——'Over Panislamisme', *Archives du Musée Teyler* (Haarlem), sér. 3, 1 (1912), 87–105.
——*The Holy War 'Made in Germany'* (New York, 1915).
Husain, Asaf, *Political Perspectives on the Muslim World* (New York, 1984).
al-Husri, Sati', 'Muslim Unity and Arab Unity', transl. by S.G. Haim, in: S.G. Haim, ed., *Arab Nationalism* (Berkeley, Calif., 1976), 147–53.
Hyman, Anthony, *Muslim Fundamentalism* (= *Conflict Studies*, 174; London, 1985).
Ibn Fayyāḍ, Zayd Ibn 'Abd al-'Azīz, *al-Waḥda al-Islāmiyya* (Riyad, 1388/1968).
Ibn Nūḥ, 'Abd Allāh, 'al-Taḍāmun al-Islāmī', in: Rābiṭat al-'ālam al-Islāmī, *Nadwat al-muḥāḍarāt* (1968), 107–16.
İbrahim, Abdürreşid, *20. asrın başlarında İslâm dünyası ve Japonya'da İslâmiyet*, i–ii (Istanbul, 1987).
Idrīs, Aḥmad, *Abū al-A'lā al-Mawdūdī ṣafaḥāt min ḥayātih wa-jihādih* (Cairo, 1979).
Iḥtiqār al-dīn al-Islāmī fī al-ṣufūf al-Fransiyya (n.p., n.d. [1916?]).

[İleri], Celal Nuri, *İttihad-ı İslâm: İslâmın mazısı, hali, istikbali* (Istanbul, 1331).
—— *İttihad-ı İslâm ve Almanya* (Istanbul, 1333H).
—— *Ittiḥād al-Muslimīn: al-Islām, māḍīh wa-ḥāḍiruh wa-mustaqbaluh. Naẓariyyāt fī muduniyyat al-ʿālam wa-madhāhibih al-siyāsiyya wa-'l-ijtimāʿiyya* (n.p. [Cairo], 1338/1920), transl. by Ḥamza Ṭāhir and ʿAbd al-Wahhāb ʿAzzām.
İmam Humeyni'nin İslam birliği ile ilgili mesaj ve nutkularından seçmeler (Ankara, 1984).
İnalcık, Halil, 'The Caliphate and Atatürk's inkılâp', *Turkish Review Quarterly Digest* (Ankara), 2/7 (Spring 1987), 25–36.
İnan, Yusuf Ziya, *Kemalist eyleme göre İslam ittihadı* (Istanbul, 1976).
Ionova, A., 'Islam i myedzhdunarodnoye ekonomichyeskoye sotrudnichyestvo', *Aziya i Afrika Syegodnya* (Moscow), 3 (1983), 15–17.
Ionova, A.I., ed., *Islam: Problyemi idyeologii, prava, politiki i ekonomiki: Sbornik statyey* (Moscow, 1985).
Iqbal, Afzal, *The Life and Times of Mohamed Ali* (Lahore, 1974).
—— *Contemporary Muslim World* (Lahore, 1985).
—— ed., *Select Writings and Speeches of Maulana Mohamed Ali* (Lahore, 1944).
Iqbal Ali Shah, Sirdar, *see* Shah
Iqbal, Javid, ed., *Stray Reflections: A Note-book of Allama Iqbal* (Lahore, 1961).
Iqbal, Sir Mohammad, *The Reconstruction of Religious Thought in Islam* (London, 1934).
L'Islam dans les relations internationales: Actes du IVe colloque franco-pakistanais Paris, 14–15 mai 1984 (Aix-en-Provence, 1986).
Islam i problyemi natsionalizma v stranakh Blidzhnyego i Sryednyego Vostoka: Sbornik statyey (Moscow, 1986).
Islam kratkiy spravochnik (Moscow, 1983).
Islam v sovryemyennoy politikye stran Vostoka: Konyets 70kh—nachala 80kh godov XX v. (Moscow, 1986).
'Islamic Development Bank', *Impact International Fortnightly* (London), 4/10 (24 May–13 June 1974), 6.
Islamic Conference of Foreign Ministers: Eleventh Session, Islamabad, 17–22 May, 1980 (Islamabad, n.d. [1980]).
'Islamic Conference: 'From Talk to Take Off', *Impact International Fortnightly*, 1/22 (14–27 Apr. 1972), 4–5.
'Islamic Conference: Realities of Muslim Statesmanship', *Impact International Fortnightly*, 5/15 (8–21 Aug. 1975), 6–8.
Islamic Council of Europe, *The Muslim World and the Future Economic Order* (London, 1979).

Islamic Council of Europe, *Muslim Communities in Non-Muslim States* (London, 1980).
—— *Islam and Contemporary Society* (London and New York, 1982).
Ismael, T.Y., and J.S. Ismael, *Government and Politics in Islam* (London, 1985).
Israeli, Raphael, 'Islam in Egypt under Nāsir and Sadat: Some Comparative Notes', in: M. Heper and R. Israeli, eds., *Islam and Politics in the Modern Middle East* (London and Sydney, 1984), 64–78.
—— 'Muslim Minorities under Non-Islamic Rule', *Current History* (Philadelphia, Pa.), 78/456 (Apr. 1980), 159–64, 184.
Jäckh, Ernst, *Die deutsch-türkische Waffenbruderschaft* (Stuttgart and Berlin, 1915).
Jamāl, Muḥammad, *Hind: bhā'iyon ko ek mushfiqānah naṣīḥat* (n.p. [Istanbul], 1917).
—— *al-Jihād fī sabīl al-ḥaqq* (n.p. [Istanbul?], 1917.
Jami'at al Islam: History—policy—program (Vienna, 1958).
al-Jammāl, Gharīb, *al-Taḍāmun al-Islāmī fī al-majāl al-iqtiṣādī* (Jedda, 1396/1976).
Jāwīsh, 'Abd al-'Azīz, *al-Khilāfa al-Islāmiyya* (Istanbul, 1334).
Johansen, Baber, *Islam und Staat: Abhängige Entwicklung, Verwaltung des Elends und religiöser Antiimperialismus* (Berlin, 1982).
al-Jundī, Anwar, *'Abd al-'Azīz Jāwīsh min ruwwād al-tarbiya wa-'l-ṣaḥāfa wa-'l-ijtimā'* (n.p. [Cairo], n.d. [1965]).
—— *al-Yaqẓa al-Islāmiyya fī muwājahat al-isti'mār: Mundh ẓuhūrihā ilā awā'il al-ḥarb al-'ālamiyya al-ūlā* (Cairo, 1398/1978).
Jung, Eugène, *L'Islam sous le joug: La Nouvelle Croisade* (Paris, 1926).
—— *Les Arabes et l'Islam en face des nouvelles croisades et Palestine et Sionisme* (Paris, 1931).
—— *Le Réveil de l'Islam et des Arabes* (Paris, 1933).
—— *L'Islam se défend* (Paris, 1934).
Justice to Islam and Turkey: Speeches Delivered at a Meeting Held at Kingsway Hall, on Thursday, the 22nd of April, 1920, to Demand Justice for Islam and Turkey (London, 1920).
'K' voprosu o Panislamizm'', *Mir' Islama*, 2/1 (1913), 1–12.
'Das Kalifat als geistiges Band', *Die islamische Welt*, 1/6 (1 May 1917), 361.
Kamel, Moustafa, 'L'Angleterre et l'Islam', *La Nouvelle Revue* (Paris), 96 (Sept.–Oct. 1895), 835–7; repr. in *Journal du Caire* (Cairo), 25 Apr. 1905.
—— 'L'Europe et l'Islam', *Le Figaro* (Paris), 12 Sept. 1901, 3.
—— 'Kaiser Wilhelm und der Islam', *Berliner Tageblatt* (Berlin), 23 Oct. 1905.

——'Patriotisme et Panislamisme', *Le Temps* (Paris), 8 Sept. 1906, 1.
Kāmil, ʿAlī, *Muṣṭafā Kāmil Bāshā fī 34 rabīʿan: sīratuh wa-aʿmāluh min khuṭab wa-aḥādīth wa-rasāʾil siyāsiyya wa-ʿumrāniyya*, i–v (Cairo, 1329/1908).
Kāmil, Muṣṭafā, *Miṣr wa-ʾl-iḥtilāl al-Injlīzī* (n.p., 1313 H).
——*Kitāb al-masʾala al-sharqiyya* (Cairo, 1898).
Kar, M., 'Khilafat and Non-cooperation Movements in Assam', in: Mushirul Hasan, ed., *Communal and Pan-Islamic Trends in Colonial India* (New Delhi, 1981), 126–40.
Kara, İsmail, ed., *Türkiye'de İslamcılık düşüncesi: Metinler/kişiler*, i–ii (Istanbul, 1986–7).
Karakaş, Ş., 'Süleyman Nazîf', *Türk Kültürü Araştırmaları* (Ankara), 25/1 (1987), 65–83.
Karal, Envèr Ziya, *Osmanlı tarihi*, viii (Ankara, 1962).
Karpat, K.H., 'The Turkic Nationalities: Turkish–Soviet and Turkish–Chinese Relations', in: W.O. McCagg and B.D. Silver, eds., *Soviet Asian Ethnic Frontiers* (New York, 1979), 117–44.
——'Pan-İslamizm ve ikinci Abdülhamid: Yanlış bir görüşün düzeltilmesi', *Türk Dunyası Araştırmaları* (Istanbul), 48 (June 1987), 13–37.
al-Kattānī, ʿAlī al-Muntaṣir, *al-Muslimūn fī al-muʿaskar al-shuyūʿī* (Mecca, n.d. [1393]).
Kazemi, Farhad, ed., *Iranian Revolution in Perspective* (=*Iranian Studies*, 13/1–4; Boston, 1980).
Kazemzadeh, F., 'Pan Movements', *International Encyclopaedia of Social Sciences* (1968), xi. 365–70.
Keddie, N.R., 'The Pan-Islamic Appeal: Afghani and Abdülhamid II', *Middle Eastern Studies*, 3/1 (Oct. 1966), 46–67.
——'Pan-Islam as Proto-nationalism', *The Journal of Modern History* (Chicago), 41/1 (Mar. 1969), 17–28.
——*Sayyid Jamāl ad-Dīn 'al-Afghānī': A Political Biography* (Berkeley, Calif., 1972).
Kedourie, Elie, 'Pan-Arabism and British Policy', in: W.Z. Laqueur, ed., *The Middle East in Transition* (New York, 1958), 100–11.
——*Afghani and ʿAbduh: An Essay on Religious Unbelief and Political Activism in Modern Islam* (London, 1966).
——'Egypt and the Caliphate, 1915–1952', in: Elie Kedourie, *The Chatham House Version and Other Middle Eastern Studies* (London, 1970), 177–207.
Kelidar, Abbas, 'The Struggle for Arab Unity', *The World Today* (London), 23/7 (July 1967), 292–300.
——'Shaykh ʿAli Yusuf: Egyptian Journalist and Islamic Nationalist', in: M.R. Buheiry, ed., *Intellectual Life in the Arab East* (Beirut, 1981), 10–20.

Kemal, Mustafa, *Nutuk*, iii (*vesikalar*) (Istanbul, 1959).
Kemal, Namık, 'İttihad-ı İslam', *İbret* (Istanbul), 1/11 (27 June 1872), 1.
Kerr, M.H., *Islamic Reform: The Political and Legal Theories of Muḥammad ʿAbduh and Rashīd Riḍā* (Berkeley, Calif., 1966).
Key, Kerim Kami, 'Jamal ad-Din al-Afghani and the Muslim Reform Movement', *The Islamic Literature*, 3/10 (Oct. 1951), 5–10.
Kh., H., *Ayyuhā al-Sāda al-Sanūsiyyūn*. (n.p., n.d. [1916–17?]).
Khadduri, Majid, 'Pan-Islamism', *Encyclopaedia Britannica*[14] (Chicago, 1966), xvii. 227–8.
Khalid, D.H. 'The Kemalist Attitude Towards Muslim Unity', *Islam and the Modern Age*, 6/2 (May 1975), 23–40.
Khālid, Detlev, 'Aḥmad Amīn: Modern Interpretation of Muslim Universalism', *Islamic Studies Journal of the Islamic Research Institute, Pakistan* (Karachi), 8/1 (Mar. 1969), 47–93.
El Khalidi, Saleb, 'Pan-Islamism', *The Spectator* (London), 99/4130 (24 Aug. 1907), 256–7.
Khan, Aga, and Zaki Ali, *L'Europe et l'Islam* (Paris, 1945).
Khan, Chaudri Nazir Ahmad, *Commonwealth of Muslim States: A Plea for Pan-Islamism* (Karachi, 1972).
—— *Thoughts on Pakistan and Pan-Islamism* (Karachi, 1977).
Khan, Inʿamullah, *Motamar-e-alam-e-Islami: World Muslim conference* (Karachi, 1951).
—— 'The Moʾtamar al-ʿAlam al-Islami: A Brief Description of its Sixth Conference Held at Mogadishu', *The Islamic Review* (Woking, Surrey, UK), 53/6 (June 1965), 27–9.
—— 'Thoughts on a Muslim Summit Conference', *The Islamic Review*, 54/4 (Apr. 1966), 3–4.
Khan, M.A., ed., *International Islamic Conference February, 1968*, i. *English Papers* (Publications of the Islamic Research Institute, 18; Islamabad, 1970).
Khan, M. Ali Asgar, 'The Turkish Nationalists and the Indian Khilafatists', *Journal of the Asiatic Society of Bangladesh* (Dacca), 19/3 (Dec. 1974), 39–50.
Khan, Rais A., 'Religion, Race, and Arab Nationalism', *International Journal* (Toronto), 34 (Summer 1979), 353–90.
Khan, Shafique Ali, 'The Khilafat Movement', *Journal of the Pakistan Historical Society* (Karachi), 34/1 (Jan. 1986), 33–73.
al-Khaṭīb, ʿAbd al-Karīm, *al-Islām fī muwājahat al-ʿaṣr wa-taḥadiyātihi aw taʿqīb ʿalā nadwat al-Ahrām (nadwa ḥurra maʿa al-Raʾīs Muʿammar al-Qadhdhāfī) fī 7.4.1972* (n..p. [Cairo], 1972).
Khorāsānī, Ḥasan Abṭaḥī, *Ittiḥād va-dūstī dar Islām* (Qum, 1382/1963).
Kidwai, Shaykh Mushir Hosain, 'Pan-Islamism', *The Morning Post* (London), 20 Aug. 1906.

——*Pan-Islamism* (London, 1908).
——*The Future of the Muslim Empire Turkey* (London, n.d. [1919]).
——*The Sword against Islam or a Defence of Islam's Standard Bearers* (London, 1919).
——*Pan-Islamism and Bolshevism* (London, n.d. [1937]).
al-Kīlānī, Najīb, *al-Ṭarīq ilā ittiḥād Islāmī* (Tripoli, Libya, 1962).
King, W.H., 'The American Treaty with Turkey at Lausanne and the Kemalist Pan-Islamic Adventure', *The Armenian Review* (Boston, Mass.), 26/103 (Autumn 1973), 3–8.
Kinross, Lord, *Ataturk: The Rebirth of a Nation* (London, 1964).
Kirimal, Edige, *Der nationale Kampf der Krimtürken mit besonderer Berücksichtigung der Jahre 1917–1918* (Emsdetten, 1952).
Kitāb maftūḥ li-Mawlāy Yūsuf Sulṭān Marrākish (n.p. [Istanbul?], n.d. [1916?]).
Klimovich, Lyutsian I., *Islam v Tsarskoy Rossii: Ochyerki* (Moscow, 1936).
Knight, E.F., *The Awakening of Turkey: A History of the Turkish Revolution* (London, 1909).
Kocabaş, Süleyman, *Avrupa Türkiyesi'nin kaybı Balkanlarda Panslavizm* (Istanbul, 1986).
Kodaman, Bayram, *Sultan II. Abdulhamid devri Doğu Anadolu politikası* (Ankara, 1987).
Kohen, Sami, 'Row Erupts in Turkey over Saudi Funding of State Clergy', *Middle East Times* (Cyprus), 29 Mar.–4 Apr. 1987, 4.
Kohn, Hans, *A History of Nationalism in the East* (New York, 1929).
Koloğlu, Orhan, *Abdülhamid gerçeği* (Istanbul, 1987).
——*Islamic Public Opinion During the Libyan War 1911–1912* (forthcoming).
Korkud, Selçuk, 'Istanbul Islamic Conference of Foreign Ministers', *Dış Politika/Foreign Policy* (Ankara), 5/4 (Apr. 1976), 18–22.
Kramer, Martin, *Political Islam* (The Washington Papers, 8; Beverly Hills and London, 1980).
——*Islam Assembled: The Advent of the Muslim Congresses* (New York, 1986).
Kreiser, Klaus, 'Der japanische Sieg über Russland (1905) und sein Echo unter den Muslimen', *Die Welt des Islams*, 21 (1981), 209–39.
——'Vom Untergang der *Ertoghrul* zur Mission Abdurrashid Efendis—Die türkisch-japanischen Beziehungen zwischen 1890 und 1915', in: Josef Kreiner, ed., *Japan und die Mittelmächte im Ersten Weltkrieg und in den zwanziger Jahren* (Bonn, 1986), 235–53.
Krieken, G.S. van, *Snouck Hurgronje en het Panislamisme* (Oosters Genootschap in Nederland, 6; Leiden, 1985).
Krishna, Gopal, 'The Khilafat Movement in India: The First Phase', *Journal of the Royal Asiatic Society* (1968), 37–53.
Kruse, Hans, 'The Politics of the Islamic Call, 1962–1966' (= app. 2 in

Kruse's *Islamische Völkerrechtslehre*[2], Bochum, 1979), 203–32.
Kudsi-Zadeh, A.A., *Sayyid Jamāl al-Dīn al-Afghānī: An Annotated Bibliography* (Leiden, 1970).
Kunitz, Joshua, *Dawn over Samarkand: The Rebirth of Central Asia* (London, 1936).
Kupferschmidt, U.M., 'The General Muslim Congress of 1931 in Jerusalem', *Asian and African Studies* (Jerusalem/Haifa), 12/1 (Mar. 1978), 123–62.
Kuran, Ahmed Bedevi, *İnkılap tarihımız ve Jön Türkler* (Istanbul, 1945).
——*İnkılâp tarihimiz ve İttihad ve Terakki* (Istanbul, 1948).
Kuran, Ercümend, 'Türk düşünce tarihinde Arap kültürlü aydın: Said Halim Paşa', in: *Türk–Arap ilişkileri: geçmişte, bugün ve gelecekte* (Ankara, 1980), 21–5.
——'Panislâmizm'in doğuşu ve gelişmesi', *Beşinci Milletlerarası Türkoloji Kongresi, İstanbul, 23–28 Eylül 1985. Tebliğler* (Istanbul, 1986), i. 395–400.
Kuttner, Thomas, 'Russian Jadīdism and the Islamic World: Ismail Gasprinskii in Cairo—1980. A Call to the Arabs for the Rejuvenation of the Islamic World', *Cahiers du Monde Russe et Soviétique* (Paris), 16/3–4 (July–Dec. 1975), 383–424.
Kutubī, Ḥasan, *Dawrunā fī zaḥmat al-aḥdāth* (Jedda, 1387/1967).
Kyrres, C.P., *Tourkía kai Valkánia* (Athens, 1986).
Landau, J.M., 'al-Afghānī's Panislamic Project', *Islamic Culture* (Hyderabad), 26/3 (July 1952), 50–4.
——*The Hejaz Railway and the Muslim Pilgrimage: A Case of Ottoman Political Propaganda* (Detroit, Mich., 1971).
——'The National Salvation Party in Turkey', *Asian and African Studies* (Jerusalem/Haifa), 11/1 (Summer 1976), 1–57.
——*Abdul-Hamid's Palestine* (Jerusalem and London, 1979).
——*Pan-Turkism in Turkey: A Study of Irredentism* (London, 1981).
——'Pan-Islam and Pan-Turkism during the Final Years of the Ottoman Empire: Some Considerations', in: Robert Hillenbrand, ed., *Union Européenne des Arabisants et Islamisants, 10th Congress Edinburgh 9–16 September 1980: Proceedings* (Edinburgh, 1982), 43–5.
——*Tekinalp, Turkish Patriot, 1883–1961* (Istanbul and Leiden, 1984).
——'Saint Priest and His *Mémoire sur les Turcs*', in: Hâmit Batu and J.-L. Bacqué-Grammont, eds., *L'Empire ottoman, la République de Turquie et la France* (Varia Turcica, 3; Istanbul and Paris, 1985), 127–49.
——'An Egyptian Petition to Abdül Ḥamīd II on Behalf of al-Afghānī', in: M. Sharon, ed., *Studies in Islamic History and Civilization in*

Honour of Professor David Ayalon (Leiden and Jerusalem, 1986), 209–19.

——'An Insider's View of Istanbul: Ibrāhīm al-Muwailiḥī's *Mā hunālika*', *Die Welt des Islams*, NS, 27/1–3 (1987), 70–81.

——'Ideologies in the Late Ottoman Empire: A Soviet Perspective', *Middle Eastern Studies*, 25/3 (July 1989), 387–8.

Lane-Poole, Stanley, 'The Caliphate', *The Quarterly Review*, 224/444 (July 1915), 162–77.

Laoust, Henri, *Le Califat dans la doctrine de Rašīd Riḍā* (Mémoires de l'Institut Français de Damas, 6; Beirut, 1938).

Larcher, M., *La Guerre turque dans la guerre mondiale* (Paris, 1926).

Larson, B.L., 'The Moslems of Soviet Central Asia: Soviet and Western Perceptions of a Growing Political Problem', Ph.D. dissertation, University of Michigan, Ann Arbor, 1983.

Layer, Ernest, *Confréries religieuses musulmanes et Marabouts, leur état et leur influence en Algérie* (Rouen, 1916).

——*Notes sur le Panislamisme et la géographie équatoriale* (Rouen, 1916).

Lazzerini, E.J., 'Ǧadīdism at the Turn of the Twentieth Century: A View from Within', *Cahiers du Monde Russe et Soviétique*, 16/2 (Apr.–June 1975), 245–77.

Le Chatelier, A., 'Les Musulmans russes', *Revue du Monde Musulman*, 1/2 (Dec. 1906), 145–68.

——'Le Pan-Islamisme et le progrès', *Revue du Monde Musulman*, 1/4 (Feb. 1907), 145–68.

——'Le Congrès musulman universel', *Revue du Monde Musulman*, 3/11–12 (Nov.–Dec. 1907), 497–502, 613–17.

——'Lettre à un conseiller d'état', *Revue du Monde Musulman*, 12/9 (Sept. 1910), 1–165.

Lee, D.E., 'A Turkish Mission to Afghanistan, 1877', *The Journal of Modern History*, 13 (Mar.–Dec. 1941), 335–56.

——'The Origins of Pan-Islamism', *The American Historical Review* (Washington, DC), 47/2 (Jan. 1942), 278–87.

Le Gall, Michel, 'The Ottoman Government and the Sanūsiyya: A Reappraisal', *International Journal of Middle East Studies* (Cambridge, UK), 21/1 (Feb. 1989), 91–106.

Lemercier-Quelquejay, Chantal, 'Islam and Identity in Azerbaijan', *Central Asian Survey*, 3/3 (1984), 29–55.

Lévi-Provençal, E., 'L'Emir Shakib Arslan (1869–1946)', *Cahiers de l'Orient Contemporain* (Paris), 4/9–10 (1947), 5–19.

Levtzion, Nehemia, *International Islamic Solidarity and its Limitations* (Jerusalem Papers on Peace Problems, 29; Jerusalem, 1979).

Lewis, Bernard, *The Middle East and the West* (London, 1964).

——'The Ottoman Empire in the Mid-Nineteenth Century', *Middle*

Eastern Studies, 1/3 (Apr. 1965), 283–95.

Lewis, Bernard, *The Emergence of Modern Turkey* (London, 1968).

—— 'The Ottoman Empire and its Aftermath', *Journal of Contemporary History* (London), 15/1 (Jan. 1980), 27–35.

—— 'The Return of Islam', in: Michael Curtis, ed., *Religion and Politics in the Middle East* (Boulder, Colo., 1981), 9–29.

Lewis, Geoffrey L., 'Pan Islamism', *Chambers Encyclopaedia* (rev. edn., 1966), x. 413.

—— 'The Ottoman Proclamation of Jihad in 1914', *Islamic Quarterly: A Review of Islamic Culture* (London), 19/1–2 (Jan.–June 1975), 157–63.

Luke, Harry, *The Old Turkey and the New: From Byzantium to Ankara* (London, 1955).

Lybyer, A.H., 'Caliphate', *Encyclopaedia of the Social Sciences*, ii (1935), 145–9.

Lyevin, Z.I., *Islam i natsionalizm v stranakh zarubyedzhnogo Vostoka (idyeyniy aspyekt)* (Moscow, 1988).

M., R., 'L'Islam et les missions', *Revue du Monde Musulman*, 16 (1911), 95–120.

MacColl, Malcolm, 'The Musulmans of India and the Sultan', *The Contemporary Review*, 71/374 (Feb. 1897), 280–94.

'Macedonian Atrocities', supplement to *Comrade* (Delhi), 13, 24, and 31 May 1913.

Madani, N.O., 'The Islamic Content of the Foreign Policy of Saudi Arabia: King Faisal's Call for Islamic Solidarity 1965–1975', Ph.D. dissertation, American University, Washington, DC, 1977.

Maddy-Weitzmann, Bruce, *Arab Politics and the Islamabad Conference, January 1980* (Shiloah's Center's Occasional Papers; Tel Aviv, 1980).

al-Maghribī, ʿAbd al-Qādir, *Jamāl al-Dīn al-Afghānī: Dhikrayāt wa-aḥādīth* (Cairo, 1948).

Maḥmūd, ʿAlī ʿAbd al-Ḥalīm, *Jamāl al-Dīn al-Afghānī* (Jedda and Riyad, 1979).

al-Makhzūmī, Muḥammad Bāshā, *Khāṭirāt Jamāl al-Dīn al-Afghānī al-Ḥusaynī*[2] (n.p. [Damascus?], 1385/1965).

Malik, Abdul Qayyum, 'Prince Saeed Halim Pasha', *al-Islam* (Karachi), 1 Aug. 1953, 72.

Malik, Hafeez, *Sir Sayyid Ahmad Khan and Muslim Modernization in India and Pakistan* (New York, 1980).

Mango, Andrew, 'Turkey in the Middle East', *Journal of Contemporary History* (London), 3/3 (July 1968), 225–36.

Manifesto of Afghan Students' Islamic Federation ([Peshawar], 1985).

Manṣūr, ʿAbd al-ʿAẓīm Ibrāhīm, *al-Jāmiʿa al-Islāmiyya naẓra fikriyya ʿanhā* ([Cairo], n.d. [1980?]).

Marchand, H., 'Un coup d'œil sur l'Islam: Panislamisme et modernisme', *Renseignemens Coloniaux et Documents* (Paris), 7 (= suppl. to *L'Afrique Française*, July 1909), 146–52.

Mardin, Şerif, *The Genesis of Young Ottoman Thought: A Study in the Modernization of Turkist Political Ideas* (Princeton, NJ, 1962).

——*Jön Turklerin siyasî fikirleri 1895–1908* (Ankara, 1964).

Margoliouth, D.S., *Pan-Islamism*, repr. from *Proceedings of the Central Asian Society* (London), 12 Jan. 1912.

——'The Caliphate Yesterday, To-day, and To-morrow', in: J.R. Mott, ed., *The Moslem World of Today* (London, 1925), 31–44.

Martin, B.G., *Muslim Brotherhoods in Nineteenth-Century Africa* (Cambridge, UK, 1976).

——'Notes on the Members of the Learned Classes of Zanzibar and East Africa in the Nineteenth Century', *African Historical Studies* (Boston), 4/3 (1981), 525–45.

Martinez Feduchy, *see* Feduchy

Marunov, Yu. V., 'Pantyurkizm i Panislamizm Mladoturok (1908–1911 gg)', *Kratkiye Soobshchyeniya Narodov Azii* (Moscow), 45 (1961), 38–56.

Massé, H., 'Le Deuxième Congrès musulman général des femmes d'Orient à Téhéran (novembre–décembre 1932)', *Revue des Etudes Islamiques* (Paris), 7 (1933), 45–141.

Massell, G.J., *The Surrogate Proletariat: Moslem Women and Revolutionary Strategies in Soviet Central Asia, 1919–1929* (Princeton, NJ, 1974).

Massignon, Louis, 'Introduction à l'étude des revendications musulmanes', *Revue du Monde Musulman*, 39 (June 1920), 1–26.

——'L'Entente islamique internationale et les deux congrès musulmans de 1926', *Revue des Sciences Politiques* (Paris), 49 (Oct.–Dec. 1926), 481–5.

Mathews, K., 'Tanzania, the Middle East and Afro-Arab Co-operation', *Taamuli, A Political Science Forum* (Dar es Salaam), 10/2 (1980), 5–30.

Maududi, S. Abu A'la, [Sayyed Abulala Maudoodi], *The Process of Islamic Revolution*^{2} (Lahore, 1955).

——*Unity of the Muslim World* (Lahore, 1967).

——[Abū al-A'la al-Mawdūdī], *Bayn al-da'wa al-qawmiyya wa-'l-rābiṭa al-Islāmiyya* (Cairo, n.d.).

May, L.S., *The Evolution of Indo-Muslim Thought after 1857* (Lahore, 1970).

Mayer, Thomas, 'Egypt and the General Islamic Conference of Jerusalem in 1931', *Middle Eastern Studies*, 18/3 (July 1982), 311–22.

Melka, R.L., 'Max Freiherr von Oppenheim: Sixty Years of

Scholarship and Political Intrigue', *Middle Eastern Studies*, 9/1 (Jan. 1973), 81–93.

Menashri, David, 'Shiite Leadership: In the Shadow of Conflicting Ideologies', in: Farhad Kazemi, ed., *Iranian Revolution in Perspective* (*Iranian Studies*, 13/1–4; Boston, 1980), 119–45.

—— 'The Shah and Khomeini: Conflicting Nationalisms', *Crossroads: International Dynamics and Social Change* (Jerusalem), 8 (Winter–Spring 1982), 53–79.

Mende, Gerhard von, *Der nationale Kampf der Russlandtürken: Ein Beitrag zur nationalen Frage in der Sovjetunion* (Berlin, 1936).

Menzie, [A.P.A.] de, *Rome sans Canossa* (Paris, n.d. [1918]).

Michaux-Bellaire, E., 'Notes sur le Gharb', *Revue du Monde Musulman*, 21 (1911–12), 1–40.

——'La Souveraineté et le califat au Maroc', *Revue du Monde Musulman*, 59 (1925), 117–45.

Miloslavskaya, T.P., and G.V. Miloslavskiy, 'Nyekotoriye voprosi tyeorii i praktiki finansogo-ekonomichyeskoy dyeyatyel'nosti OIK', in: A.I. Ionova, ed., *Islam: Problyemi idyeologii, prava, politiki i ekonomiki: Sbornik statyey* (Moscow, 1985), 59–67.

Minault, Gail, 'Islam and Mass Politics: The Indian Ulama and the Khilafat Movement', in: D.E. Smith, ed., *Religion and Political Modernization* (New Haven and London, 1974), 168–82.

——*The Khilafat Movement: Religious Symbolism and Political Mobilization in India* (New York, 1982).

Mīrzā, *see* Qājār, Abū al-Ḥasan Mīrzā

Mitchell, R.P., *The Society of Muslim Brothers* (London, 1969).

Moazzam, Anwar, 'Jamāl al-Dīn al-Afghānī in India', *Bulletin of the Institute of Islamic Studies* (Aligarh), 4 (1960), 84–95.

Mohammed [pseud.], 'La Mission d'Enver Pacha en Chine et le rapprochement turco-chinois', *Revue du Monde Musulman*, 2/5 (Mar. 1907), 51–2.

Mohammedan History: the Pan-Islamic Movement (London, 1920).

Moinuddin, Hasan, *The Charter of the Islamic Conference and Legal Framework of Economic Co-operation among its Member States* (Oxford, 1987).

Le Monde islamique: Information basique sur les pays membres du secrétariat islamique (Jedda, n.d. [1974]).

Mortimer, Edward, *Faith and Power: The Politics of Islam* (London, 1982).

Motamar al-Alam al-Islami, *Proceedings of the World Muslim Conference* (Karachi, 1962).

——*Proceedings of the World Muslim Conference* (Karachi, 1965).

Muʿāraḍat wujahāʾ al-Jazāʾir min al-Muslimīn ḍidd al-qānūn al-ahlī (n.p., n.d. [1916?]).

al-Mubārak, Muḥammad, *al-Mujtamaʿ al-Islāmī al-muʿāṣir* (Beirut, 1971).
Muhammad, Shan, ed., *Unpublished Letters of the Ali Brothers* (Delhi, 1979).
al-Mujahid, Sharif, 'Pan-Islamism', in: Muhammad Husain *et al.*, eds., *A History of the Freedom Movement* (Pakistan Historical Society Publication, 19; Karachi, 1961), iii. 88–117.
——'Muslim Nationalism: Iqbal's Synthesis of Pan-Islamism and Nationalism', *American Journal of Islamic Social Sciences* (Silver Spring, Md.), 2/1 (1985), 29–40.
Mujeeb, M., *The Indian Muslims* (London, 1967).
al-Munajjid, Ṣalāḥ al-Dīn, *al-Taḍāmun al-Marksī wa-'l-taḍāmun al-Islāmī* (Beirut, 1967).
Munawwar, Muhammad, 'Khilafat Movement: A Pathway to Pakistan', *Iqbal Review: Journal of the Iqbal Academy Pakistan* (Lahore), 27/3 (Oct.–Dec. 1986), 81–96.
Munaẓẓamat al-shuʿūb al-Islāmiyya. al-Markaz al-raʾīsī fī Karātchī, Pakistan. Farʿ Lubnān, *Ahdāf* (Beirut, 1952).
Muqarrāt al-muʾtamar al-Islāmī al-ʿāmm fī dawratih al-ūlā (Jerusalem, n.d. [1932]).
Muslehuddin, Muhammad, 'Two Needs of Islamic World: 1. A Commonwealth of Islamic Countries. 2. A Muslim World Bank', *The Islamic Review and Arab Affairs* (London), 58/12 (Dec. 1970), 25–32.
'Der muslimische Kongreß von Europa Genf September 1935', *Die Welt des Islams* (Berlin), 17/3–4 (Feb. 1936), 99–104.
Muṣṭafā, Shākir, 'Jamāl al-Dīn al-Afghānī al-thāʾir al-sharīd al-muftarā ʿalayh', *al-ʿArabī* (Kuwait), 301 (Dec. 1983), 15–20.
'Myedzhdunarodniye Musul'manskiye organizatsii', in: Akadyemiya Nauk SSSR, Institut Vostokovyedyeniya, *Islam: Kratkiy spravochnik* (Moscow, 1983), 127–33.
Nadvi, Rais Ahmad Jafri, ed., *Selections from Mohammad Ali's Comrade* (Lahore, 1965).
al-Nahḍa al-Islāmiyya bi-Dār al-Khilāfa al-ʿUlyā (Istanbul, n.d.)
Nakkash, A. Kh., 'Doktrina "Islamskogo ekonomiki" v izlodzhyenii tyeologov Visshyego Sovyeta po Dyelam Islama', in: A.I. Ionova, ed., *Problyemi idyeologii, prava, politiki i ekonomiki: Sbornik statyey* (Moscow, 1985), 248–52.
Nallino, C.A., *Notes on the 'Caliphate' in general and on the Alleged 'Ottoman Caliphate'*, transl. from the 2nd edn. (Rome, 1919).
——'Panislamismo', *Enciclopedia italiana* (Rome, 1935), xxvi. 196.
Narain, Prem, 'Political Views of Sayyid Ahmad Khan: Evolution and Impact', *Journal of Indian History* (Trivandrum, Kerala), 53/1 (Apr. 1973), 105–53.

Narayan, B.K., *Pan-Islamism Background and Prospects; With a Supplement on Imam Ali and His Doctrines* (New Delhi, 1982).
Naṣr, Naṣr al-Dīn ʿAbd al-Ḥamīd, *Miṣr wa-ḥarakat al-Jāmiʿa al-Islāmiyya min ʿām 1882 ilā ʿām 1914* (n.p. [Cairo], 1984).
Nielsen, Alfred, 'The International Islamic Conference at Jerusalem', *The Moslem World*, 22/4 (Oct. 1932), 340–54.
Niemeijer, A.C., *The Khilafat Movement in India 1919–1924* (The Hague, 1972).
Nikitine, B., 'Le Bolchevisme et l'Islam. II. Hors de Russie', *Revue du Monde Musulman*, 52 (Dec. 1922), 1–53.
Nimr, ʿAbd al-Munʿim, *Kifāḥ al-Muslimīn fī taḥrīr al-Hind* (Cairo, 1384/1964).
Nissman, David, 'Iran and Soviet Islam: The Azerbaijan and Turkmenistan SSRs', *Central Asian Survey*, 2/4 (Dec. 1983), 45–60.
'Non-Talks in Jeddah: The Inside Story', *Impact International Fortnightly*, 5/5 (14–27 Mar. 1974), 5–6.
Norman, Henry, *All the Russias: Travel and Studies in Contemporary European Russia, Finland, Siberia, the Caucasus, and Central Asia* (London, 1902).
'Not Everyone is for *wahdah*', *Crescent International* (Willowdale, Ont.), 16–31 Oct. 1983, repr. in: Kalim Siddiqui, ed., *Issues in the Islamic Movement 1982–1984* (London, 1985), 85–7.
Nuri, 'Teşyid-i revabit', *İbret*, 1/14 (2 July 1872), 1–2.
Nuri, Celal, *see* [İleri], Celal Nuri
Ochsenwald, William, *The Hijaz Railroad* (Charlottesville, Va., 1980).
Office Musulman International, *Statuts* (Lausanne, 1916).
Öklem, Necdet, *Hilâfetin sonu* (Izmir, n.d. [1983]).
Okyar, Osman, 'Un témoignage français sur Union et Progrès et la défense de la Tripolitaine (1911–1912)', in: Abdeljelil Temimi, ed., *Mélanges Professeur Robert Mantran* (Zaghouan, 1988), 173–8.
Olcott, M.B., 'Soviet Islam and World Revolution', *World Politics* (Princeton, NJ), 34 (Oct. 1981–July 1982), 487–504.
O'Leary, De Lacy, *Islam at the Cross Roads: A Brief Survey of the Present Position and Problems of the World of Islam* (London, 1923).
Ölümünün ellinci yılında Yusuf Akçura sempozyumu tebliğleri (11–12 mart 1985) (Ankara, 1987).
'Organisation of the Islamic Conference, Instrument of Islamic Solidarity', *The Middle East* (London), Aug. 1983, 2–15.
Organisation of the Islamic Conference, *International Coordination of Labour Exchange and Social Security: A Collection of International Instruments Involving the Member States of the Organisation of the Islamic Conference* (Ankara, 1982).

Les Organisations Islamiques internationales (Rome, 1984).
Orhan, 'İttihad-ı İslam', *Türk* (Cairo), 2/58 (8 Dec. 1904), 2.
Ortaylı, İlber, *İkinci Abdülhamit döneminde Osmanlı İmparatorluğunda Alman nüfuzu* (Ankara, 1981).
Osman, A.A., 'The Ideological Development of the Sudanese Ikhwan Movement', *Proceedings of the 1988 International Conference on Middle Eastern Studies* (Leeds, 1988), 387–430.
Ozankaya, Özer, 'Cumhuriyet'i hazırlayan düşünce ortamı', *Siyasal Bilgiler Fakültesi Dergisi* (Ankara), 42/1–4 (1987), 129–42.
Özön, Mustafa Nihat, *Namık Kemal ve İbret gazetesi* (Istanbul, 1938).
Page, David, 'Prelude to Partition: All India Moslem Politics 1920–1932', D.Phil. dissertation, Oxford University, 1974.
Pahlen, K.K., *Mission to Turkestan 1908–1909*, transl. by N.J. Couriss, ed. by R.A. Pierce (London, 1964).
Pakadaman, Homa, *Djamal-ed-Din Assad Abadi dit Afghani* (Paris, 1969).
Palgrave, W.G., *Essays on Eastern Questions* (London, 1872).
'Panislamism', *Asiatic Review* (London), 23/73–6 (Jan.–Oct. 1927), 209–15.
'Panislamism and the Caliphate', *The Contemporary Review*, 43 (Jan. 1883), 57–68.
'Panislamism and the Caliphate', *The Times* (London), 19 Jan. 1882, 8.
'Panislamism in Africa', *The Times* (London), 22 Mar. 1900, 5.
'Le Panislamisme turc en Afrique et en Arabie et la presse arabe', *Bulletin du Comité de l'Asie Française* (Paris), 71 (Feb. 1907), 59–61.
'Panislamismo', *Enciclopedia Universal Illustrada Europeo-Americana* (Madrid, 1958), xli. 794–5.
'Panislamizm', *Bol'shaya Sovyetskaya Entsiklopyediya* (Moscow, 1939), xliv, cols. 62–3.
'Panislamizm', *Bol'shaya Sovyetskaya Entsiklopyediya*[2] (Moscow, 1955), xxxii. 3–4.
'Panislamizm', *Istorichyeskaya Entsiklopyediya* (Moscow, 1967), x, cols. 787–8.
'Panislamizm', *Bol'shaya Sovyetskaya Entsiklopyediya*[3] (Moscow, 1975), xix. 146.
'Panislamizm' i Pantyurkizm'', *Mir' Islama*, 2/8 (1913), 556–71; 2/9 (1913), 596–619.
Pears, Edwin, *Life of Abdul Hamid* (London, 1917).
—— 'Turkey, Islam, and Turanianism', *The Contemporary Review*, 114/634 (Oct. 1918), 371–9.
Peters, Rudolph, *Islam and Colonialism: The Doctrine of Jihad in Modern History* (The Hague, 1979).

Pierce, R.A., *Russian Central Asia 1867–1917: A Study in Colonial Rule* (Berkeley, Calif., 1960).
Pinon, René, *L'Europe et l'Empire ottoman: Les Aspects actuels de la question d'Orient* (Paris, 1909).
—— *L'Europe et la Jeune Turquie: Les Aspects nouveaux de la question d'Orient* (Paris, 1911).
Pipes, Daniel, 'This World is Political!: The Islamic Revival of the Seventies', *Orbis: A Journal of World Affairs* (Philadelphia, Pa.), 24/1 (Spring 1980), 9–51.
—— *In the Path of God: Islam and Political Power* (New York, 1983).
Pipes, Richard, 'Muslims in the Soviet Union', in: Jaan Pennar, ed., *Islam and Communism* (Munich, 1960).
—— *The Formation of the Soviet Union: Communism and Nationalism 1917–1923* (rev. edn, Cambridge, Mass., 1964).
Piscatori, J.P., *Islam in a World of Nation-states* (Cambridge, UK, 1986).
Polonskaya, L.R., 'Sovryemyenniye Musul'manskiye idyeyniye tyechyeniya', in: A.I. Ionova, ed., *Islam: Problyemi idyeologii, prava, politiki i ekonomiki: Sbornik statyey* (Moscow, 1985), 6–25.
Ponomariyev, Yu. A., *Istoriya Musul'manskoy ligi Pakistana* (Moscow, 1982).
Pouwels, R.L., *Horn and Crescent: Cultural Change and Traditional Islam on the East African Coast, 800–1900* (Cambridge, UK, 1987).
Pozzi, Jean, *Le Khalifat et les revendications arabes* (Paris, 1917).
Prasad, Amba, 'Pan-Turanian and Pan-Islamic Movements and India 1908–1922', *Indian Historical Records Commission; Proceedings of the Forty-fourth Session* (New Delhi), 44 (Feb. 1976), 246–59.
Programma Musul'manskoy parlamyentskoy fraktsii v' gosudarstvyennoy dumi (St Petersburg, 1907).
Programmniye dokumyenti Musul'manskikh politichyeskikh partiy 1917–1920 gg. (Society for Central Asian Studies reprint ser. 2; Oxford, 1985).
'The Proposed Annual Islamic Congress at Mecca', *The Islamic Review*, 42/11 (Nov. 1954), 26–7.
Proyekt': 'Muslimān ittifākı' cemiyetinin niẓāmnāmesī (St Petersburg, n.d. [1906]).
Proyekt': Ustav' obshchyestva 'Musul'manskoy partii' (St Petersburg, n.d. [1906]).
Putra, Tunku Abdur Rahman, 'Islamic Unity: That Problem of Ego and Indifference', *Impact International Fortnightly*, 1/12 (12–25 Nov. 1971), 4–5.
Qā'imī, 'Alī, *Tafriqa mas'ala-i rūz-i mā* (Tehran, 1358).
Qājār, Abū al-Ḥasan Mīrzā (Shaykh al-Ra'īs), *Ittiḥād-i Islām* (Bombay, 1312/1894, 2nd edn., Tehran, 1363/1984).

Qāsim, Jamāl Zakariyā, 'Mawqif Miṣr min al-ḥarb al-Ṭarāblusiyya', *al-Majalla al-Ta'rīkhiyya al-Miṣriyya* (Cairo), 13 (1967), 308–44.

Ọṣuntokun, Jide, 'The Response of the British Colonial Government in Nigeria to the Islamic Insurgency in the French Sudan and the Sahara during the First World War', *ODU: A Journal of West African Studies* (Ile-Ife, Nigeria), NS, 10 (July 1974), 98–107.

Qunaybir, Salīm 'Abd al-Nabī, *al-Ittijāhāt al-siyāsiyya wa-'l-fikriyya wa-'l-ijtimā'iyya fī al-adab al-'Arabī al-mu'āṣir: 'Abd al-'Azīz Jāwīsh 1872–1929* (Beirut, 1968).

Qureshi, M.N., 'The Khilafat Movement in India, 1919–1924', Ph.D. dissertation, University of London, 1973.

——'The Indian Khilafat Movement (1918–1924)', *Journal of Asian History* (Wiesbaden), 12/2 (1978), 152–68.

——'Bibliographic Soundings in Nineteenth Century Pan-Islam in South Asia', *The Islamic Quarterly* (London), 24/1–2 (1980), 22–34.

——'Mohamed Ali's Delegation to Europe, 1920', *Journal of the Pakistan Historical Society* (Karachi), 28/2 (Apr. 1980), 79–117; 28/3 (July 1980), 157–85 (= Pakistan Historical Society Publ., 71).

——'The Ali Brothers: A Study in Political Participation', in: Ahmad Hasan Dani, ed., *Founding Fathers of Pakistan* (Islamabad, 1981), 109–36.

——'The Rise of Atatürk and its Impact on Contemporary Muslim India, the Early Phase', in Boğaziçi Üniversitesi, *Atatürk konferansı* (Istanbul, 1981), iii/55. 1–9.

R., R., 'Un Uléma de Pékin à Constantinople: Deux envoyés du Sultan à Pekin', *Revue du Monde Musulman*, 3/11–12 (Nov.– Dec. 1907), 612–17.

R., Z., 'Pan-Islam', in: Yaacov Shimoni and Evyatar Levine, eds., *Political Dictionary of the Middle East in the Twentieth Century* (London, 1972), 312–14.

Rābī, Ḥāmid, *al-Islām wa-'l-quwwa al-awwaliyya* (Cairo, 1981).

Rābiṭat al-'ālam al-Islāmī, *Qarār al-majlıs al-ta'sīsī—al-dawra al-ḥādiya 'ashra* (Mecca, n.d. [1969–70]).

Rado, Şevket, *Ahmet Midhat Efendi* (Istanbul, 1986).

Raḍwān, Fatḥī, *Mashhūrūn mansiyyūn* (n.p. [Cairo], n.d. [1970]).

——*Muṣṭafā Kāmil* (Cairo, 1974).

al-Rāfi'ī, 'Abd al-Raḥmān, *Muṣṭafā Kāmil bā'ith al-ḥaraka al-qawmiyya*[3] (Cairo, 1369/1950).

——*Muḥammad Farīd ramz al-ikhlāṣ wa-'l-taḍḥiya*[3] (Cairo, 1381/1962).

Rahman, F., 'Internal Religious Developments in the Present Century Islam', *Journal of World History* (Paris), 2/1 (1954), 862–79.

Rakowska-Harmstone, Teresa, *Russia and Nationalism in Central*

Asia: The Case of Tadzhikistan (Baltimore, Md., 1970).
Rakowska Harmstone, Teresa 'Soviet Moslem Nationalism in Comparative Perspective', in: C. Lemercier *et al.*, eds., *Turco-Tatar Past Soviet Present: Studies Presented to Alexandre Bennigsen* (Louvain and Paris, 1986), 469–80.
Ramaḍān, Muṣṭafā Muḥammad, *al-ʿĀlam al-Islāmī fī al-taʾrīkh al-ḥadīth al-muʿāṣir*, i (Cairo, 1985).
Ramsaur, E.E., *The Young Turks: Prelude to the Revolution of 1908* (Beirut, 1965).
Réby, J., 'Un cri d'alarme russe', *Revue du Monde Musulman*, 10/1 (Jan. 1910), 106–7.
Redhouse, J.W., *A Vindication of the Ottoman Sultan's Title of 'Caliph'; Shewing its Antiquity, Validity, and Universal Acceptance* (London, 1877).
Refik [Altınay], Ahmed, *Abdülhamid-i sani ve devr-i saltanatı*, iii (Istanbul, 1327/1911).
Reichardt, Thomas, *Der Islam vor der Toren* (Leipzig, 1939).
Reid, Anthony, 'Nineteenth Century Pan-Islam in Indonesia and Malaysia', *The Journal of Asian Studies*, 26/2 (1966–7), 267–83.
Reissner, J., 'Internationale islamische Organisationen', in: W. Ende and U. Steinbach, eds., *Der Islam in der Gegenwart* (Munich, 1984), 539–47.
Rémond, Georges, *Aux camps turco-arabes: Notes de route et de guerre au Tripolitaine et en Cyrenaïque* (Paris, 1917).
Report on Islamic Summit 1974 Pakistan Lahore, February 22–24, 1974 (Islamabad, n.d.).
'Resolutions and Recommendations of the World Conference of Islamic Organizations', *Impact International Fortnightly*, 4/10 (24 May–13 June 1974), 6.
Rey, A. A., *La Question d'Orient devant l'Europe: Constantinople et les Détroits: Vues historiques et diplomatiques*, i–ii (Paris, 1917).
—— *Le Réveil de l'Islam est-il possible? La politique des alliés* (Paris, 1917).
Riḍā, Muḥammad Rashīd, *Taʾrīkh al-ustādh al-imām*, i–iii (Cairo, 1907–28).
—— 'Madrasat al-daʿwa wa-'l-irshād', *al-Manār* (Cairo), 14/1 (30 Jan. 1911), 52–3.
—— *al-Khilāfa aw al-imāma al-ʿuẓmā* (Cairo, 1341 H, repr. from *al-Manār*, 42–3).
—— *Kitāb al-waḥda al-Islāmiyya wa-'l-ukhuwwa al-dīniyya*² (Cairo, 1346).
al-Rīs, Muḥammad Ḍiyāʾ al-Dīn, *al-Islām wa-'l-Khilāfa fī al-ʿaṣr al-ḥadīth. Naqd kitāb al-Islām wa-uṣūl al-ḥukm* (Beirut, 1393/1973).

Risāla-i inqilāb-i Islāmī-i Īrān dar tavḥīd-i kalima (Tehran, 1403 H).
Rıza, Ahmad, 'Le Panislamisme', *Mechveret* (Paris monthly supplement, in French), 11/179 (1 Sept. 1906), 3–4.
——*La Crise de l'Orient*, part 1 (Paris, 1907).
Rondot, Pierre, 'Perspectives du congrès islamique mondial', *Signes du Temps* (Paris), NS, 2 (Nov. 1963), 31–2.
Rose, J.H., '1815 and 1915', *The Contemporary Review*, 107 (Jan. 1915), 12–18.
Rosenthal, E.I.J., *Islam in the Modern National State* (Cambridge, UK, 1965).
Rothmann, Lothar, 'Ägypten im Exil (1914–1918)—Patrioten oder Kollaborateure des deutschen Imperialismus?', *Asien in Vergangenheit: Beiträge der Asienwissenschaftler der DDR zum XXIX. Internationalen Orientalistenkongress 1973 in Paris* (Berlin-Ost, 1974).
Rouire, [A.M.], 'La Jeune-Turquie et l'avenir du Panislamisme', *Questions Diplomatiques et Coloniales*, 28/301 (1 Sept. 1909), 257–70.
Roumani, Jacques, 'From Republic to Jamahiriya: Libya's Search for Political Community', *The Middle East Journal*, 37/2 (Spring 1983), 151–68.
Roy, Gilles, *Abdul-Hamid le Sultan rouge* (Paris, 1936).
Royal Academy for Islamic Civilization Research (Āl Albait Foundation), *Beginnings* (n.p. [Amman], 1401/1981).
Ruthven, Melise, *Islam in the World* (London, 1984).
Rywkin, Michael, *Moscow's Muslim Challenge: Soviet Central Asia* (London, 1982).
Saab, Hassan, *The Arab Federalists of the Ottoman Empire* (Amsterdam, 1958).
Sabaheddine, 'The Sultan and the Panislamic Movement', *The Times* (London), 13 Aug. 1906, 6.
Sadiq, Mohammad, 'The Ideological Legacy of the Young Turks', *International Studies* (New Delhi), 18/2 (Apr.–June 1979), 177–207.
——*The Turkish Revolution and the Indian Freedom Movement* (Delhi, 1983).
Salem, Elie, 'Nationalism and Islam', *The Muslim World*, 32/3 (July 1962), 277–87.
Salem (El Arafati), Mahmoud Ben, *La Coordination des forces alliées* (Paris, 1916).
——*Le Congrès islamo-européen de Genève (août 1933)* (Paris, 1933).
Salih, 'Ittihad-ı İslam', *Osmanlı* (Istanbul), 1/53 (3 Feb. 1881), 1.
Salmoné, H.A., 'Is the Sultan of Turkey the True Caliph of Islam?',

The Nineteenth Century, 39 (Jan. 1896), 173–80.
Samardžiev, Božidar, 'Traits dominants de la politique d'Abdülhamid II relative au problème des nationalités (1876–1885)', *Études Balkaniques* (Sofia), 4 (1972), 57–79.
Samné, Georges, *Le Khalifat et le Panislamisme* (Paris, 1919).
Samra, M.D., 'Pan Islamism and Arab Nationalism: A Study of the Ideas of Syrian Muslim Writers 1860–1918', *Faculty of Arts Journal* (Amman), 3/2 (Aug. 1972), 5–32.
Sanhoury, A., *Le Califat: Son évolution vers une société des nations orientales* (Paris, 1926).
Saray, Mehmet, '1874'de Kaşgar'a gönderilen Türk subayları', *Türk Kültürü Araştırmaları*, 17–21/1–2 (1979–83), 244–51.
—— *Rus işgalı devrinde Osmanlı devleti ile Türkistan hanlıkları ve arasındaki siyasi münasebetler (1775–1875)* (Istanbul, 1984).
—— *Türk dünyasında eğitim reformu ve Gaspıralı İsmail Bey (1851–1914)* (Ankara, 1987).
Sareen, T.R., *Indian Revolutionary Movement Abroad (1905–1921)* (New Delhi, 1979).
Saroop, Vijay, 'The Islamic Summit', *The World Today*, 30/4 (Apr. 1974), 138–40.
Saunders, Lloyd, 'The Sultan and the Caliphate', *The North American Review* (New York), 176/4 (15 Apr. 1906), 546–53.
Sawant, Ankush B., 'Nationalism and National Interest in Egypt', *International Studies* (New Delhi), 22/2 (1985), 135–51.
al-Ṣawwāf, Fā'iq Bakr, *al-ʿAlāqāt bayn al-dawla al-ʿUthmāniyya wa-iqlīm al-Ḥijāz fī al-fatra bayn 1293–1334 H (1876–1916 M)* (n.p. [Mecca?], 1398/1978).
Schäfer, Rich., *Der deutsche Krieg, die Türkei, Islam and Christentum*², (Leipzig, 1915).
—— *Islam und Weltkrieg* (Leipzig, 1915).
Schimmel, Annemarie, *Gabriel's Wing: A Study into the Religious Ideas of Sir Muhammad Iqbal* (supplements to *Numen*, 6; Leiden, 1963).
Schmitz, Paul, *All-Islam! Weltmacht von Morgen?* (Leipzig, 1937).
Schulze, Reinhard, 'Die Islamische Weltliga (Mekka) 1962–1987', *Orient* (Hamburg), 29/1 (Mar. 1988), 58–67.
—— *Islamischer Internationalismus in 20. Jahrhundert* (Leiden, 1988).
—— 'Die Politisierung des Islam im 19. Jahrhundert', *Die Welt des Islams* (Bonn), NS, 22/1–4 (1982), 103–16.
—— 'Regionale Gruppierungen und Organisationen', in: U. Steinbach and R. Robert, eds., *Der nahe und mittlere Osten: Politik, Gesellschaft, Wirtschaft, Geschichte, Kultur* (Opladen, 1988), ii. 469–76.
Secretariat of the Motamar al-Alam al-Islami (World Muslim Con-

gress), compiler, *Studies on Commonwealth of Muslim Countries* (Karachi, n.d. [1964–5]).

Sékaly, Achille, *Le Congrès du Khalifat (le Caire, 13–19 mai 1926) et le congrès du monde musulman (la Mekke, 7 juin–5 juillet 1926)* (Collection de la Revue du Monde Musulman; Paris, 1926).

Shah, Sirdar Ikbal Ali, 'Ferments in the World of Islam', *Journal of the Central Asian Society*, 14/2 (1927), 130–46.

——'Les États-Unis d'Islam', *Revue Politique et Littéraire (Revue Bleue)* (Paris), 67/5 (2 Mar. 1929), 146–7.

Shakir, Moin, *Khilafat to Partition: A Survey of Major Political Trends among Indian Muslims During 1919–1947*[2] (Delhi, 1983).

al-Sharabāṣī, Aḥmad, *Rashīd Riḍā ṣāḥib al-Manār ʿaṣruh wa-ḥayātuh wa-maṣādir thaqāfatih* (n.p. [Cairo], 1970).

al-Sharīf, Ṣāliḥ, *Sharḥ dasā'is al-Fransīs ḍidd al-Islām wa-'l-Khalīfa* (n.p., n.d. (1915]).

Sharipova, R.M., 'Ekonomichyeskiye kontsyeptsii idyeologov Ligi Islamskogo Mira', in: A.I. Ionova, ed., *Islam: Problyemi idyeologii, prava, politiki i ekonomiki: Sbornik statyey* (Moscow, 1985), 68–79.

——*Panislamizm syegodnya: Idyeologiya i praktika Ligi Islamskogo Mira* (Moscow, 1986).

——'Dvidzhyeniye Islamskoy solidarnosti: Osnovniye tyeoryetichyeskiye kontsyeptsii', in: *'Islamskoy faktor' v myedzhdunarodnikh otnoshyeniyakh v Azii* (Moscow, 1987).

——and T.P. Tikhonova, 'Liga Islamskogo Mira: Ot traditsionalizma k ryeformatorstvu', *Narodi Azii i Afriki*, 2 (1984), 30–9.

Shaw, S.J., and E.K. Shaw, *History of the Ottoman Empire and Modern Turkey*, ii (Cambridge, UK, 1977).

Shāwīsh, *see* Jāwīsh

Shaykh al-Ra'īs, *see* Qājār

al-Shaykha, Ḥasan, *Aqlām thā'ira* (Cairo, 1963).

Shepard, William, *The Faith of a Modern Muslim Intellectual: The Religious Aspects and Implications of the Writings of Ahmad Amin* (New Delhi, 1982).

Shibli, A.R., *Muslim World Economic Collaboration* (Lahore, n.d. [1975–6]).

Shukla, R. L., 'The Pan-Islamic Policy of the Young Turks and India', *Indian History Congress: Proceedings of the Thirty-Second Session Jabalpur, 1970* (New Delhi), 2 (1970), 302–7.

Sid-Ahmed, Mohamed, 'Shifting Sands of Peace in the Middle East', *International Security* (Cambridge, Mass.), 5/1 (Summer 1980), 53–79.

Siddiqui, Kalim, *Beyond the Muslim Nation-states* (London, 1980).

——ed., *Issues in the Islamic Movement 1983–1984* (London, 1985).

Simon, Rachel, *Libya between Ottomanism and Nationalism: The Ottoman Involvement in Libya During the War with Italy (1911–1919)* (Berlin, 1987).

Sindi, Abdullah Mohamed, 'The Muslim World and its Efforts in Pan-Islamism', Ph.D. dissertation, University of Southern California, 1978.

——'King Faisal and Pan-Islamism', in: W.E. Beling, ed., *King Faisal and the Modernisation of Saudi Arabia* (London, 1980).

Singh, Kushwant, 'Pax Islamica: (A Study of Pan-Islamic Movements)', *Journal of the Panjab University Historical Society* (Lahore), 9 (1946), 27–42.

Singh, Sangat, *Pakistan's Foreign Policy: An Appraisal* (New York, 1970).

Sırma, İhsan Süreyya, 'Fransa'nın Kuzey Afrika'daki sömürgeciliğine karşı Sultan II. Abdülhamid'in Panislamist faaliyetlerine ait bir kaç vesika', *Tarih Enstitüsü Dergisi* (Istanbul), 7–8 (1976–7), 157–84.

——'Ondokuzuncu yüzyıl Osmanlı siyâsetinde büyük rol oynayan tarîkatlara dâir bir vesîka', *Tarih Dergisi* (Istanbul), 31 (Mar. 1977), 183–98.

——'Pekin Hamidiyye Üniversitesi', *Atatürk Üniversitesi İslâmî İlimler Fakültesi Tayyib Okiç Armağanı* (Ankara, 1978), 159–69.

——'Sultan II. Abdülhamid'in Uzak Doğu'ya gönderdiği ajana dair', *I. Milli Türkoloji Kongressi, 6–9 Şubat 1978* (Istanbul, 1978), 323–5.

——'II. Abdülhamid'in Çin Müslümanlarını sünnî mehzebine bağlama gayretlerine dâir bir belge', *Tarih Dergisi*, 32 (Mar. 1979), 559–62.

——'Quelques documents inédits sur le rôle des confréries Tarîqat dans la politique Panislamique du Sultan Abdulhamid II', *Atatürk Üniversitesi İslâmi İlimler Fakültesi Dergisi* (Erzurum), 3/1–2 (1979), 283–93.

——'Sultan II. Abdülhamid ve Çin müslümanları', *İslam Tetkikleri Enstitüsü Dergisi* (Istanbul), 7/3–4 (1979), 199–205.

——*II. Abdül Hamid'in İslam birliği siyaseti* (Istanbul, 1985).

'Sixth World Muslim Conference', *The Islamic Review*, 39/6 (June 1951), 24–36.

S(lousch), N., 'Une opinion allemande sur le Panislamisme', *Revue du Monde Musulman*, 1/3 (Jan. 1907), 406–7.

Smirnov, N.A., 'Turyetskaya agyentura pod flagom Islama', in: Akadyemiya Nauk SSSR—Institut Istorii, *Voprosi istorii ryeligii i atyeizma: Sbornik statyéy* (Moscow, 1950), 11–63.

Smith, W.C., *Modern Islam in India: A Social Analysis* (London, 1946).

Smoor, Pieter, 'Krachtlijnen tussen de polen Islām en nationalisme in

Egypte', *Internationale Spectator* (The Hague and Brussels), 23/12 (22 June 1969), 1110–31.

Snyder, L.L., *Macro-nationalisms: A History of Pan-movements* (Westport, Conn., 1984).

Sokol, E.D., *The Revolt of 1916 in Russian Central Asia* (Johns Hopkins Studies in Historical and Political Science, ser. 71, no. 1; Baltimore, Md., 1954).

Soulié, G.J.-L., 'Le Monde musulman à la recherche de son unité', *Revue de Défense Nationale* (Paris), 23 (Feb. 1967), 231–41; (Mar. 1967), 468–75; (Apr. 1967), 633–40.

Srour, Hani, *Die Staats- und Gesellschaftstheorie bei Sayyid Ǧamaladdīn 'al-Afghānī' als Beitrag zur Reform der islamischen Gesellschaften in der zweiten Hälfte des 19. Jahrhunderts* (Freiburg, 1977).

'Statement by Muslim World League', *Muslim World League: A Monthly Magazine* (Mecca), 3/5 (Oct.–Nov. 1965), 4–5.

'Statuts du Congrès musulman universel', *Revue du Monde Musulman*, 4/2 (Feb. 1908), 399–403.

Stead, Alfred, 'The Problem of the Near East', *The Fortnightly Review*, 478 (1 Oct. 1906), 575–601.

Steinbach, Udo, '"Re-Islamisierung" und die Zukunft des Nahen Ostens', *Aus Politik und Zeitgeschichte: Beilage zur Wochenzeitung Das Parlament* (Bonn), 24 Jan. 1979, 23–37.

—— and Rüdiger Robert, eds., *Der nahe und mittlere Osten: Politik, Gesellschaft, Wirtschaft, Geschichte, Kultur*, i–ii (Opladen, 1988).

Steiner, M.J., *Inside Pan-Arabia* (Chicago, 1947).

Stepaniants, Marietta, 'Development of the Concept of Nationalism: The Case of the Muslims in the Indian Subcontinent', *The Muslim World*, 69/1 (Jan. 1979), 27–41.

Steppat, Fritz, 'Nationalismus und Islam bei Muṣṭafā Kāmil', *Die Welt des Islams*, NS 4 (1956), 241–341.

—— 'Kalifat, *Dār al-Islām* und die Loyalität der Araber zum Osmanischen Reich bei Ḥanafitischen Juristen des 19. Jahrhunderts', *Correspondance d'Orient* (Brussels), 11 (1970), 443–62.

Stoddard, Lothrop, *The New World of Islam* (London, 1921).

Streater, N.M., 'Pan-Islamism and Pan-Turanianism', Seminar Paper, Columbia University of New York, n.d.

Stuermer, Harry, *Zwei Kriegsjahre in Konstantinopel: Skizzen deutsch-jungtürkischer Moral und Politik* (Lausanne, 1917).

Sultan-Galiyev, M.S., *Stat'i* (Society for Central Asian Studies, repr. ser. 1; Oxford, 1984).

Swietochowski, Tadeusz, *Russian Azerbaijan, 1905–1920: The Shaping of National Identity in a Muslim Community* (Cambridge, UK, 1985).

Tabibi, Abdul Hakim, *The Political Struggle of Sayid Jamal ad-Din al-Afghani* (Kabul, 1977).

al-Taḍāmun al-Islāmī fī 'ālam al-yawm (n.p. [Jedda], n.d. [1977]).

Tahir, Bursalı Mehmed, *Osmanlı müellifleri*, i–iii (Istanbul, 1333–42 H).

Tahir, Mahmud, 'Abdurrashid Ibragim', *Central Asian Survey*, 7/4 (1988), 135–40.

Tahsin Paşa, *Abdülhamit Yıldız hatıraları* (Istanbul, 1931).

Ṭāliqānī, Maḥmūd, *Vaḥdet ve-āzādī* (Tehran, 1361).

Tamura, Amiri, 'Ethnic Consciousness and its Transformation in the Course of Nation-building: The Muslim and the Copt in Egypt, 1906–1919', *The Muslim World*, 75/2 (Apr. 1985), 102–14.

[Tansel], Fevziye Abdullah, *Mehmet Akif* (Istanbul, 1945).

——'Mizancı Mehmed Murad Bey', *Tarih Dergisi*, 2/3–4 (Sept. 1950–Mar. 1951), 67–88.

al-Tawba, Ghāzī, *al-Fikr al-Islāmī al-mu'āṣir: Dirāsa wa-taqwīm* (n.p. [Beirut?], 1389/1969).

Third Islamic Summit Spotlight on Muslim World Problems (Islamabad, n.d.).

Tibi, Bassam, *Nationalismus in der Dritten Welt am arabischen Beispiel* (Frankfurt a/M., 1971).

Tlili, Béchir, 'Au seuil du nationalisme tunisien: Documents inédits sur le Panislamisme au Maghreb', *Africa* (Rome), 28/2 (June 1973), 211–36.

Torre, R.G., 'Panislamismo, Panarabismo y acción ibérica', *Nuestro Tiempo* (Madrid), 25/315 (Mar. 1925), 257–72.

Toynbee, A.J., *Civilization on Trial* (London, 1948).

——'The Ineffectiveness of Panislamism', in: A.J. Toynbee, *A Study of History* (London, 1954), viii. 692–5.

——ed., *Survey of International Affairs 1925*, i. *The Islamic World Since the Peace Conference* (London, 1927).

Trumpener, Ulrich, *Germany and the Ottoman Empire 1914–1918* (Princeton, NJ, 1968).

Tschudi, Rudolf, *Der Islam und der Krieg* (Hamburg, 1914).

Tunaya, T.Z., *İslâmcılık cereyanı* (Istanbul, 1962).

——*Türkiye'de siyasal partiler*^{2}, i–ii (n.p. [Istanbul], 1984–6).

al-Tūnisī, Ṣāliḥ al-Sharīf, *Ākhir ghadr li-'l-Inglīz li-'l-sharq wa-ahlih* (n.p., n.d. [1915]).

——*Bayān tawaḥḥush Fransā fī al-quṭr al-Tūnisī al-Jazā'irī wa-'l-istinjād ilayh* (n.p., n.d.).

——*Brīṭāniyā al-'uẓmā wa-hadhayānuhā al-fi'lī wa-'l-qawlī fī ḥaqq al-Khilāfa al-Islāmiyya* (n.p., n.d.).

al-Turabi, Hassan, 'The Islamic State', in: J.L. Esposito, ed., *Voices of Resurgent Islam* (New York, 1983), 241–51.

Turgut, 'İttihad-ı İslam meselesi', *Türk* (Cairo), 2/57 (1 Dec. 1904), 2.
'Turkey, Russia and Islam', *The Round Table* (London), 8/29 (Dec. 1917), 100–38.
'Turks and Arabs', *The Egyptian Gazette* (Cairo), 26 Apr. 1913.
Tütsch, H.E., *From Ankara to Marrakesh: Turks and Arabs in a Changing World* (London, 1964).
Tyan, E., 'Djihād', *EI*[2] ii (1965), 538–40.
Ular, Alexander, and Enrico Insabato, *Der erlöschende Halbmond: Türkische Enthüllungen* (Frankfurt a/M., 1909).
Ülken, Hilmi Ziya, *Türkiye'de çağdaş düşünce tarihi* (Iṣtanbul, 1979).
Unité Islamique (Paris, 1871).
al-ʿUrwa al-wuthqā (repr. Cairo 1957).
ʿUways, ʿAbd al-Ḥalīm, 'al-Mu'tamar al-Islāmī', *al-Waʿy al-Islāmī* (Kuwait), 114 (June 1974), 94–8.
Vacca, Virginia, 'Il congresso dei Musulmani d'Europa a Ginevra', *Oriente Moderno* (Rome), 15/10 (Oct. 1935), 501–3.
——'Critiche palestinesi al congresso dei Muslumani d'Europa a Ginevra', *Oriente Moderno*, 15/10 (Oct. 1935), 503–4.
——'Seguito dei lavori del congresso dei Muslumani d'Europa a Ginevra', *Oriente Moderno*, 15/11 (Nov. 1935), 563–5.
——'L'Emiro Arslan dichiara che il congresso musulmano Europeo non ha fatto propaganda per la Spagna e per l'Italia', *Oriente Moderno*, 15/11 (Nov. 1935), 565–7.
Vaidyanath, R., *The Formation of the Soviet Central Asian Republics: A Study in Soviet Nationalities Policy, 1917–1936* (New Delhi, 1967).
Vambéry, Arminius, *Western Culture in Eastern Lands* (London, 1906).
——'Pan-Islamism', *The Nineteenth Century and After* (London), 60 (July–Dec. 1906), 547–58; 61 (Jan–June 1907), 860–72.
——'Pan-Islamism and the Sultan of Turkey', *The Imperial and Asiatic Quarterly Review and Oriental and Colonial Record* (London), 3rd ser. 23/45–6 (Jan.–Apr. 1907), 1–11.
Vatikiotis, P.J., 'Muḥammad ʿAbduh and the Quest for a Muslim Humanism', *Arabica Revue d'Études Arabes* (Leiden), 4 (1957), 55–72.
——'Islam as a World Force', *New Society* (London), 23 Sept. 1965, 10–12.
Voll, J.O., *Islam Continuity and Change in the Modern World* (Boulder, Colo., 1982).
Vollers, K., 'Ueber Panislamismus', *Preussische Jahrbücher* (Berlin), 170 (July–Sept. 1904), 18–40.
Vucinich, W.S., *Russia and Asia: Essays on the Influence of Russia on the Asian Peoples* (Stanford, Calif., 1972).

Waardenburg, Jacques, 'Islam as a Vehicle of Protest', in: Ernest Gellner, ed., *Islamic Dilemmas: Reformers, Nationalists and Industrialization: The Southern Shore of the Mediterranean* (*Religion and Society* 25; Berlin and New York, 1985), 22–48.
Waddy, Charis, *The Muslim Mind* (London, 1976).
Wahby Bey, Behdjet, 'Pan Islamism', *The Nineteenth Century*, 61/363 (May 1907), 860–72.
Walker, Dennis, 'Muṣṭafā Kāmil's Party: Islam, Pan-Islamism and Nationalism', *Islam and the Modern Age*, 11/3 (Aug. 1980), 230–93; 11/4 (Nov. 1980), 329–88; 12/2 (May 1981), 79–113.
War Office, *Anti-Moslem Germany: Measures to Stop the Spread of Islam* (London, 1916).
Wasti, S. Razi, 'The Khilafat Movement in the Indo-Pakistan Subcontinent', *India Quarterly Journal of Bazm-i Iqbal* (Lahore), 19/3 (Jan.–Mar. 1972).
—— 'The Role of the Aga Khan in the Muslim Freedom Struggle', in: Ahmad Hasan Dani, ed., *Founding Fathers of Pakistan* (Islamabad, 1981), 101–8.
Watt, W.M., 'Thoughts on Islamic Unity', *Islamic Quarterly* (London), 3/3 (Oct. 1956), 188–95.
Weinbaum, M.G., and Gautam Sen, 'Pakistan Enters the Middle East', *Orbis*, 22/3 (Fall 1978), 595–612.
Wilson, H.A. 'The Moslem Menace: One Aspect of Pan-Islamism', *The Nineteenth Century and After* (London), 367 (Sept. 1907), 378–87.
Wilson, S.G., *Modern Movements among Muslims* (London, 1916).
Wirth, Albrecht, 'Panislamismus', *Deutsche Rundschau* (Berlin), 163 (Apr.–June 1915), 429–40.
X, 'Les Courants politiques dans la Turquie', *Revue du Monde Musulman*, 21 (1912), 158–221.
—— 'Le Panislamisme et le Panturquisme', *Revue du Monde Musulman*, 22 (1913), 179–220.
—— 'Les Courants politiques dans le milieu arabe', *Revue du Monde Musulman*, 25 (1913), 236–81.
XX, 'La Solidarité islamique et l'Angleterre', *Orient et Occident* (Paris), 1/3 (Mar. 1922), 366–72; 1/4 (Apr. 1922), 525–41; 2/5 (May 1922), 66–74; 2/6 (June 1922), 215–26.
Yakan, Fatḥī, *Islamic Movement: Problems and Perspectives*, transl. by Maneh al-Johani (Indianapolis, Ind., 1984).
Young, George, 'Pan-Islamism', *Encyclopaedia of the Social Sciences*, xi. 542–4.
Young, T.C., 'Pan-Islamism in the Solidarity and Conflict among Muslim Countries', in: J.H. Proctor, ed., *Islam and International*

Relations (London, 1965), 194–221.

Yūsuf, 'Alī, *Bayān fī khuṭṭat al-Mu'ayyad tijāh al-Dawla al-'Uthmāniyya al-'Aliyya* (n.p. [Cairo], 1909).

Zarevand, *United and Independent Turania: Aims and Designs of the Turks* (Leiden, 1971).

Zeine, Z.N., *Arab–Turkish Relations and the Emergence of Arab Nationalism* (Beirut, 1958).

Zenkovsky, S.A., *Pan-Turkism and Islam in Russia* (Cambridge, Mass., 1967).

Zürcher, E.J., *The Unionist Factor: The Rôle of the Committee of Union and Progress in the Turkish National Movement 1905–1926* (Leiden, 1984).

Zürrer, Werner, *Kaukasien 1918–1921: Der Kampf der Großmächte um die Landbrücke zwischen Schwarzem und Kaspischem Meer* (Düsseldorf, 1978).

Zwemer, S.M., 'Present-day Journalism in the World of Islam', in: J.R. Mott, ed., *The Moslem World of Today* (London, 1925).

——*Across the World of Islam: Studies in Aspects of the Mohammedan Faith and in the Present Awakening of the Moslem Multitudes* (New York, n.d. [1929]).

Zyevyelyev, A.I. *et al.*, *Basmachyestvo: Pravda istorii i vimisyel fal'sifikatorov* (Moscow, 1986).

Index

Conjunctions, prepositions and articles have not been considered in the alphabetical sequence.

ʿAbbās Ḥilmī 38, 47, 107, 124, 127, 132, 243
Abbassids 337
Abbott, G.F. 136–7
ʿAbd al-ʿĀlim, Muḥammad Ibn ʿAlī 123
ʿAbd Allāh (Shaykh) 47
ʿAbd al-ʿAzīz (Sultan of Morocco) 139
ʿAbd al-Ḥāfiẓ (Sultan of Morocco) 139
ʿAbd al-Nāṣir, Jamāl 251, 255–6
ʿAbd al-Qādir (Emir) 103, 324
ʿAbd al-Rāziq, ʿAlī 21
ʿAbd al-Rāziq, Muṣṭafā 16
ʿAbdallāh (King) 270, 272
ʿAbduh, Muḥammad 3, 16, 25, 48, 63, 125, 219
  and Pan-Islam 25–7, 367–8
Abdülaziz (Ottoman Sultan) 11, 12, 22, 40
Abdul Bari, Maulvi Qayyamul Din Muhammad 199, 201–3, 205, 350
Abdul Hamid (Judge) 375 n.
Abdul Wali, Hakim 350
Abdülhamid II (Ottoman Sultan) 11, 13, 33, 35–6, 73, 127, 188, 195, 331, 368
  and al-Afghānī 19–20, 31
  and Caliphate 31–2, 36–7, 44, 58–9
  deposition of 88, 101, 124
  and Islamic fraternities 51–4, 321–5, 328–9
  and Pan-Islam 1, 10, 22, 36, 38–49, 54–6, 86, 104, 139, 155, 171, 173, 176, 187, 308–9, 322, 364–6
Abdülmecid (Ottoman Sultan) 11
Abdur Rab 204 n.
ʿAbdurrahman (Emir of Afghanistan) 44
al-ʿĀbid, ʿIzzat Pasha 70
ʿAbīdī, Āl-i Aḥmad 275
Abū al-Hudā (Shaykh), *see* al-Ṣayyādī, Abū al-Hudā
Abū Zahra, Muḥammad 262
al-ʿAdl 109
Afghan Students' Islamic Federation 264–5, 380–1
al-Afghānī, Jamāl al-Dīn 13, 48, 75, 122, 150, 183–5, 190, 242, 261
  and ʿAbduh 16, 25
  and Abdülhamid II 19–20, 31
  and Pan-Islam 13–21, 171–2, 318–20, 326–7, 366–7
  and *al-ʿUrwa al-wuthqā* 3, 16
Afghanistan 217
  Ottoman interest in 19, 20, 40, 44, 65
  Pan-Islam in 69, 91, 103–4, 195–7, 212, 231, 234, 239, 240, 264–5, 337, 339
  Soviet war in 168, 174, 289, 291, 297, 301, 380–1
Africa 302
  Muslims in 33, 41, 42, 67, 116, 124, 287, 330, 333
  Pan-Islam in 140–2
Aga Khan 185, 194, 210
Ağaoğlu (Agayev), Ahmed 150, 156
al-Ahibba 274–5, 277–8, 283
Ahmad, Mirza Mahmud 205
Ahmad, Rafiüddin 188
Ahmadis 205
al-Ahrām 123, 235
Akaba, conflict over 122
Akçura (Akchurin), Yusuf 150, 156
Akhbār al-ʿĀlam al-Islāmī 286
Âkif (Ersoy), Mehmed 75–7
Akun, Nur Muhammad 333
al-ʿĀlam al-Islāmī 109, 110, 132
Albania, Albanians 38, 49, 57, 76, 81, 231–2, 337, 339

Albanian League, *see* League of Prizrend
Algeria 12, 36, 41, 57, 66, 76–7, 117, 339, 358
Ali (Shaykh) 103
Ali, Mohsin 301
ʿAlī, Muḥammad (of Egypt) 84
Ali, Muhammad (of India) 191, 193–4, 199, 201–2, 205, 207–10, 212, 239, 240, 307, 350
ʿAlī, Muḥammad Kurd 109
Ali, Sayyid Ameer 186, 198, 210
Ali, Shaukat 195, 199–202, 205, 207, 210, 239, 240, 307, 350
ʿAlī Pasha 103
Aligarh Institute Gazette 194
Alkın, Giray 157
All-Africa Muslim Congress 287
All-India Muslim League 185, 188, 194, 197, 204, 206
Alsagoff, Syed Ibrahim Bin Omar 269
Americas, Muslims in 287
Amin, Muhammad Anwar 298
al-Amīrī, Bahā' al-Dīn 251
al-ʿĀmirī, Sayyid Qāsim Ibn Sayyid al-Shamākhī 62
Anjumān-i Islām 189, 197
Anjuman-i Khuddam-i Kaʿba 198–203, 213, 346–50
Ansari, Mukhtar Ahmad 193, 205
Aperçu sur l'illégitimité du Sultan turc en tant que Khalife 177
al-Aqṣā Mosque 288
al-ʿArabī 251
Arab League 252, 269–70, 278, 289
Arabic (language) 57, 75, 81, 90, 118, 123, 148, 274, 281, 294, 302, 304, 345, 371
Arabs 54, 57, 70, 76, 78–9, 244, 257
  and Pan-Islam 60–3, 74, 81, 236, 309, 337, 339
  and Turks 38, 118, 230, 243, 328
ʿArafāt 222
al-ʿArafātī, Maḥmūd Sālim 222–3, 243
ʿArjūn, Muḥammad al-Ṣādiq 260–1
Arnold, T.W. 177
Arslān, Shakīb 92–3, 107, 109, 135, 243–6, 344–5
Arusî, Shaykh Mihriddin, *see* Hilmi, Şehbenderzade Ahmed
Asʿad, Aḥmad (Shaykh) 71, 322–3, 365
Asʿad, Ṣāḥib Zādeh (Shaykh) 113
aşiret mektebi 57
Asociación Pan-Islamismo 232
Association of Islamic Brotherhood 280
Astrakhan 155
Atatürk, *see* Kemal, Mustafa
atheism 161, 164, 167
Australia, Muslims in 287
Avn-ül-Tevfik (Emir) 323
ʿAwwād, Muḥammad Ḥasan 267
Aya Sofia Mosque 188
al-Ayyām 246
Azad, Abul Kalam 190, 192, 194, 205–6, 214
Azerbaijan, Pan-Islam in 150, 153, 164
al-Azhar 104, 120, 122, 127, 221, 225, 237, 251, 256, 260
al-ʿAẓm, Rafīq (Azmzadeh, Refik) 86 n.
ʿAzzām, ʿAbd al-Wahhāb 226

Baghdad Pact 254, 270
Bahadur II (Emperor) 182
Bahaeddin (Dr) 151, 330 n.
Baku 94
  Pan-Islamic convention in 154, 158
al-Balāgh 75, 190
Balbo (Marshal) 246
Balkan wars (1912–13) 92, 125, 192, 195–6, 198, 345
Baluchistan 19
Bamford, P.C. 193
Bandung meeting 273
Bangladesh 254
al-Bannā, Ḥasan 223–4
Barakatullah, Muhammad 195–6
al-Barāwī, Rāshid 271
Bareilles, Bertrand 22
al-Bārūdī, Sāmī 42
Başak, Hasan Tahsin 181
Basiret 3, 57
Bayur, Yusuf Hikmet 151
al-Bazzāz, ʿAbd al-Raḥmān 251
Behram Agha 324
Benevolent Islamic Society (*Cemiyet-i Hayriye-yi İslamiye*) 92, 104, 108, 114, 137, 308, 344–5
Benevolent Islamic Society (St Petersburg) 153
Ben Nabi, Malik 273

Bennigsen, Alexandre 144
Berbers 225
Berlin, as centre of Pan-Islam 140, 229, 230
Bethmann Hollweg (Reichskanzler) 105
Beyazıt Mosque 76
Bhutto, Zulfikar 254 n.
Bismarck, Otto von 47
Blunt, Wilfrid Scawen 41, 187
Boer war (1899–1902) 67
Bolsheviks 159, 162, 167
Bosnia, Bosnians 38, 81, 95
Bouteiller, Georges de 284
British Commonwealth 268, 272, 273
Brotherhood 107
Browne, E.G. 1
Buchanan, George 165, 166
Buddhism 305
Bukhara 160, 162, 171, 231
Bulgaria, Muslims (Pomaks) in 89, 301
Bulletin de la Société Endjouman Terekki İslam, see *Encümen-i Terekki-yi İslam*
Bulletin of the Muslim People's Organisation 277
Būrgība, Ḥabīb 279
Burkina Faso 289 n.
Burrows, R.A.B. 269
Bury, G.W. 6
Byelyayev, Ye. A. 173

Caffarel, E. 70–1, 321–5
Calcutta 188, 197–8
Caliph, Caliphate 172, 177, 185, 191, 196, 221–3, 228, 235, 241, 277, 321, 323, 335–41
  abolition of 176, 180, 210–11, 213, 216, 218, 222, 231, 236, 291
  alternatives to Ottoman 38, 136, 139, 237, 240, 255
  attacks on 127, 129, 131, 187
  authority of 98, 115, 137, 171, 208, 220
  Khilafat movement and 204, 206
  as rallying point 5, 24, 31–3, 59, 79, 82, 86, 116, 123, 125, 157, 198, 224, 275, 317, 328, 369
Cambodia, Muslims in 292
Cameron, William 307
Cameroon 289 n.
Catholic Church 290
  priests 339
Caucasus 11, 103, 151, 158, 169, 173, 234, 244, 330
Cavid, Mehmet 92
Celâlettin Pasha, Hasan Enver 43
Cemal Pasha 104, 229, 231
Cemiyet-i Hayriye-yi İslamiye, see Benevolent Islamic Society
Central Islamic Committee 72, 334, 369
Cevdet, Ahmed 182
Chad 41, 301
Charmes, Gabriel 2, 364
China 11, 44, 47–8, 66
Chirol, Valentine 35
Chotani, Seth 210
Christianity, Christians 191, 208, 223, 262, 305
  Christianization 144–5, 339
  Copts 133
  in Ottoman Empire 39, 47, 90, 316, 338
  threat to Islam 12, 15, 35, 36, 38, 63, 75, 77, 83–4, 112–13, 130, 189, 206, 261, 284, 328, 365
Committee for the rights of the Turco-Tatar Muslims in Russia 106
Committee of Union and Progress, *see* Young Turks
Communism 159, 161–2, 167–8, 171, 173, 175, 255, 263
  Muslim Communism 162, 163, 167
Communist party (Soviet Union) 168
Communité des Pays d'Expression Française 268
Comrade 191–3
Congo 141–2
Congress of Berlin (1878) 49
Congress of Nationalities (Lausanne) 156
Constans 333
Copts, *see* Christians
The Crescent Versus the Cross 27
Crete, Cretans 88, 188
Crimean war 11, 145, 184, 185, 365
Cromer (Lord) 129
Cumhuriyet 285
Cyprus 187, 253, 289, 292

al-Dajānī, Aḥmad Ṣidqī 251
al-Dānā, ʿAbd al-Qādir 68
Danishway Incident 27, 122
Darfur 142
Dār al-ḥarb (Domain of war), *see* Islam
Dār al-Islām (Domain of Islam), *see* Islam
al-Dawālibī, Maʿrūf 280
Daʿwat al-ḥaqq 286
Dawn Overseas Weekly 301
Derviş Pasha 49
Detering, H.W.A. 231
Deutsch-Islamisch Gesellschaft für Islamkunde 106
El-Djabri, Ihsan 244
Dost Muhammad 366
Dufferin (Lord) 60
Duma, Muslims in 153, 154, 158
al-Dustūr 123
Dutch East Indies 212, 237

Eastern Question 139
Ebu-El-Toeme (Shaykh) 342
Echos de l'Islam 209, 232
Edib, H. Eşref 75, 179
education 82, 126, 145, 148, 228, 252, 275, 294–5, 300, 302, 310, 343, 363, 374
  schools 33, 37–8, 75, 89, 147, 320, 341
  tribal schools 57
Egypt 11, 36, 42–3, 62, 69, 96, 108, 111, 116, 139, 187, 217, 237, 239, 242, 246, 249, 254, 267, 270, 279, 289, 339
  and Pan-Islam 121–34, 226, 253, 255–6
Encümen-i Terekki-yi İslam (Society for the Progress of Islam) 94, 130, 131
Enver Pasha 89, 93–5, 100, 104, 106, 108, 110, 114, 135, 140, 193, 229, 231
Erbakan, Necmettin 264
Eritrea 142, 371
Esat 79
Essays on Pan-Islam and Pan-Turkism in Russia 169
Etherton, P.T. 162
Ethiopia, Muslims in 246, 292
Evren, Kenan 299 n.

Faḍl, Sayyid 71, 322, 323
Faḥṣ, Hānī 261
al-Faraed 108
Farīd, Muḥammad 130–2
al-Fārikhī, Sayyid Muṣṭafā Ibn Ismāʿīl al-ʿAmrī 62
fatwā (fetva) 88, 91, 99, 100, 108, 194, 203
Fayṣal (King) 254–5, 261, 267, 275, 288, 298
fetva, see *fatwā*
Le Figaro 128
The First Year of the War 195
France 41, 53, 59, 63, 81, 88, 111, 121, 139, 165, 209, 245, 247, 358–9
fraternities, *see* Islamic fraternities
Frémont, P. 41
Friendly Advice to Indian Brothers 195
Fu'ād (King) 237

Gandhi, Mahatma Mohandas 205, 210 n.
Gasprinskiy (Gasprali), İsmail 146–50, 152, 154–6, 235, 307
General Islamic congress 279
Geographical Society of Algiers 55, 333
Germany 129, 133, 140
  Muslims in 242, 285
  Nazi 246, 247
  as Ottoman ally 45–8, 59, 103–7, 111, 121, 171–2, 194, 202, 366
Ghaussy, A. Gh. 265
al-Ghayātī, ʿAlī (Shaykh) 120, 121
Ghayret 187
al-Ghiṭā', Muḥammad al-Ḥusayn al-Kāshif 241
Gibb, H.A.R. 15
Gladstone, W.E. 187, 192
Goethe, W. von 166
Gökalp, Ziya 74, 80
Gorchakov (Prince) 316
Gordlyevskiy, V. 157
Gordon-Polonskaya, L.R. 172, 173
Grand National Assembly (in Turkey) 178–81, 236
Great Britain 21, 59, 81, 87, 111, 121, 123, 131, 165, 172, 187, 194, 196, 207, 213, 220, 245–7, 366–7
  Muslims in 136

and Palestine Mandate 240, 241
Great Soviet Encyclopaedia 170, 173
Greece 88
Grey, Sir Edward 27–8, 78
Grey (Major) 62
Guignard, François Emmanuel Comte de Saint-Priest 10
Gümülçineli, İsmail Bey 344
Güngörge, Mustafa Talip 181–2

Habibullah Khan (Emir of Afghanistan) 104
Ḥabl-ol-Matīn 91, 193
al-Hacaik 108
El-Hadj Abdullah (lieutenant) 117, 121, 358 n.
Hakkı, İsmail (adjutant) 55, 333–4, 344
Hakkı, İsmail (emissary) 72, 333
Haleem, A.B.A. 282, 283
Halid, Halil 27, 132, 181
Halil Bey [Kut] 93
Halil Pasha 22
Ḥalīm, ʿAbbās 344
Ḥalim Pasha, Said 84–5, 93, 132, 344
Ḥamāda, Fārūq 249–50
Ḥamba, ʿAlī Baş 93, 106
Ḥamdard 193, 194
Hamsa, Abdul Malik 110, 111
Hanotaux, Gabriel 26, 63
Haq, Inamul 272, 283
al-Haqq yaʿlū 133
al-Ḥarakān, Muḥammad ʿAlī 287
Ḥasan II (King) 288, 299
Ḥasan, Nazir-uddin 350
Hasbi, Süleyman 24
Haushofer, Karl 304
Ḥaydar Pasha, Sharīf ʿAlī 93, 344
Ḥaydar, Sharīf Jaʿfar 344
al-Hidāya 75, 201
Hijaz 52, 69, 81, 82, 211, 237, 321
Hijaz railway 55, 70, 189, 197, 333, 365
Ḥijāzī, Miskīn 250
Hikmet 78
al-Hilāl 190, 192, 194
El-Hilal el-Osmani 133
Hilmi, Şehbenderzade Ahmed 77–9, 335–41
Hindus 205, 206, 210, 214
Historical Encyclopaedia 173
Hitler, Adolf 247
Ḥizb al-Iṣlāḥ 125
Ḥoly pilgrimage, *see* Pilgrimage
Holy war, see *Jihad*
Hürriyet 3
Ḥusayn (King of Jordan) 279
Ḥusayn (Sharīf of Mecca) 38, 109, 111, 120, 236, 344, 365
Husayn, Sayyid Ali 228
Husayn, Vajahat 195
al-Ḥusaynī, Amīn 239–41, 247, 280, 283
Hüseyin (General) 41
al-Ḥuṣrī, Sāṭiʿ 219
Hyderabad 371
  Nizam of 188, 213

Ibn ʿAlawī Ibn Sahl, Faḍl 67
Ibn Fayyāḍ, Zayd Ibn ʿAbd al-ʿAzīz 252
Ibn Saʿūd, ʿAbd al-ʿAzīz (King) 211, 236, 238, 239, 268
Ibn Sumayṭ, Aḥmad 67–8
İbrahim (Ibragimov), Abdürreşid 29–30, 103, 150, 153, 154, 156, 247
İbret 3, 25
İkdam 193
India 253, 254, 371
  Muslims in 11, 19, 50, 57, 60, 66, 69, 76, 77, 94, 101–2, 105, 108, 124, 132, 136, 171, 226, 239, 247
  Mutiny 184, 185
  Pan-Islam in 176–215, 306, 308, 309
İnönü, İsmet 210
Inter-Departmental Committee on Eastern Unrest 179
Iqbal, Muhammad 214, 215 n., 226–7, 241
Iran 20, 44, 65, 71–2, 76, 89, 91, 150, 179, 191 n., 217, 226, 233–4, 240, 254, 257–8, 270, 337
  Islamic Republic of 174, 258–60
Iran–Iraq war 289, 292, 301
Iraq 217, 236–7, 251, 270
ISESCO, *see* Islamic Educational, Scientific and Cultural Organization
Ishaq, Wares 270
Islam 22, 80, 127, 129, 144, 262, 282, 304–5, 319, 336
  as bond 82, 84–5, 111, 157

Islam (*cont.*)
Dār al-ḥarb and *Dār al-Islām* 4, 187, 304, 357
Holy Places 199–201, 204, 208, 211, 219, 225, 238, 240, 244, 263 n., 290, 298, 346, 348, 372
religion and politics 4, 6, 11–12, 17, 219, 223, 248, 251, 253–4, 257–8, 299
sects 317
Shiʿa v. Sunna 15, 16, 32, 50, 81, 91, 125, 164, 241, 257, 259, 260, 316–17, 340
universality of 4, 166, 259
al-Islām 212
L'Islam aujourd'hui 295
Islam Today 295, 301
al-Islām al-yawm 295
Islam and Contemporary Society 295
Islamic Academy for Jurisprudence 295
Islamic conference 279
Islamic Council for Europe 295, 301
Islamic Development Bank 293, 298, 300
Islamic Educational, Scientific and Cultural Organization (ISESCO) 294–5, 299, 301
Islamic fraternities 12, 51–4, 65, 105, 134, 140, 203, 328–9, 358, 365
Naqshbandiyya 113
Qādiriyya 137
Rifāʿiyya 52–3, 71, 324, 365
Sanūsiyya 52–3, 66, 77, 98, 118, 141, 233, 321, 324, 354
Shādhiliyya-Madaniyya 52–4, 71, 323, 328, 365
The Islamic Fraternity 196
The Islamic Guidance Association 225
Islamic News 207
Islamic News Agency 295
Islamic Solidarity Fund 300
Islamic Solidarity in Today's World 266
Islamic World Centre 256
Die Islamische Welt 109–11, 132
Islamism 9, 22, 37–8, 73, 84, 124, 310
see also Islam; Pan-Islam
İslam Mecmuası 75
Ismāʿīl (Khedive) 69
Ismailis 194
Israel 253, 256, 288, 291, 301
Muslims in 292 n.
Istanbul 145
as centre of Pan-Islam 68–72, 140, 230
Italy 93, 101, 133, 136, 165, 209, 246
İtilaf ve Hürriyet 344
İttifak-ul-Muslimin see Union of Muslims
İttihad-ı İslam (term) *see* Union of Islam
İttihad-ı İslam (weekly) 77
Ittiḥād ol-Islām (Iranian association) 91
al-Ittiḥād al-Maghribī 139
İttihad ve Terakki 344

Jackson, J.B. 119
Jaʿfar (Sharīf) 344
Jamāʿa Khayriyya 139
Jamaʿat-i Islami 249
Jamʿiyyat al-Islam 277
al-Jammāl, Gharīb 265
Japan 43, 66, 247
Jarīdat al-sharq 109
al-Jawāʾib 41, 60–2
Jedda 278
Jedidism 148, 149, 166, 171
Jerusalem 225, 279, 285, 288, 289
Muslim university plan 241, 242
Jews 223, 226
in Ottoman Empire 90, 338
Jihad 6, 11, 171, 188
Israel–Arab conflict as 291
Libyan war as 135, 136
Soviet invasion of Afghanistan 380
World War I as 94, 98–103, 108, 110, 112–14, 116, 118, 119, 141–2, 156, 194–5, 352–5, 358, 365
World War II as 247
Jihad for Right 195
Jihān-i Islām 93, 108
Jinnah, Muhammad Ali 204, 253–4, 268
Johansen, Baber 9
Jordan (Kingdom) 256, 270, 279
see also Transjordan

Kaʿba 199, 263 n., 346–50
see also Islam, Holy Places

Kāmil, Muṣṭafā 127–9, 307
Kâmil Bey 44
Kâmil Pasha (grandson of Shaykh Ali) 103
Kâmil Pasha (Grand Vizir) 324, 325
Karachi 280, 281
Karpat, K.H. 87
Kāshānī (Ayatollah) 258
Kāshānī, Muṣṭafā 258
Kashgar 41, 202, 322
Kashmir 283, 371
al-Kattānī, ʿAlī al-Muntaṣir 285
al-Kawākibī, ʿAbd al-Raḥmān 34, 222, 368
Kayhān 257
Kazan 145
  Muslim convention in 159
Kazanskiy Tyelyegraf 157
Keddie, N.R. 15, 18
Kemal (Atatürk) Mustafa 104, 178–81, 210–13, 229, 235
Kemal, Namık 3, 24–5, 84
Khaliquzzaman, Chaudry 269, 276, 277
Khan, Ayyub 254 n.
Khan, Azim 366
Khan, Chaudri Nazir Ahmad 273–5, 277, 278
Khan, Inʿamullah 281, 283
Khan, Khudayar 364
Khan, Sayyid Ahmad 185
Khan, Sultan Ahmad 233
Khan, Zafar Ali 191, 193, 200, 201
Khan, Zafrullah 270, 271
Khaṭīb, Muḥibb al-Dīn 225
Khayr al-Dīn (Khayreddin) Pasha 70, 324, 344
Khilafat Bulletin 207
Khilāfat movement 171, 203–16, 226, 239, 308
Khilāfat-i ʿUthmāniyya 207
Khomeyni, Rōḥallāh (Ayatollah) 259–60
Khorasānī, Ḥasan Abṭaḥī 258
Khuddam-i Kaʿba see *Anjuman-i Khuddam-i Kaʿba*
Khuddāmul Kaʿba 202
Kidwai, Mushir Husain 50, 187 n., 189–90, 199–201, 207, 209, 214 n., 350
al-Kīlānī, Najīb 263–4
King ʿAbd al-ʿAzīz university 300
al-Kiyānī, Muḥammad 137
Klimovich, L. 169
Koran 17, 18, 64, 79, 127, 224, 257–8, 260, 267, 273, 284–5, 297, 300, 318, 320, 354, 364
Kramer, Martin 7
Küçük Kaynarca Ottoman–Russian treaty (1774) 10
Kuran, Ahmed Bedevi 151
Kurds 57, 76, 92, 244, 337
Kutubī, Ḥasan 263

Lacau (French Consul) 53, 328–9
Lausanne Peace treaty (1923) 181
Layard, A.A. 40
League of Nations 222
League of Pan-Islam 342–3
League of Prizrend (Albanian League) 48–9
Lebanon 253, 277
Lenin, V.I. 161, 167
Lewis, Bernard 4, 305
Liberalism, Islamic 185
Libya 36, 41, 101, 237, 246, 321, 328–9
  in Chad 301
  Pan-Islam in 298
  1969 revolution 256
  war in 123, 125, 133–8, 198
Libyan war (1911–12), *see* Ottoman–Italian war
Lie, Trygve 253
Little Scientific Atheist Dictionary 173
al-Liwā' 132
Lloyd George, David 204, 208
London, as centre of Pan-Islam 207
Lowther, Gerald 78
Lutfi Bey 107
Lutfullah (Prince) 27

MacDonnell, Anthony 188
Madaniyya, *see* Islamic fraternities
Madrid, as center for Pan-Islam 140
al-Maghribī, ʿAbd al-Qādir 109, 184
al-Mahdī, Muḥammad 53
Mahmud II (Ottoman Sultan) 11
Majallat rābiṭat al-ʿālam al-Islāmī 286
al-Makhzūmī, Muḥammad 15
Maldive Islands 285
Malumat 59–60
Manapir-Zade, Mustafa Nuri 3
al-Manār 125–6, 219, 220

al-Mardīnī, ʿĀrif Bey 344
Marianne 246
Marunov, Yu. V. 169, 170
Marxism, *see* Communism
Massignon, Louis 4
Matthews, Basil 182
al-Maududi, Abu Aʿla 226–8, 249, 254, 271–2, 284, 302
Majallat al-ḥajj 262
Mecca 199, 236, 247, 284, 321, 323
  see also pilgrimage
Medina 199, 247, 275, 321
Mehdi, Shaykh-Omer-Ejel Rubini 342
Mehmed V Reşad (Ottoman Sultan) 108, 130, 196, 339
Mehmet Bey 68
Menzie, de 216
The Message 195
Midhat, Ahmed 59
Mihriddin Arusî, *see* Hilmi, Şehbenderzade Ahmed
minority groups 156, 161, 285–6, 289, 292, 294, 302
  in Ottoman Empire, *see* Christians; Jews; Kurds; etc.
Mir'āt 150
Mir Islama 167
Mizan 33
modernism, modernization 11, 74, 87, 90, 144, 181, 249
moguls 182, 184
Mohammedanische Gesellschaft, *see* Office Musulman International
Le Monde 114
Montagu, Edwin 208
Montaẓerī (Ayatollah) 260
Montet, Edouard 120
Morgenthau, Henry 120
Morocco 66–7, 96, 101, 116, 117, 217, 225, 232n., 233, 237, 337, 358
  Pan-Islam in 138–40, 298–9
Mosalmānān-i Mojāhid 258
Mōsavī, Ḥoseyn 260
Moscow, Muslim convention in 158–9
The Moslem Chronicle 188–93
Mount ʿArafāt 82
Moẓaffar al-Dīn Shah 45
Moẓafferī 34
al-Mu'ayyad 26, 124, 139
Muḥammad (Prophet) 60, 76, 115, 121, 139, 176, 183, 236, 253, 255, 258, 267, 273, 337, 338
al-Mubārak, Muḥammạd 264
Muhammadan Union (*Ittihad-ı Muhammedi Fırkası*) 50–1
Muhtar Bey 344
Muhyi al-Din 324
al-Mujāhid 212, 219, 232
Mumcu, Uğur 285
al-Munajjid, Ṣalāḥ al-Dīn 263
Münif Pasha 44–5
Murad, Mehmed 33, 222
Murad Efendi (Franz von Werner) 1
Musa (Shaykh) 323
Muscat, Sultan of 331
Muslehuddin, Muhammad 272
Muslim Brethren 223–6, 250, 256, 280
Muslim Communities in Non-Muslim States 295
Muslim Outlook 207
Muslim People's Organization 269, 276–7
The Muslim Revival in the Seat of the Supreme Caliphate 112
The Muslim Standard 207
The Muslim World: A Weekly Review of the Motamar 281
Muslim World Congress 280–3, 286, 296, 308, 371–2
Muslim World Islamic Quarterly 269
Muslim World League 7, 280, 283–7, 296, 302, 308, 309
Muslim World League Journal 286
The Muslim World and the Future Economic Order 295
Mussolini, Benito 246, 247
Muştak, İsmail 344
al-Mustaqbal 121
al-Muwayliḥi, Ibrāhīm 31

al-Nahḍa 155
Najaf 91, 100, 225
Nallino, C.A. 1, 4, 177
Naqshbandiyya, *see* Islamic fraternities
Narayan, B.K. 7
Nāṣir (Sharīf) 344
Nāṣir, *see* ʿAbd al-Nāṣir
Nāṣir al-Dīn Shah 44, 367
La Nation Arabe 244
National Congress (India) 210

The National Demarcation of Central Asia 168
National Islamic Front 250
National Salvation Party 256–7, 264
nationalism 86, 149, 160, 167–8, 172–3, 203, 215, 219, 225, 227, 261, 304, 306, 311
  Arab nationalism 41, 101, 104, 125, 196, 252, 255; *see also* Pan-Arabism
  clash with Ottomanism 43, 73
  clash with Pan-Islam 15, 76, 81, 126–7, 212, 214, 217, 223, 248–51, 262, 273, 281
  Egyptian nationalism 129, 134, 224
  Syrian nationalism 220
  Turkish nationalism 179, 181, 235
Nationalist party (Egypt) 127, 129, 130, 133
Nazîf, Süleyman 84
Nazım (Dr) 151, 330n.
Nazis, Nazism 168, 172
Netherlands 165
Nibrās al-Mashāriqa wa-'l-Maghāriba 62
Nicholas II (Tsar) 153, 165
Nicolson, A. 166
Nigeria, Muslims in 141, 289
Nijni Novgorod, Pan-Islamic congress in 152, 153
Nikolay (Grand Duke) 40
non-aligned states 263, 268
La Nouvelle Revue 128
Novoye Vryemya 167
Nuri [İleri], Celal 25, 80–4, 133, 307

Office Musulman International (Islamische Gesellschaft) 107
Okhrana 157, 166
Oppenheim, Max von 93, 96–8, 105
Organization of African Unity 289
Organization of the Islamic Conference 280, 287–95, 297–9, 301, 308–10
Organization of the Muslim Peoples 279–80, 296, 373–4
L'Orient Journal de Défense des Intérêts de l'Empire Ottoman 63–4
Orientalism 183
Osmanlı 58
Ottoman Bank 324
Ottoman Empire 9–13, 47, 145, 148–9, 221
  and Arabs 243, 328, 339
  break up of 73, 79, 81, 217
  and Caliphate 136, 185, 335, 340
  government of 133, 151
  and Indian Muslims 187–8, 192, 194, 201
  and Pan-Islam 19, 20, 26, 40, 49, 83, 86, 144, 150, 166, 170, 172, 184, 228, 306, 308, 315–17, 330
  and press 58, 59
  and World War I 110, 112, 126–7, 156, 202
  see also Turkey
Ottoman–Italian war (1911–12) 125–6, 134–8, 140, 192–3, 198
Ottomanism 9, 73–4, 77, 87, 90, 124, 130, 181
Ottomano 108

Pahlavī, Moḥammed Rezā Shah 257, 270
Pahlavī, Rezā Shah 257
Pahlen, K.K. (Count) 65
Pakistan 214
  and Pan-Islam 174, 249, 253–5, 267, 270, 275–7, 279, 297–8, 307–9
Pakistan–Bangladesh war 291
Palestine 121, 225, 233, 237, 239–42, 245, 285, 288–91, 371
Palestine Liberation Organization 251, 288–9, 291
Palestine Mandate 240, 241
Pan-Arabism 218, 220, 224, 252, 304–5
  see also nationalism, Arab
Pan-Germanism 2, 6, 304, 305
Pan-Hellenism 2, 6, 177, 304
Pan-Islam
  conventions 34, 131, 152–60, 232, 234–47, 295–303, 360–3; Cairo 237–9; Geneva 242–6; Jerusalem 240–3, 280; Mecca (1924) 236; Mecca (1926) 238–40, 280; Sivas 235; Stockholm 235
  definition 5–8
  early developments 9–35
  in Egypt 121–34
  emissaries 64–8, 139, 184, 308, 322–5

Pan-Islam (*cont.*)
financing 55–6, 198, 206
ideology 13–35, 73–86, 218–28, 248–67, 304–7
in India 176–215
intelligence gathering 54–5, 333–4
opposition to 23
organization of 35–72, 101, 137, 197–203, 267–303, 307–9
origins 1–4
political parties 154, 160
press 32–3, 57–63, 73–6, 97, 123–5, 135, 192–3
propaganda 24, 29, 30, 36, 38, 39, 56–64, 69, 90, 97, 104–21, 134, 212, 230, 308
in Russia and Soviet Union 143–75, 360–3
as solidarity 260–7
and Young Turks 73–142
Pan-Islam (association) 50
Pan-Islam (periodical) 50
Pan-Islamic League 231, 233
Pan-Islamic Society (London) 49
Pan-Italianism 309
Pan-Romanianism 309
Pan-Slavism 2, 6, 144, 217, 304, 305
Pan-Turkism 6, 9, 73, 74, 89–90, 130, 144–6, 149, 150, 152, 156, 163, 165–8, 181, 304, 305
Pears, Edwin 52
Peïk Islām 60
Persia, *see* Iran
Philby, St John 268
Philippines 66, 289, 292
Phillips, J.E. 141
pilgrimage 56, 145, 183, 200, 217, 222–3, 238, 262, 284–5, 333
Pan-Islam and 57, 118–19, 233, 236
as unifying factor 24, 26, 32, 35, 82, 198
Piscatori, J.P. 7, 254
Poland 237
Pomaks, *see* Bulgaria
Provisional Central Bureau of Russian Muslims 158
Putra, Tunku Abdur Rahman 289 n.

Qadhāfī, Muʿammar 256, 298
Qādiriyya, *see* Islamic fraternities
Qā'imī, ʿAlī 258
Qājār, Prince Abū al-Ḥasan Mīrzā (Shaykh al-Ra'īs) 31–2
Qajars 257
Qanātābādī, Shams 258
Qārūn 257
Quṭb, Sayyid 260, 277

al-Ra'y al-ʿĀmm 109, 113
Red Army 162
Red Crescent Society 55–6, 193, 200
Renan, Ernest 24
Research Centre for Islamic History, Art and Culture 299
Resulzade, Mehmet 360
Revue des Deux Mondes 2, 364
Revue du Monde Musulman 155
Riḍā, Muḥammad Rashīd 125–6, 219–20, 239, 241
Rifā'iyya, *see* Islamic fraternities
al-Rīs, Muḥammad Ḍiyā' al-Dīn 278
Riskulov, Turar 162
Riyad Pasha 366
Rıza, Ahmad 29
Rıza, Ali 44
Rous-Keppel 104
Royal Academy for Islamic Civilization 279
Royal Dutch/Shell 234
Ruhi Efendi 367
Russia 40, 59, 61–2, 65, 95, 187, 319, 342
Muslims in 29, 77, 89, 94, 117, 343, 360–3
Pan-Islam in 143–61, 165–7, 306, 308–9
Russification 143–5, 147, 148
see also Soviet Union
Russian revolution 158, 160, 171, 173
Russo-Japanese war 146

Sabaheddine (Prince) 27–8
al-Sādāt, Anwar 256
Sadık Bey 39, 44
Ṣadīqī, Mīr 257
Sahara, French 141
Saheb Mullah Aftab Hussain, *see* Kidwai, Mushir Hussain
Said Pasha 323
Saint-Priest, Comte de, *see* Guignard
Salah, Khualdiyya 334
Ṣalāḥiyya school (Jerusalem) 104
Salih 187
Samarkand 65, 149
Sanhoury, A. 221–2

al-Sanūsī, Ibrāhīm 66, 365
al-Sanūsī, Idrīs 237
Sanūsī, Sayyid Shaykh Aḥmad 233
Sanūsiyya, *see* Islamic fraternities
Sarikat al-Islām 212
Saudi Arabia 217, 236, 252–5, 267, 279, 284, 298, 300, 307, 308
see also Hijaz
al-Ṣayyādī, Shaykh Abū al-Hudā 30–1, 71, 322, 324–5, 365
and al-Afghānī 31
Sayyid ʿAlī Ibn Ḥamūd, *see* Ibn Ḥamūd
Sayyid Khalīfa (Sultan of Zanzibar) 68, 142
Sazonov, Syergyey Dimitrovich 165
Schnee (German governor) 142
schools, *see* education
Schulze, Reinhard 7
Sebilür Reşad 74, 179
Selim I (Ottoman Sultan) 10, 366
Senegal 287
şeriat, see *sharīʿa*
Service des Affaires Musulmanes et Sahariennes 53
Şeyhülislam (Shaykh al-Islām) 26, 88, 99, 108, 322, 329, 344
al-Shaʿb 93
Shadhiliyya, *see* Islamic fraternities
sharīʿa (şeriat) 34, 49, 51, 85, 121, 184, 223, 265, 282, 284, 293, 295, 300
Sharīʿatmadārī (Ayatollah) 259
al-Sharīf (al-Tūnisī), Ṣāliḥ 93, 106, 114–15, 140, 229
al-Sharīʿī Pasha 344
Shatwān Bey, Yūsuf 93, 108, 344–5
Shāwīsh, ʿAbd al-ʿAzīz 93, 109–11, 116, 131–3, 135, 193, 200, 212, 229, 235, 344
Shiʿa, Shiites 15, 32, 100, 135, 205, 226, 257, 260, 367
see also Islam
al-Shidyāq, Aḥmad Fāris 60–1
Shinketi, Shaykh Muhammad 322
Shir Ali 366
Shīrāzī, Mīrzā ʿAlī Agā 34
Sid al-Mahdi 321
Siddiqui, Kalim 252
Sierra (Italian Consul) 92, 93
Sī Ḥamza, *see* Ẓāfir, Ḥamza
Sī Qāsim 54
Sirāj ol-akhbār 197
Sırat-ı Müstakim 74, 190
Sırma, İ.S. 36
Smirnov, N.A. 169
socialism 160
Société Progrès de l'Islam, see *Encümen-i Terekki-yi İslam*
Society of Islamic Brotherhood 225, 226
South Africa, Muslims in 136, 237, 294
Soviet Academy of Sciences
Institute of History 169
Institute of Peoples of Asia 169
Soviet Encyclopaedic Dictionary 173
Soviet Union 263
Pan-Islam in 160–4, 167–75, 229, 230, 239, 245, 261, 306
see also Russia
The Standard 195
Stalin, J.V. 167, 171
Stolipin, Pyotr Arkad'yevich 166
Strempel, Captain von 67
Sudan 11, 66, 141–2, 233, 250
Sufis 51–2
Suhrawardy, Abdullah al-Mamun 50
Sultan-Galiyev, Mir Said 162, 164, 168
Sunnites 15, 182, 184, 205, 257, 260
see also Islam
Supreme Muslim Council 239
Switzerland 242
Syeutov, S. 174, 302
Syria 11, 52, 61, 242, 256, 270

Ṭabāṭabā'ī, Ḍiyā' al-Dīn 241
al-Taḍāmun al-Islāmī 261–2
Tahir, Mehmed 59
al-Taḥrīr 256
Talat Bey 229, 231, 344
Ṭāliqānī, Maḥmud 259
Tatars 89, 144–5, 147–8, 153–4, 159, 162, 174, 302
Teâli-yi İslâm cemiyeti 228
Tercüman 149, 154, 155
Teşkilat-ı Mahsusa 104
Tevfik Pasha 55, 333
Thailand, Muslims in 292
Topchibashyev 154
Transjordan 236, 242
see also Jordan (Kingdom)
Tripolitania, *see* Libya
al-Tuhamī, Ḥasan 290 n.
Tunisia 12, 36, 41–2, 66, 70, 92, 141, 237, 256, 339

al-Turabī, Ḥasan 250
Turco-Greek war (1897) 60, 122
Turco-Russian war (1877–8) 186
Türk 33
Turkestan 65, 100, 124, 157, 162, 166, 277
  East 12
  1916 uprising 157
Turkey 209, 210, 220, 229, 235, 254, 256–7, 285 n.
  and Pan-Islam 177–82, 212, 234, 270, 299
  see also Ottoman Empire
Türkische Skizzen, *see* Murad Efendi
Turkism, *see* Pan-Turkism

Uganda 141, 142, 289 n.
Ulema 37, 47, 65, 88, 100–1, 106–7, 127, 148, 185, 187, 202–4, 221, 225, 237, 284
  Meeting in Najaf 91
ʿUmar (Caliph) 11
al-Umma 123
Union of Islam (*İttihad-ı İslam*) 2–3, 23, 230
Union of Muslims (*İttifak-ul-Muslimin*) 154–5, 171
Union of Muslims of Russia (*Russiya Müsülmanlarının İttifakı*) 153
Unité Islamique 23, 315–17
United Arab Federation 270
United Arab Republic 270
United Nations 279, 284, 289–91, 371, 372
United Nations' Educational, Scientific and Cultural Organization (UNESCO) 294
United States 172, 263
  Muslims in 294
A Universal Proclamation to All the People of Islam 119, 351–7
ʿUrābī, Aḥmad 42
ʿUrābī uprising (1881–2) 25, 42–3, 71, 323
al-ʿUrwa al-wuthqā 3, 16, 21, 25, 366, 367

Vaḥdat-i Islāmī 260
Vaḥdeti 50
Vakıt 41
Validi, Cemal 157
Vambéry, Arminius 1, 12, 49, 64
Vatican 291
Versailles Peace Treaty 216
The Voice of the Crescent 232
Voice of Islam 256
Volkan 50–1

Waday, Sultan of 329
Wahba, Ḥāfiẓ 201
Waliullah, Shah 183
Wangenheim, Hans von 98
waqf 147, 363
Warsaw Pact 272
Wilhelm II (Kaiser) 46–7, 67, 95–6, 98, 104, 121, 129, 243, 247
Wilson, Woodrow 107, 156
World Council of Churches 291
World Development Bank 293
World War I 94–121, 156–7, 171, 202, 345
World War II 168, 172–3, 247

Yakub Beg 40
Yemen 237, 240, 270, 320
  war 254
Young Bukharans 160
Young India 205
The Young Maghrib 139
Young Men's Muslim Association 225
Young Ottomans 2, 146
Young Turks 24, 51, 155, 165, 167, 178, 190, 210, 217, 219, 229, 235
  Congresses of 88–9
  and Pan-Islam 73–142, 151, 166, 173, 306, 308, 330
Yugoslavia 174, 371
Yūsuf, ʿAlī 124–5, 139
Yūsuf (Sultan of Morocco) 117
Yusuf İzzedin (Ottoman Crown Prince) 93, 344

Ẓāfir, Ḥamza 324, 328, 365
Ẓāfir, Shaykh Muḥammad 53–4, 71, 321–4, 328, 365
Zakariya, Haji Nur Muhammad 197–8
zakāt 275
Zamīndār 191, 193, 195
Zanzibar 67–8
Zaytūna Mosque (Tunis) 114
al-Zayyāt, Aḥmad Ḥasan 277
Zia-ul Haq 254
Zionism 241, 245, 255, 279, 304
Ziya Pasha 3